Intermittent and Nonstationary Drying Technologies

Principles and Applications

Advances in Drying Science & Technology

Series Editor: Arun S. Mujumdar

PUBLISHED TITLES

Intermittent and Nonstationary Drying Technologies: Principles and Applications
Azharul Karim and Chung-Lim Law

Handbook of Drying of Vegetables and Vegetable Products
Min Zhang, Bhesh Bhandari, and Zhongxiang Fang

Computational Fluid Dynamics Simulation of Spray Dryers: An Engineer's Guide
Meng Wai Woo

Advances in Heat Pump-Assisted Drying Technology
Vasile Minea

Intermittent and Nonstationary Drying Technologies

Principles and Applications

Edited by
M. Azharul Karim and Chung-Lim Law

CRC Press is an imprint of the
Taylor & Francis Group, an informa business

CRC Press
Taylor & Francis Group
6000 Broken Sound Parkway NW, Suite 300
Boca Raton, FL 33487-2742

Printed on acid-free paper

International Standard Book Number-13: 978-1-1387-4629-9 (Paperback)
International Standard Book Number-13: 978-1-4987-8409-2 (Hardback)

Visit the Taylor & Francis Web site at
http://www.taylorandfrancis.com

and the CRC Press Web site at
http://www.crcpress.com

Contents

Preface

The first book on intermittent drying (ID), *Intermittent and Nonstationary Drying Technologies: Principles and Applications* covers key topics ranging from the technology, effect of operating parameters and mathematical modelling to the energy and quality aspects of ID. This book is an important reference for food engineers, chemical product engineers, pharmaceutical engineers, plant design engineers, researchers, students and technologists in food processing, food engineering, food science, chemical engineering and pharmaceutical engineering. The readers are expected to

- Understand the basic principles and fundamentals of ID
- Understand how to use ID to minimize energy consumption of a drying process
- Understand how to use ID to produce better quality dried products
- Understand the variants of ID dryers
- Understand the modelling of ID and its application in designing ID dryers

Drying is a very energy-intensive process and typically consumes about 20%–25% of the total energy used in the food processing industry. Energy efficiency and product quality are two key factors in the processing and drying of bio-products, chemicals, pharmaceuticals, etc.

This book demonstrates the benefits of ID, which has been considered as one of the most promising solutions for improving energy efficiency and product quality without increasing the capital cost of the drier and therefore has received much attention recently. ID is achieved by applying energy intermittently which is very different from continuous application of energy in the conventional drying method. Intermittency can also be achieved by varying the mode of heat input (e.g. convection, conduction, radiation or microwave/radio frequency heating). Multiple heat inputs can be used to remove both surface and internal moisture simultaneously, which can reduce the drying time substantially.

Continuous use of energy during drying results in wastage of huge amounts of energy and tends to produce poor-quality products. Therefore, ID is useful in minimizing energy usage, which allows the drying product to relax during the tempering period. With regard to ID, the right level of intermittency and the proper selection of processes are vital in achieving optimum energy efficiency while producing good-quality dehydrated products.

M. Azharul Karim

Chung-Lim Law

The MathWorks, Inc.
3 Apple Hill Drive
Natick, MA, 01760-2098 USA
Tel: 508-647-7000
Fax: 508-647-7001
E-mail: info@mathworks.com
Web: www.mathworks.com

Series Preface

ADVANCES IN DRYING SCIENCE AND TECHNOLOGY

Series Editor: Professor Arun S. Mujumdar

It is well known that the unit operation of drying is a highly energy-intensive operation encountered in diverse industrial sectors, ranging from agricultural processing, ceramics, chemicals, minerals processing, pulp and paper, pharmaceuticals, coal polymer, food, forest products industries as well as waste management. Drying also determines the quality of the final dried products. The need to make drying technologies sustainable and cost effective via application of modern scientific techniques is the goal of academic as well as industrial R&D activities around the world.

Drying is a truly multi- and interdisciplinary area. Over the last four decades, the scientific and technical literature on drying has seen exponential growth. The continuously rising interest in this field is also evident from the success of numerous international conferences devoted to drying science and technology.

The establishment of this new series of books entitled Advances in Drying Science and Technology is designed to provide authoritative and critical reviews and monographs focusing on current developments as well as future needs. It is expected that books in this series will be valuable to academic researchers as well as industry personnel involved in any aspect of drying and dewatering.

The series will also encompass themes and topics closely associated with drying operations, for example mechanical dewatering, energy savings in drying, environmental aspects, life cycle analysis, technoeconomics of drying, electrotechnologies, control and safety aspects, and so on.

Series Editor

Professor Arun S. Mujumdar is an internationally acclaimed expert in drying science and technologies. He is the founding chair in 1978 of the International Drying Symposium (IDS) series and editor-in-chief of *Drying Technology: An International Journal* since 1988. The fourth enhanced edition of his *Handbook of Industrial Drying* published by CRC Press has just appeared. He is the recipient of numerous international awards, including honorary doctorates from Lodz Technical University, Poland, and University of Lyon, France.

Please visit www.arunmujumdar.com for further details.

Editors

Dr. M. Azharul Karim is a senior lecturer in mechanical engineering at Queensland University of Technology, Australia. He earned his BSc in mechanical engineering from Chittagong University of Engineering and Technology, Bangladesh, and masters (by research, mechanical) from National University of Singapore. He received his PhD from Melbourne University in 2007. Through his scholarly, innovative, high-quality research, he has established a national and international standing. His excellence in research has been demonstrated by the development of many innovative, new products; 2 international patents; 142 high-quality refereed publications (including 57 journal papers, 2 books and 12 book chapters); 13 research grants amounting to A$3.1 million, including one highly competitive Advanced Queensland fellowship; being invited by reputed universities for seminars and, finally, by the establishment of national and international research collaboration. Dr Karim's innovative product development capability is well recognized. He has developed ultrasonic washing machines and ultrasonic dishwashers (both patented), highly efficient v-grove solar thermal collectors for drying applications, stratified thermal storage for air-conditioning systems, intermittent microwave-assisted convective dryers and intermittent microwave heat pump dryers. Dr Karim's papers are highly cited. In last 5 years, 12 of his papers have attracted approximately 1200 citations. Dr Karim serves as an assistant editor for *Drying Technology*, a journal published by Taylor & Francis. He is also an editorial board member of the *International Journal of Food Properties*.

Professor Chung-Lim Law, PhD, is professor of chemical and process engineering at the University of Nottingham Malaysia Campus. He serves as head of department in the Department of Chemical and Environmental Engineering. He received his bachelor's and master's degrees from the National University of Malaysia and his PhD from the same university under the supervision of Professor Wan Ramli Wan Daud. He has worked on fluidized bed drying technology for many years and has extended his interest in drying research to heat pump-assisted drying and low-temperature drying. His recent research activities include drying of herbal products using heat pump-assisted solar drying and heat pump-assisted drying and edible birdnest drying using low-temperature drying technique. He has recently completed an industrial project in fabricating and installing industrial-scale heat pump drying systems for the drying of medical-grade filter paper. Chung Lim serves as an associate editor for *Drying Technology*, a journal published by Taylor & Francis. He is an assessor in accrediting chemical engineering programmes in Malaysia. He also serves the Honorary Secretary of IChemE (Institution of Chemical Engineers) Malaysian Board and the Chair of the Malaysia Professional Formation Forum.

Chung Lim has more than 17 years of experience in R&D, especially in the areas of industrial drying and fluidized bed. He has extensive experience in various areas related to these fields, namely, diffusion, product quality such as colour change, physical appearance, textural properties, retention of bio-active ingredients, product safety, powder characteristics, handling of powder, powder mixing and extraction. Further, he has designed a number of industrial dryers, including fluidized bed dryers, low-temperature dryers and intermittent dryers for various types of products. Some of the dryers have been applied in the food processing industry.

Contributors

A.M. Nishani Lakmali Abesinghe
Department of Animal Science
Uva Wellassa University of Sri Lanka
Badulla, Sri Lanka

Yehya Baradey
Department of Mechanical Engineering
International Islamic University Malaysia
Kuala Lumpur, Malaysia

Shu-Hui Gan
Department of Chemical and Environmental Engineering
University of Nottingham Malaysia Campus
Selangor, Malaysia

Sami Ghnimi
Food Science Department
United Arab Emirates University
Abu Dhabi, United Arab Emirates

Y.T. Gu
Queensland University of Technology
Brisbane, Queensland, Australia

MNA Hawlader
Department of Mechanical Engineering
International Islamic University Malaysia
Kuala Lumpur, Malaysia

Meftah Hrairi
Department of Mechanical Engineering
International Islamic University Malaysia
Kuala Lumpur, Malaysia

Ahmad Faris Ismail
Department of Mechanical Engineering
International Islamic University Malaysia
Kuala Lumpur, Malaysia

S.V. Jangam
Department of Chemical and Biomolecular Engineering
National University of Singapore
Singapore, Singapore

Mohammad U.H. Joardder
Rajshahi University of Engineering and Technology
Rajshahi, Bangladesh

M. Azharul Karim
Queensland University of Technology
Brisbane, Queensland, Australia

Md. Imran H. Khan
Queensland University of Technology
Brisbane, Queensland, Australia

and

Dhaka University of Engineering and Technology
Gazipur, Bangladesh

Stefan Jan Kowalski
Poznań University of Technology
Poznań, Poland

Chandan Kumar
School of Chemistry, Physics and Mechanical Engineering
Queensland University of Technology
Brisbane, Queensland, Australia

Chung-Lim Law
Department of Chemical and Environmental Engineering
University of Nottingham Malaysia Campus
Selangor, Malaysia

Md. Mahiuddin
Queensland University of Technology
Brisbane, Queensland, Australia

M.H. Masud
Rajshahi University of Engineering and Technology
Rajshahi, Bangladesh

Arun S. Mujumdar
Department of Chemical & Biochemical Engineering
Western University
London, Ontario, Canada

and

Bioresource Engineering
McGill University
Montreal, Quebec, Canada

Scott Muller
Innovative Climate Control (ICCON)
Archerfield, Queensland, Australia

Andrzej Pawłowski
Poznań University of Technology
Poznań, Poland

Tony Petley
Innovative Climate Control Pty Ltd
Archerfield, Queensland, Australia

Nghia Duc Pham
Department of Chemistry, Physics and Mechanical Engineering
Queensland University of Technology
Brisbane, Queensland, Australia

and

Department of Food Engineering
Vietnam National University of Agriculture
Hanoi, Vietnam

Aditya Putranto
School of Chemistry and Chemical Engineering
Queen's University Belfast
Belfast, United Kingdom

M.M. Rahman
Queensland University of Technology
Brisbane, Queensland, Australia

S.C. Saha
Queensland University of Technology
Brisbane, Queensland, Australia

Justyna Szadzińska
Poznań University of Technology
Poznań, Poland

Yuchuan Wang
School of Food Science and Technology
Jiangnan University
Wuxi, Jiangsu, People's Republic of China

R. Mark Wellard
School of Chemistry, Physics and Mechanical Engineering
Queensland University of Technology
Brisbane, Queensland, Australia

Min Zhang
School of Food Science and Technology
Jiangnan University
Wuxi, Jiangsu, People's Republic of China

1 Developments in Intermittent and Non-Stationary Drying Technologies

S.V. Jangam and Arun S. Mujumdar

CONTENTS

1.1 INTRODUCTION

Drying is a commonly used industrial unit operations where a solid, semi-solid or liquid feed material is converted into a solid product by evaporation of liquid using heat. Drying is widely used in numerous industries. For example, the food, pharmaceuticals, specialty chemicals, wood and paper industries use drying as their final production steps. As a result of its widespread applications, the study of drying has attracted significant attention from both industry and academia. This can be seen from a huge number of research articles published and from the success of conferences such as the *International Drying Symposium* (IDS) and the dedicated journal *Drying Technology* (DRT).

One of the most important facets of drying is the selection of an appropriate drying technique for a given application. For the drying operation, although seemingly simple, there are a number of ways one can remove the moisture from the wet material. It has been reported in the literature that there are over 400 types of dryers available. Therefore, selecting an appropriate dryer for a given application is itself a big task. Traditionally, most techniques have relied on adding thermal energy to vaporize water within the substrate. The heat may be supplied by convection (direct dryers), by conduction (indirect dryers), radiation or volumetric heating using microwave (MW).

The indirect dryers are found to be the most energy efficient as less amount of heat is lost with the exhaust gas. The examples of some commonly used dryers are spray, belt, fluidized bed, tray and rotary dryers. Dehydration can also be induced by freezing a drying substrate and then allowing the frozen liquid to sublime. This is the idea behind freeze drying, where water is allowed to sublime after freezing at its triple point. This can enhance the product quality substantially but with considerable additional cost. There are several advanced techniques available and the choice solely depends on the user's requirements. However, one should always remember that the wrong choice of dryer will result in compromise in energy efficiency and final product quality (Jangam, 2011; Mujumdar, 2014).

Drying is a complex process involving transient heat and mass transfer. The wrong choice of drying conditions can result in unacceptable product quality. The drying of wet material occurs by two simultaneous processes: removal of surface moisture by evaporation and transfer of internal moisture to surface. The rate at which surface moisture is removed depends on external conditions such as temperature, humidity and surface area available for drying. On the other hand, the internal moisture diffusion is a function of temperature, instantaneous moisture content and the physical nature of the solids. Hence, the drying conditions used largely determine physical and structural changes to the dried material (such as case hardening, shrinkage, puffing, crystallization and glass transition). The slower drying rates mean long exposure of wet material to the drying conditions. This may damage colour (chemical and biochemical reactions) and nutritional properties, especially in food, pharmaceutical and bioproducts. In case of most of the food products, these quality changes are common as most of the moisture present is bound moisture; hence, the drying occurs in a falling rate period. This happens as a result of excessive moisture gradient from the interior to the surface, especially during final stages of drying. This results in overdrying and shrinkage, which cause structural damage.

There have been a large number of attempts to tackle the aforementioned difficulties in drying. Some researchers have tried to improve the existing dryers, either by modifying dryer design to improve energy efficiency, by changing the operating conditions in existing dryers to improve product quality or by using an appropriate pre-treatment for quality improvement (especially in case of food products). For example, the commonly used drying systems in food and agriculture sector, the solar dryer systems, have greatly evolved over the years, including improved design of solar collectors (Du et al., 2012; Labed et al., 2012; Fudholi et al., 2015a,b; Hong et al., 2015), better designs of solar dryers (Bennamoun, 2013; Schiavone et al., 2013), improving the efficiency of solar drying systems using hybrid techniques, such as use of rotary desiccant (Kabeel and Abdelgaied, 2016) and heat pump with solar drying system (Wang et al., 2012; Sevik, 2014; Qiu et al., 2016). In case of fluidized bed drying, which is the most commonly used technique for particulate drying because of high heat and mass transfer rates, the developments have been the changes in the design of drying chamber for better mixing (Li et al., 2009; Park et al., 2010; Yuan et al., 2011), improvements in gas distribution system for efficient and uniform drying (Tao et al., 1999; Patel et al., 2008; Jangam et al., 2009; Liu et al., 2010), use of hybrid systems for enhancing the drying rates (Wang et al., 2002; Jangam, 2011; Sivakumar et al., 2016) or use of immersed heat exchangers and other techniques for improving energy efficiency (Agraniotis et al., 2009; Aziz et al., 2010, 2011;

Lechner et al., 2014). Spray drying is another important industrial drying system for drying of liquid materials. The amount of energy used per kg of moisture removed in spray drying is very high. This is because of very high volume of gas and high gas temperatures used in spray drying. There have been numerous studies on improvement of spray drying energy efficiency as well as product quality. This is done either through improved designs of spray drying chambers (Southwell et al., 2001; Huang et al., 2003; Huang and Mujumdar, 2005, 2006), or by improved design of atomizer (Huang et al., 2006; Lampa and Fritsching, 2011; Verma and Singh, 2015).

On the other hand, some researchers have proposed a number of innovative and advanced drying techniques. Some examples are heat pump drying system (Perera and Rahman, 1997; Jangam and Mujumdar, 2012; Minea, 2013), superheated steam drying (Karimi, 2010; Devahastin and Mujumdar, 2014; Romdhana et al., 2015), swell drying (Mounir et al., 2011, 2012) and hybrid drying systems (Kowalski and Mierzwa, 2011; Kudra, 2012; Zhang and Jiang, 2014). The main goal in each of these techniques is to improve energy efficiency and produce final product with the best quality possible. Although some of these techniques can be very expensive, but used for some products which can attract a good market price (special food products and many pharmaceutical and biological products).

The use of intermittent drying is another recent innovative idea of great interest. By varying the airflow, drying air temperature, humidity, operating pressure, or the mode of heat input individually or together, the operating condition in drying process can be controlled, in order to reduce the operating cost, for example thermal and power input. The concept of intermittency can be used in any existing drying system; hence, various intermittent drying operations are tested at different scales (laboratory, pilot and industrial scales). The objectives of using intermittency is to improve the drying rate to reduce energy consumption and improve the product quality by reducing the exposure of product beyond its permissible temperature limit, while maintaining high moisture removal rate.

There are two ways of applying intermittent heat input profiles, subjecting the drying materials to intermittent heat input: time-varying flow of drying medium or use of cyclically varying operating pressure in the drying chamber. The purpose is to allow internal moisture to migrate to the material surface during non-active phase of drying, often termed the tempering period. Intermittent drying consists of two distinctive drying periods, namely, active drying and non-active drying. During the active period, heat input is supplied by the drying medium, while during the passive drying period, heat input is stopped. The two distinctive periods are carried out in an intermittent mode. Since water content on the surface is increased during the tempering period, the drying rate during the subsequent active drying is increased noticeably, which helps enhance the drying kinetics. This also evens out the moisture resulting in product with uniform moisture content. However, since the rate of drying is almost negligible during the passive period, the overall drying time may be longer, but it is often offset by the reduction in energy consumed and the better product quality achieved by avoiding exposure to high temperature and overdrying of the external surface of the product.

Another intermittent drying strategy is to apply stepwise change of operating conditions in order to minimize energy requirement. In general, the drying is faster during the initial phase (especially when the surface moisture is present) and any improvement in external factors will enhance removal of this surface moisture. As the surface moisture is removed completely, the overall drying rate is controlled by the internal moisture diffusion and the external drying conditions have limited effect on the drying rate. Typically, in case of food products mostly the internal moisture is present and the drying process is controlled by internal moisture diffusion. As such, one possible way to reduce energy loss is to gradually reduce the heat input to the materials along the drying process. In addition to this, the product being dried is exposed to high drying temperatures for a shorter time; hence, this results in less product shrinkage and less physical product damage. Figure 1.1 shows the possible ways of using intermittency. As mentioned earlier, one can also vary the mode of heat input (e.g. convection, conduction, radiation or MW/radio frequency heating) to achieve the intermittency. Multiple heat inputs can be used to remove both surface and internal moisture simultaneously, which can reduce the drying time substantially. It should be noted that intermittent heat input is not relevant to removal of surface moisture that is available for evaporation without any internal resistance to heat and mass transfer. For heat-sensitive materials like most bioproducts, as the moisture content drops it is necessary to lower the energy input to avoid overheating.

The concept of intermittent drying can be applied to various products, using different drying methods, including the use of a tray dryer, conveyor dryer, vibrating bed dryer and fluidized bed dryer. For batch drying, the intermittency can be cyclic or stepwise as discussed before, while for continuous dryers the use of intermittency could be along the length of the dryer. One should note that some continuous dryers are inherently intermittent in nature and this intermittency cannot be controlled. Figure 1.2 shows various possibilities of intermittency in commonly used batch and continuous dryers.

It is important to note that several commonly used dryers have built-in intermittency, although it is not a design variable. Solar drying is inherently intermittent since solar energy input varies with time. Rotary dryers operate with time-varying heat input; axially flowing drying medium provides heat to the falling wet particles,

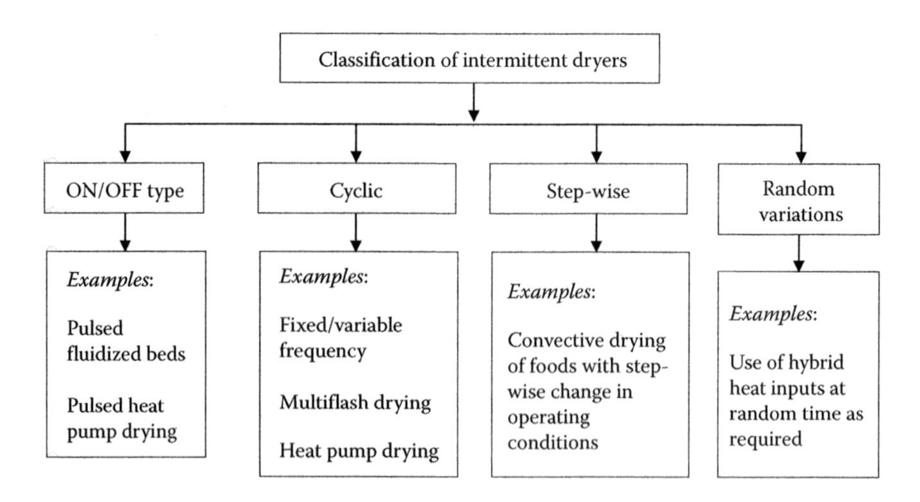

FIGURE 1.1 Classification of type of intermittency.

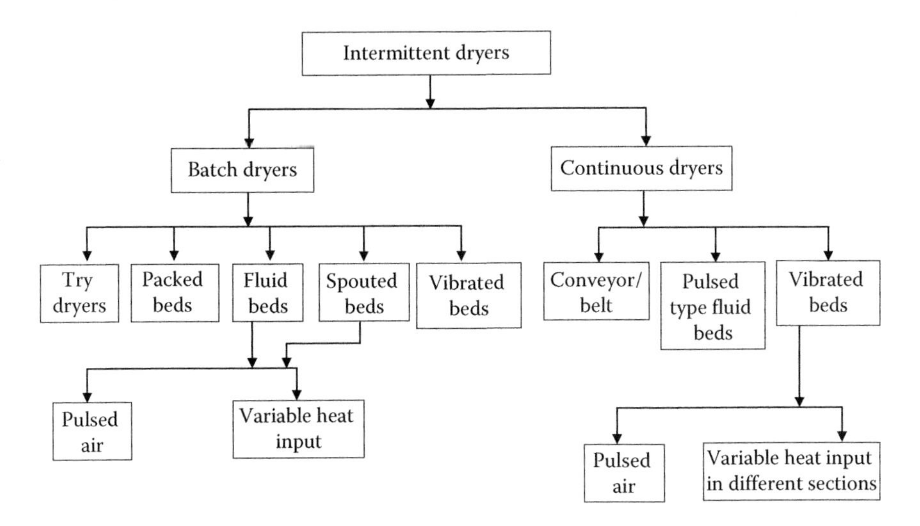

FIGURE 1.2 Types of intermittent modes for batch and continuous dryers.

while there is practically no heat input as bed of particles is carried to the top of the drum by internal flights. In spouted bed dryers, most of the heat input occurs in the spout region, while the duration of the particles in the downcomer zone acts as a "tempering" period during which internal moisture migrates to the surface before the particles are lifted upwards by entrainment in the spouting gas.

In addition to the changes in operating conditions and/or heat inputs, the intermittency can also include the flipping of the drying object itself to improve uniformity of drying for better product quality. The following sections will discuss the concepts and application of intermittency to various drying techniques and specific products.

1.2 USE OF INTERMITTENCY FOR BETTER PRODUCT QUALITY

The product quality is of utmost importance for food and pharmaceutical products. Almost all food, pharmaceutical and biological products are highly heat-sensitive. The use of high temperature in case of fruits and vegetables can result in degradation of colour, texture and nutritional value; in case of herbs and spices the use of high temperatures can result in loss of aroma; for pharmaceutical products the use of inappropriate drying conditions can result in degradation of active pharmaceutical ingredient or undesired product (polymorphism, glass transition); while, in case of biological products high temperature and longer drying times can result in degradation of useful quality attributes. Hence, it is necessary to maintain the product temperature and reduce the exposure of material to hot drying medium. The homogenous moisture content is also an important aspect of drying in the aforementioned products to reduce local microbial attack and also to attract good market value. Besides the food and pharmaceutical products, the uniformity of drying and heat sensitivity is also important in case of products like ceramic and wood. The use of intermittency—as discussed before—can help overcome many of these issues. There is a considerable amount of work done to understand the idea of intermittency. Table 1.1 is

TABLE 1.1
Selected Applications of Intermittent Drying for Quality Improvement

Product Dried with Method	Detailed Observations	Reference
Intermittent drying of rough rice	Various tempering intervals and drying intervals were considered for deep-bed drying of rough rice. A shorter drying time interval and a longer tempering time interval are found to be a preferable choice in recirculating-type rice dryer.	Shei and Chen (1999)
Intermittent drying of carrots in a batch vibrated fluid bed	The kinetics of degradation of β-carotene for various temperature conditions was studied. Tempering-intermittent drying was found to reduce the degradation of β-carotene and shorten the drying time.	Pan et al. (1999)
Drying kinetics of olive cake using intermittent heat input in a tray dryer	The effect of air temperature and intermittent heat input on drying rate (diffusion rates) and material moisture content was studied in small-scale tray dryer for both continuous and intermittent (on/off) modes. Providing longer tempering resulted in more moisture levelling, higher initial moisture removal in each active drying period, and the most effective energy utilization as a consequence.	Jumah et al. (2007)
Drying of banana chips using intermittent low pressure superheated steam and vacuum drying	The effect of intermittent heat input and vacuum on drying kinetics and quality parameters was studied. Although the overall drying rates in intermittent drying were not significantly different than the continuous method, the net drying time in intermittent conditions was shorter, resulting in energy savings. In general, the intermittently dried material had comparable or better quality attributes compared to continuous one.	Thomkapanich et al. (2007)
Non-stationary drying conditions for convective drying of carrot in a cabinet dryer	The effect of different frequencies and amplitudes of the periodically changeable drying air temperature on the quality of dried carrot (colour change, water activity, retention of β-carotene) was investigated. The non-stationary drying conditions helped improve the quality attributes.	Kowalski et al. (2013)
Beetroot drying using convective dryer	The effect of stepwise change of air temperature and pre-treatment methods such as osmotic dehydration and blanching on the material quality was investigated. The intermittent conditions resulted in shorter drying time, higher retention of betanin, better colour preservation and less water activity.	Kowalski and Szadzińska (2014)

(Continued)

TABLE 1.1 (*Continued*)
Selected Applications of Intermittent Drying for Quality Improvement

Product Dried with Method	Detailed Observations	Reference
Mathematical modelling of intermittent microwave convective drying of food materials	A mathematical modelling of intermittent microwave convective drying was carried out and validated with the experimental data. The results showed that the interior temperature of the material was higher than the surface and the temperature fluctuated and re-distributed due to intermittency of microwave power. This significantly improves the product quality.	Kumar et al. (2016)
Deep bed wheat drying	A fixed bed drying method featuring swing temperature and alternating flow was investigated. The wheat drying uniformity with tempering was higher by 2.55%, with an improvement rate of 58.7% compared to one without tempering. The wheat drying uniformity was found to improve with increasing air velocity in both drying and tempering processes.	Jia et al. (2016)

the compilation of the use of intermittent drying for quality improvements. Chua et al. (2003) have published a comprehensive review on the intermittent drying of bioproducts. They have discussed several dryer types using intermittency for energy reduction and better product qualities. Kumar et al. (2014) and Jangam (2011) have discussed in detail the application of intermittent drying of foods. Kumar et al. (2014) have focused on energy efficiency, quality improvement and modelling aspect of intermittent drying and they have covered many applications in food sector. Barbosa de Lima et al. (2016) have published a book chapter which also describes in detail the fundamentals of intermittent drying with more focus on mathematical modelling part.

1.3 USE OF INTERMITTENCY FOR REDUCING ENERGY

As discussed earlier, the other important objective of using intermittent drying conditions and heat inputs is to reduce the energy consumption in drying, which, in general, is very high. Many industrial dryers are poorly designed or the operating conditions used in the drying process may not be appropriate and necessary; hence, there could be a huge scope to alter the drying conditions, which can save a lot of energy. Considering the importance of sustainability in recent years, any reduction in energy consumption will be attractive. This section discusses the use of intermittency in commonly used dryers.

1.3.1 Intermittency in Fluidized Bed Dryers

Fluidized beds are widely used for several applications in chemical and process industries. The fluidized beds are considered to be the most effective way of drying the particulate materials. According to the Geldart classification (Yang, 2003), the conventional fluidized beds are best suited for drying of Type A and B particles (particles size in the range of 50 μm to 2 mm). However, some industries may need the drying of Type C (ultrafine) and Type D (larger than 2 mm) particles and the existing fluidized beds are not capable of handling these types of particles. In addition, many applications may need handling a wide particle sizes in a single fluidized bed dryer. The use of conventional fluidization methods may result in non-uniform drying or longer drying times. As observed in other dryers, fluidized bed dryers also need improvements to reduce the energy consumption. Although some attempts have been made to improve the fluidization quality in such scenarios, for example use of mechanical agitation, however, the concept of intermittency can be effectively used to address some of the aforementioned issues. The pulsed fluidized beds work by stopping and starting the fluidizing gas flow. The pulsed flow can be either applied to all the flow or part of the flow, also termed as partially pulsed (Ireland et al., 2016). There are various ways how the pulsation can be achieved; readers can refer to the relevant literature for the detailed discussion on type of pulsation and the design of pulsation devices (Ireland et al., 2016).

It has been observed in several studies that the fluidization quality improves substantially as a result of using pulsed flow compared to continuous flow in fluidized beds (Nitz and Taranto, 2007, 2009). The heat and mass transfer also improved in case of pulsed fluid beds. Several authors have demonstrated in their study that heat transfer coefficients for pulsed flow are significantly higher than that for conventional flow at the same superficial fluidizing gas velocity (Bhattacharya and Harrison, 1976; Zhang and Koksal, 2006).

There have been several applications of pulsed fluidized beds. Gawrzynski and Glaser (1996) studied pulsed fluid bed drying technology for drying of heat-sensitive materials. They investigated the flow and heat and mass transfer during drying of materials such as granulated sugar, beans, seeds, wheat, carrot and onion. De Souza et al. (2010) used the pulsed fluidized bed dryer for sodium acetate and found that the intermittent flow helps to break the agglomerates and provides better contact between the drying gas and particles. The drying rates were higher in pulsed fluidized bed compared to conventional fluidization (where the energy consumption was 2.5 times higher). Niamnuy et al. (2011) studied the hydrodynamic characteristics of pulsed spouted bed for several food products (soybean, peanut and French bean). It was observed that the maximum spouting height in pulsed conditions was more than conventional spouted bed, which can help with the drying process. Reyes et al. (2010) studied the atmospheric freeze drying of carrot particles in pulsed fluidized bed with variable temperature and found reduction in drying time. In general, pulsed fluidized bed dryers can help improve the product quality as well as reduce energy consumption as the drying time is reduced.

1.3.2 Intermittency in Heat Pump Drying

Heat-pump-assisted dryer is one of the most important novel drying techniques used for quality improvement as well as for reducing energy usage. It consists of two main components: a heat pump system (mechanical/chemical or other type) and a convective dryer. A detailed discussion on heat pump drying can be found elsewhere (Jangam and Mujumdar, 2012). Although this drying technique is essentially developed for quality improvement and reducing energy usage, further enhancement can be possible by varying the operating conditions. For the heat pump drying systems, the intermittent drying process can be categorized into four variations: airflow, air temperature, air humidity and external heat inputs (e.g. infrared and microwave). All of this can be achieved in single equipment with simple modifications and control system. One such system is proposed by Islam et al. (2003). They carried out computational analysis of varying heat input in convective dryers and suggested various routes to enhance the drying. The drying system with multiple modes of heat transfer and option for variable operating conditions is shown in Figure 1.3 (Islam and Mujumdar, 2008; Jangam and Mujumdar, 2012).

Among the possible variations listed before, the intermittent regulation of air temperature is considered to have the most substantial influence on the product drying kinetics and various quality parameters. Fatouh et al. (1998) carried out intermittent drying of onions in heat pump dryer and compared the performance with continuous heat pump drying. They found 30% reduction in energy consumption when intermittency was used. Ong and Law (2011a,b) carried out constant and intermittent heat pump drying of salak fruit. It was observed that the intermittent mode can maintain the product quality with reduced energy consumption. Zhao and Yang (2012) studied the heat pump intermittent drying of cabbage seeds and found that the use of intermittency provides the cabbage seeds with more uniform moisture content and nearly 50% reduction in energy consumption. In another study, Zhu et al. (2016) studied the effectiveness of intermittent drying with constant and time-variant intermittency ratio compared with continuous drying. Intermittent drying was accomplished by periodically changing the heat input and air velocity. The specific moisture extraction rates of intermittent drying were 1.21–4.94 and 0.93–5.40 times more than the

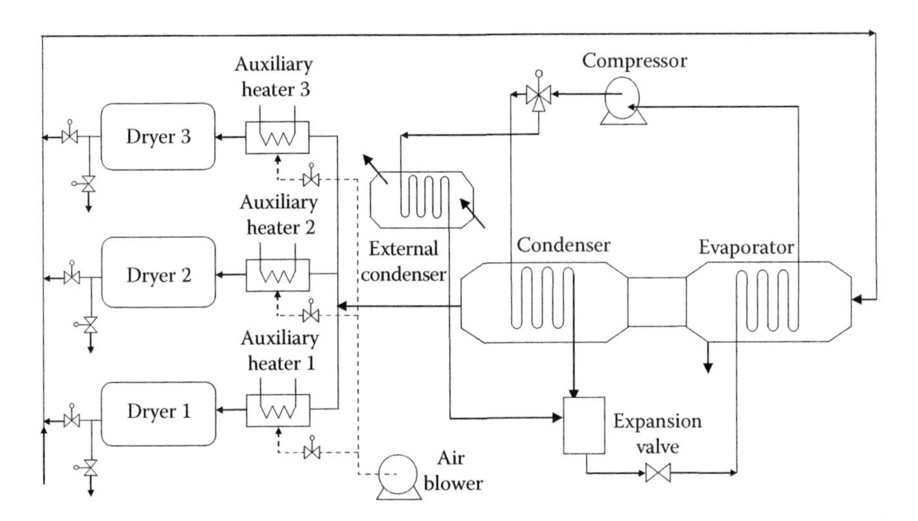

FIGURE 1.3 Intermittent heat pump drying system with arrangement for all possible variations. (From Jangam and Mujumdar, 2014.)

continuous drying for different initial moisture contents (23% and 28%), respectively. In one more important example, Gan et al. (2017) studied the influence of temperature and relative humidity (RH) during intermittent heat pump drying of edible bird's nest. The results showed the intermittent drying at optimized conditions greatly reduced effective drying time by 84.2% and colour change compared to fan drying, and retained good energy efficiency.

1.3.3 Intermittent Microwave Drying

As discussed before, during drying of any material it is very important to enhance the drying rate in order to prevent thermal damage of the product. It was also discussed in previous sections that it is necessary to have a dried product with uniform moisture content for longer shelf life and also to get better market value for the product. The use of MW is the most appropriate option to address these issues as MW provides a volumetric heating resulting in homogenous dried product (Mujumdar, 2014; Deepika and Sutar, 2015). The internal heat generation in MW drying rapidly generates vapours inside the material. This creates the pressure difference, resulting in the mass transfer (Mujumdar, 2014). The higher the moisture content, the greater is the pressure difference, leading to very rapid drying without causing any surface overheating of the product. The intermittent use of MW heat source in convective drying can significantly help address some of the difficulties discussed before.

Chen et al. (2001) carried out a theoretical study of MW heating patterns on batch fluidized bed drying of porous material. The intermittent MW heating was found to have high energy consumption but shortest drying time. Wang et al. (2002) also made similar observations in MW-assisted fluidized bed drying of carrots. They observed that changing the MW input pattern from uniform to intermittent mode can prevent material from overheating under the same power density. It was also observed that supplying more MW energy at the beginning of drying can increase the utilization of MW energy while keeping temperature low within the particle. Chua and Chou (2005) compared the intermittent MW and infrared (IR) drying of potato and carrot. The experimental comparison of the drying kinetics showed intermittent MW drying to be an effective method with shorter drying time to reach a certain moisture content compared with intermittent IR drying and convective-MW drying. A comparison of the colour degradation of the product showed that step variation of the IR intensity during intermittent drying reduced the colour change of the product less than constant radiative intermittent IR drying.

In another study, Soysal et al. (2009) studied the intermittent MW convective drying of red pepper compared with convective drying. The comparison was done based on the drying rate, physical properties and sensory qualities. Although the drying time for intermittent MW drying was longer than the continuous MW drying, the intermittent method resulted in good product quality compared to convective air drying. Recently, Zhao et al. (2014) showed the used of intermittent MW drying of carrot as it showed the lowest drying time with relatively low energy consumption and provided the best quality of final products with the best colour appearance, highest rehydration ratio and highest α- and β-carotene contents; while Lv et al. (2015) successfully used intermittent MW fluidized bed drying for ginger. They observed that the damage of the microstructure during microwave fluidized drying (MFD) was severe; however, in the intermittent drying the MW's interference on polar molecules was weakened, and the products' microstructure was better than that from continuous MFD. In general, the intermittent use of MW in convective drying can substantially help to reduce the energy consumption and improve the quality attributes.

1.3.4 Variable Pressure Methods

Another way of using the concept of intermittent drying is varying the pressure in the drying system. Drying can be carried out by using either a controlled instantaneous pressure drop method or a cyclic pressure drop method (Mounir et al., 2012; Allaf and Allaf, 2014). In a standard controlled instantaneous pressure drop (Déetente Instantanée Contrôlée, "DIC") treatment, the product is subjected to high pressure, generally using gas or steam, while temperature is maintained by one of the various means available. This is followed by an instantaneous drop in pressure, usually assisted by vacuum technologies, which leads to auto-vaporization of water and sometimes volatiles. This process can be easily controlled and results in very good product quality, especially with respect to the product expansion ratio and porosity and also in the enhancement of product texture. Swell drying is a combination of hot air drying and the DIC texturizing process. The reduction in drying time, using swell drying, results in the significant enhancement of product quality. A second stage of air drying can be used to increase both performance and final product quality (Albitar et al., 2011). This air drying stage can be operated intermittently. In a dehydration process, followed by successive decompression (Déshydratation par Détentes Successives, "DDS"), the thermally sensitive product undergoes a series of cycles, during which it is placed at a particular pressure for a defined time, followed by an instantaneous drop in processing pressure (Allaf, 2009; Allaf and Allaf, 2014). The product is maintained at lower pressures for some time, before the following cycle begins; each decompression step results in partial removal of water by auto-vaporization. The amount of water removed depends on various factors, such as operating conditions and the type of water present in the material dried.

The major advantage of these techniques is the considerable reduction in drying time that results compared to normal vacuum drying or hot air drying techniques used for food products. The reduction in drying time always helps control the physical properties of dried products. The techniques of instantaneous and cyclic pressure drop have been successfully applied to various food products, including grains, fruits, vegetables and fish. The DIC method is specifically used to tackle the problem of product shrinkage.

1.4 FLIPPING OF DRYING OBJECT—A NEW CONCEPT

Apart from improving the drying rate, uniformity of drying is an equally important criterion for the drying process. While many options discussed so far mainly make use of varying the operating conditions and/or modes of heat transfer to improve the drying rates and quality, only few of them focus on uniformity of drying. Most drying processes involve blowing of the drying air unidirectionally, mostly

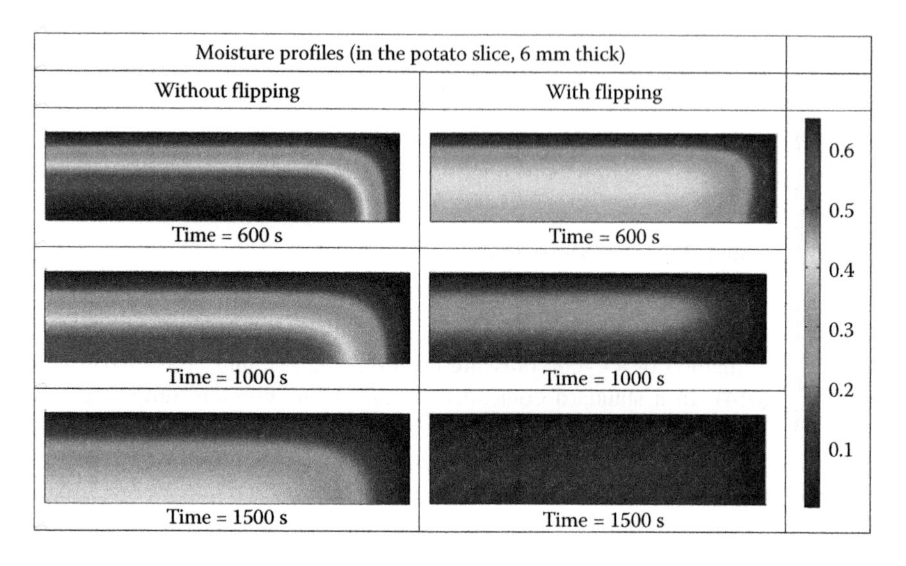

FIGURE 1.4 Selected results for drying of potato slice with "flipping."

perpendicular to the surface. As such, there is a possibility that non-uniform drying occurs as only one side of the surface is dried, while the inside and bottom of the material remain wet as compared to the surface exposed to the drying medium. This may result in shrinkage of the surface or product degradation, which may not be ideal for the drying process. In order to improve the drying process such that more uniform drying can be achieved, one can use the concept of "flipping" the drying object itself. This can be done with different time cycles so that each time an alternate surface is exposed to the drying air. This process can ensure that the dried product has more uniform moisture content. In our recent numerical study on the impingement drying of potato slices, the idea of flipping was successfully implemented (Tan et al., 2017). The results (Figure 1.4) show a substantial improvement in uniformity of moisture content in drying with flipping compared to that without flipping at same operating conditions. Although promising, this idea needs more research to implement in real dryers.

1.5 CONCLUDING REMARKS

In this chapter, a short introduction to the novel concept of intermittent drying is provided. The common idea behind using intermittency is to improve the drying kinetics, reduce the energy consumption per unit kg of moisture removed and improve the product quality. Intermittency in batch drying maybe introduced by changing the operating conditions (such as flow rate of drying gas, drying temperature, humidity) or by varying heat input (convection, conduction, radiation or volumetric) at selected time intervals in order to achieve the required targets. The use of different heat transfer modes can be applied sequentially or simultaneously.

In this chapter, we have discussed various examples of intermittent drying, including applications to various products for quality improvement, and examples of specific drying techniques such as pulsed fluidized beds, heat pump drying, microwave drying and swell drying. In general, the research carried out so far demonstrates that intermittent drying is an effective method; however, because of the complexity of the drying process the understanding about intermittent drying process cannot be generalized. There is still a lot of scope to carry out experimental and numerical study on the application of intermittent drying concept to different products and in different drying systems.

REFERENCES

Agraniotis, M., Stamatis, D., Grammelis, P., and E. Kakaras. 2009. Numerical investigation on the combustion behaviour of pre-dried Greek lignite. *Fuel* 88(12):2385–2391.

Albitar, N., Mounir, S., Besombes, C., and K. Allaf. 2011. Improving the drying of onion using the instant controlled pressure drop technology. *Drying Technology* 29(9):93–101.

Allaf, K. 2009. The new instant controlled pressure-drop DIC technology. In *Essential Oils and Aromas: Green Extraction and Applications*, ed. F. Chemat, pp. 85–121. Har Krishan Bhalla & Sons, New Delhi.

Allaf, T. and K. Allaf. 2014. *Instant Controlled Pressure Drop (DIC) in Food Processing: From Fundamental to Industrial Applications*. Springer, New York.

Aziz, M., Fushimi, C., Kansha, Y., Mochidzuki, K., Kaneko, S., and A. Tsutsumi. 2010. Innovative high energy efficiency brown coal drying based on self-heat recuperation technology. *27th Annual International Pittsburgh Coal Conference 2010*, Pittsburgh, PA, Vol. 2, pp. 1450–1459.

Aziz, M., Kansha, Y., and A. Tsutsumi. 2011. Self-heat recuperative fluidized bed drying of brown coal. *Chemical Engineering and Processing: Process Intensification* 50(9):944–951.

Barbosa de Lima, A.G., Delgado, J.M.P.Q., Neto, S.R.F., and C.M.R. Franco. 2016. Intermittent drying: Fundamentals, modeling and applications. In *Drying and Energy Technologies*, Vol. 63, eds. Barbosa de Lima, A.G. and J.M.P.Q. Delgado, pp. 19–41. Springer International Publishing AG, Cham, Switzerland.

Bennamoun, L. 2013. Improving solar dryers' performances using design and thermal heat storage. *Food Engineering Reviews* 5(4):230–248.

Bhattacharya, S.C. and D. Harrison. 1976. Heat transfer in a pulsed fluidised bed. *Transactions of the Institution of Chemical Engineers* 54:281–286.

Chen, G., Wang, W., and A.S. Mujumdar. 2001. Theoretical study of microwave heating patterns on batch fluidized bed drying of porous material. *Chemical Engineering Science* 56(24):6823–6835.

Chua, K.J., Mujumdar, A.S., and S.K. Chou. 2003. Intermittent drying of bioproducts—An overview. *Bioresource Technology* 90(3):285–295.

Chua, K.J. and S.K. Chou. 2005. A comparative study between intermittent microwave and infrared drying of bioproducts. *International Journal of Food Science and Technology* 40(1):23–39.

Deepika, S. and P.P. Sutar. 2015. Microwave assisted hybrid drying in food and agricultural materials. In *Drying Technologies for Foods: Fundamentals and Applications (Part I)*, eds. Nema, P.K., Kaur, B.P., and A.S. Mujumdar, pp. 121–154. New India Publishing Agency, New Delhi, India.

De Souza, L.F.G., Nitz, M., Lima, P.A., and O.P. Taranto. 2010. Drying of sodium acetate in a pulsed fluid bed dryer. *Chemical Engineering and Technology* 33(12):2015–2020.

Devahastin, S. and A.S. Mujumdar. 2014. Superheated steam drying of foods and biomaterials. In *Modern Drying Technology*, Vol. 5, eds. Tsotsas, E. and A.S. Mujumdar, pp. 57–84. Wiley-VCH Verlag GmbH & Co., Weinheim, Germany.

Du, J., Yang, S., Liu, G., and Z. Guo. 2012. Improved design of air collector in solar energy forage drying equipment. *Transactions of the Chinese Society of Agricultural Machinery* 43(Suppl. 1):227–230.

Fatouh, M., Abou-Ziyan, H.Z., Metwally, M.N., and H.M. Abdel-Hameed. 1998. Performance of a series air-to-air heat pump for continuous and intermittent drying. *American Society of Mechanical Engineers. Advanced Energy Systems Division (Publication) AES* 38:435–442.

Fudholi, A., Sopian, K., Bakhtyar, B., Gabbasa, M., Othman, M.Y., and M.H. Ruslan. 2015a. Review of solar drying systems with air based solar collectors in Malaysia. *Renewable and Sustainable Energy Reviews* 51:1191–1204.

Fudholi, A., Sopian, K., Gabbasa, M., Bakhtyar, B., Yahya, M., Ruslan, M.H., and S. Mat. 2015b. Techno-economic of solar drying systems with water based solar collectors in Malaysia: A review. *Renewable and Sustainable Energy Reviews* 51:809–820.

Gan, S.H., Ong, S.P., Chin, N.L., and C.L. Law. 2017. A comparative quality study and energy saving on intermittent heat pump drying of Malaysian edible bird's nest. *Drying Technology* 35(1):4–14.

Gawrzynski, Z. and R. Glaser. 1996. Drying in a pulsed-fluid bed with relocated gas stream. *Drying Technology* 14(5):1121–1172.

Hong, L., Yuan, G., Li, X., Xu, L., Li, Z., and Z. Wang. 2015. Thermal performance investigation and improvement of solar air collector. *Acta Energiae Solaris Sinica* 36(2):467–472.

Huang, L., Kumar, K., and A.S. Mujumdar. 2003. Use of computational fluid dynamics to evaluate alternative spray dryer chamber configurations. *Drying Technology* 21(3):385–412.

Huang, L., Kumar, K., and A.S. Mujumdar. 2006. A comparative study of a spray dryer with rotary disc atomizer and pressure nozzle using computational fluid dynamic simulations. *Chemical Engineering and Processing: Process Intensification* 45(6):461–470.

Huang, L. and A.S. Mujumdar. 2005. Development of a new innovative conceptual design for horizontal spray dryer via mathematical modeling. *Drying Technology* 23(6):1169–1187.

Huang, L. and A.S. Mujumdar. 2006. Numerical study of two-stage horizontal spray dryers using computational fluid dynamics. *Drying Technology* 24(6):727–733.

Ireland, E., Pitt, K., and R. Smith. 2016. A review of pulsed flow fluidisation; the effects of intermittent gas flow on fluidised gas–solid bed behaviour. *Powder Technology* 292:108–121.

Islam, Md.R., Ho, J.C., and A.S. Mujumdar. 2003. Convective drying with time-varying heat input: Simulation results. *Drying Technology* 21(7):1333–1356.

Islam, Md.R. and A.S. Mujumdar. 2008. Heat pump assisted drying. In *Drying Technologies in Food Processing*, eds. Chen, X.D. and A.S. Mujumdar, pp. 190–224. Blackwell Publishing Ltd., West Sussex, U.K.

Jangam, S.V. 2011. An overview of recent developments and some R&D challenges related to drying of foods. *Drying Technology* 29(12):1343–1357.

Jangam, S.V. and A.S. Mujumdar. 2012. Heat pump assisted drying technology—Overview with focus on energy, environment and product quality. In *Modern Drying Technology*, Vol. 4, eds. E. Tsotsas and A.S. Mujumdar, pp. 121–162. Wiley-VCH Verlag GmbH & Co, Weinheim, Germany.

Jangam, S.V., Mujumdar, A.S., and B.N. Thorat. 2009. Design of an efficient gas distribution system for a fluidized bed dryer. *Drying Technology* 27(11):1217–1228.

Jia, C., Wang, L., Guo, W., and C. Liu. 2016. Effect of swing temperature and alternating airflow on drying uniformity in deep-bed wheat drying. *Applied Thermal Engineering* 106:774–783.

Jumah, R., Al-Kteimat, E., AL-Hamad, A., and E. Telfah. 2007. Constant and intermittent drying characteristics of olive cake. *Drying Technology* 25(9):1421–1426.

Kabeel, A.E. and M. Abdelgaied. 2016. Performance of novel solar dryer. *Process Safety and Environmental Protection* 102:183–189.

Karimi, F. 2010. Applications of superheated steam for the drying of food products. *International Agrophysics* 24(2):195–204.

Kowalski, S.J. and D. Mierzwa. 2011. Hybrid drying of red bell pepper: Energy and quality issues. *Drying Technology* 29(10):1195–1203.

Kowalski, S.J. and J. Szadzińska 2014. Kinetics and quality aspects of beetroots dried in non-stationary conditions. *Drying Technology* 32(11):1310–1318.

Kowalski, S.J., Szadzinska, J., and J. Łechtanska. 2013. Non-stationary drying of carrot: Effect on product quality. *Journal of Food Engineering* 118:393–399.

Kudra, T. 2012. Energy performance of convective dryers. *Drying Technology* 30(11–12):1190–1198.

Kumar, C., Joardder, M.U.H., Farrell, T.W., Millar, G.J., and M.A. Karim. 2016. Mathematical model for intermittent microwave convective drying of food materials. *Drying Technology* 34(8):962–973.

Kumar, C., Karim, M.A., and M.U.H. Joardder. 2014. Intermittent drying of food products: A critical review. *Journal of Food Engineering* 121(1):48–57.

Labed, A., Moummi, N., and A. Benchabane. 2012. Experimental investigation of various designs of solar flat plate collectors: Application for the drying of green chili. *Journal of Renewable and Sustainable Energy* 4(4):043116.

Lampa, A. and U. Fritsching. 2011. Spray structure analysis in atomization processes in enclosures for powder production. *Atomization and Sprays* 21(9):737–752.

Lechner, S., Merzsch, M., and H.J. Krautz. 2014. Heat-transfer from horizontal tube bundles into fluidized beds with Geldart A lignite particles. *Powder Technology* 253:14–21.

Li, Q.H., Zhang, Y.G., and A.H. Meng. 2009. Design and application of novel horizontal circulating fluidized bed boiler. *Proceedings of the 20th International Conference on Fluidized Bed Combustion*, Xian, China, pp. 206–211.

Liu, W., Gao, W.-Y., Yang, S.-X., Li, F., Wang, S.-F., and Y.L. Zhang. 2010. Drying intensity of fluidized bed with novel corrugated gas distributor. *Journal of Nanjing University of Science and Technology* 34(6):818–820.

Lv, W., Han, Q., Li, S., Zhang, X., Xu, T., and L. Sun. 2015. Thermal dynamic and physical qualities of ginger (*Zingiber Officinale*) slices in intermittent microwave fluidized drying. *International Agricultural Engineering Journal* 24(1):39–46.

Minea, V. 2013. Heat-pump-assisted drying: Recent technological advances and R&D needs. *Drying Technology* 31(10):1177–1189.

Mounir, S., Allaf, T., Mujumdar, A.S., and K. Allaf. 2012. Swell drying: Coupling instant controlled pressure drop dic to standard convection drying processes to intensify transfer phenomena and improve quality—An overview. *Drying Technology* 30(14):1508–1531.

Mujumdar, A.S. 2014. *Handbook of Industrial Drying*, 4th Ed. Taylor & Francis Group, Boca Raton, FL.

Niamnuy, C., Kanthamool, W., and S. Devahastin. 2011. Hydrodynamic characteristics of a pulsed spouted bed of food particulates. *Journal of Food Engineering* 103(3):299–307.

Nitz, M. and O.P. Taranto. 2007. Drying of beans in a pulsed fluid bed dryer: Drying kinetics, fluid-dynamic study and comparisons with conventional fluidization. *Journal of Food Engineering* 80:249–256.

Nitz, M. and O.P. Taranto. 2009. Drying of a porous material in a pulsed fluid bed dryer: The influences of temperature, frequency of pulsation, and airflow rate. *Drying Technology* 27:212–219.

Ong, S.P. and C.L. Law. 2011a. Drying kinetics and antioxidant phytochemicals retention of Salak fruit under different drying and pretreatment conditions. *Drying Technology* 29(4):429–441.

Ong, S.P. and C.L. Law. 2011b. Microstructure and optical properties of salak fruit under different drying and pretreatment conditions. *Drying Technology* 29(16):1954–1962.

Pan, Y.K., Zhao, L.J., Dong, Z.X., Mujumdar, A.S., and T. Kudra. 1999. Intermittent drying of carrot in a vibrated fluid bed: Effect on product quality. *Drying Technology* 17(10):2323–2340.

Park, J., Shun, D., Bae, D.-H., Lee, S., Seo, J.H., and J. Park. 2010. Drying kinetics of low rank coal in multi-chamber fluidized bed. *27th Annual International Pittsburgh Coal Conference 2010, PCC 2010*, Pittsburgh, PA, Vol. 3, pp. 2477–2480.

Patel, A.K., Waje, S.S., Thorat, B.N., and A.S. Mujumdar. 2008. Tomographic diagnosis of gas maldistribution in gas–solid fluidized beds. *Powder Technology* 185(3):239–250.

Perera, C.O. and M.S. Rahman. 1997. Heat pump dehumidifier drying of food. *Trends in Food Science and Technology* 8(3):75–79.

Qiu, Y., Li, M., Hassanien, R.H.E., Wang, Y., Luo, X., and Q. Yu. 2016. Performance and operation mode analysis of a heat recovery and thermal storage solar-assisted heat pump drying system. *Solar Energy* 137: 225–235.

Romdhana, H., Bonazzi, C., and M. Esteban-Decloux. 2015. Superheated steam drying: An overview of pilot and industrial dryers with a focus on energy efficiency. *Drying Technology* 33(10):1255–1274.

Schiavone, D.F., Teixeira, A.A., Bucklin, R.A., and S.A. Sargent. 2013. Design and performance evaluation of a solar convection dryer for drying tropical fruit. *Applied Engineering in Agriculture* 29(3):391–401.

Sevik, S. 2014. Experimental investigation of a new design solar-heat pump dryer under the different climatic conditions and drying behavior of selected products. *Solar Energy* 105:190–205.

Shei, H.-J. and Y.-L. Chen. 1999. Thin-layer models for intermittent drying of rough rice. *Cereal Chemistry* 76(4):577–581.

Sivakumar, R., Saravanan, R., Elaya Perumal, A., and S. Iniyan. 2016. Fluidized bed drying of some agro products—A review. *Renewable and Sustainable Energy Reviews* 61:280–301.

Southwell, D.B., Langrish, T.A.G., and D.F. Fletcher. 2001. Use of computational fluid dynamics techniques to assess design alternatives for the plenum chamber of a small spray dryer. *Drying Technology* 19(2):257–268.

Soysal, Y., Ayhan, Z., Eştürk, O., and M.F. Arikan. 2009. Intermittent microwave-convective drying of red pepper: Drying kinetics, physical (colour and texture) and sensory quality. *Biosystems Engineering* 103(4):455–463.

Tan, Y.M.K., Jangam, S.V., and A.S. Mujumdar. 2017. Innovative ways to improve the performance of impingement drying. Submitted to *Asia Pacific Drying Conference 2017*, to be held in China.

Tao, X., Chen, Q., Luo, Z., Liang, C., and G. Yang. 1999. Distribution regularity of moisture on coal surface and its effects on separation by an air-solid fluidized bed. *Journal of China University of Mining and Technology* 28(4):326–330.

Thomkapanich, O., Suvarnakuta, P., and S. Devahastin. 2007. Study of intermittent low-pressure superheated steam and vacuum drying of a heat-sensitive material. *Drying Technology* 25(1):205–223.

Verma, A. and S.V. Singh. 2015. Spray drying of fruit and vegetable juices—A review. *Critical Reviews in Food Science and Nutrition* 55(5):701–719.

Wang, Q., Wang, D., Du, J., and G. Zhai. 2012. Design and experiment of heat pump assisted solar energy heat-storage drying equipment for herbage seed. *Transactions of the Chinese Society of Agricultural Machinery* 43(Suppl. 1):222–226.

Wang, W., Thorat, B.N., Chen, G., and A.S. Mujumdar. 2002. Simulation of fluidized-bed drying of carrot with microwave heating. *Drying Technology* 20(9):1855–1867.

Yang, W.-C. 2003. *Handbook of Fluidization and Fluid-Particle Systems*. Marcel Dekker, New York.

Yuan, L.-Y., Zheng, Y.-P., Yang, A.-S., Sun, Q., and R. Cheng. 2011. Drying of group C particles in circulating fluidized bed dryer. *Chemical Engineering (China)* 39(6):50–54.

Zhang, D. and M. Koksal. 2006. Heat transfer in a pulsed bubbling fluidized bed. *Powder Technology* 168:21–31.

Zhang, M. and H. Jiang. 2014. Recent food drying R&D at Jiangnan University: An overview. *Drying Technology* 32(15):1743–1750.

Zhao, D., An, K., Ding, S., Liu, L., Xu, Z., and Z. Wang. 2014. Two-stage intermittent microwave coupled with hot-air drying of carrot slices: Drying kinetics and physical quality. *Food and Bioprocess Technology* 7(8):2308–2318.

Zhao, H. and Z. Yang. 2012. Variation of moisture content in cabbage seeds with heat pump intermittent drying. *Transactions of the Chinese Society of Agricultural Engineering* 28(11):261–267.

Zhu, Z., Yang, Z., and F. Wang. 2016. Experimental research on intermittent heat pump drying with constant and time-variant intermittency ratio. *Drying Technology* 34(13):1630–1640.

2 Effect of Periodic Drying on Energy Utilization, Product Quality, and Drying Time

Yehya Baradey, MNA Hawlader, Ahmad Faris Ismail, and Meftah Hrairi

CONTENTS

2.1 INTRODUCTION

The quality and safety of food materials, especially those rich in nutritional value such as fruits and vegetables, have recently received great attention from researchers worldwide and have become the most significant research topic in industries. This can be attributed to the harmful effects likely to occur to human life due to consumption of spoilt food stuff (Rahman, 1999, p. 19). Approximately one third of global food production is wasted annually due to the lack of proper food preservation processes (Kumar et al., 2014). According to Karim and Hawlader (2005), about 30%–40% of total yearly production of fruits and vegetables in Bangladesh is lost due to improper food preservation processing. Baini and Langrish (2007) reported that between 40% and 45% of annual production of banana is lost due to mold growth and postharvest spoilage in India and Brazil. Dehydration technologies are considered the most widely used techniques for food preservation. The most common method in ancient times was solar drying. Nowadays, different types of drying are available for commercial use in the market place, such as freeze drying, osmotic dehydration, convective drying, and vacuum microwave drying. The most widely used dehydration process is hot air drying due to its low comparative costs and the quality of its products. The main drawback of this technology is that a long time period is required to complete the drying process, as well as the hardening and darkening problems that occur to the dried product due to the high operating air temperature. Some researchers (Kowalski and Pawlowski, 2011; Golmohammadi et al., 2015) have reported that energy consumption related to this method needs to be minimized to make it more efficient and competitive. In order to overcome such difficulties, researchers have considered intermittent convective drying as an alternative and viable solution.

The applications of drying technologies are not only exclusive to food but are also extended to other materials that need to be dried such as wood, clay, cartons, and clothes. Drying technologies for food and other materials usually require a large amount of energy that sometimes accounts for up to 15% of all industrial energy usage, often with relatively low thermal efficiency in the range of 20%–25%, according to Chua et al. (2001). In this chapter, the effects of traditional hot air drying and intermittent drying on energy consumption, drying time, and product quality are summarized and compared.

2.2 CONVECTIVE DRYING METHOD

Convective or hot air drying refers to product dehydration in order to obtain specific properties from dried materials. It is a traditional drying method widely used in industry, because it is simple and economically feasible compared to other methods. Chemical and thermal treatments are usually required for the product prior to the drying process. A temperature range of 50°C–80°C is maintained for this kind of drying. It is easily operated under steady state conditions but usually it is not efficient and not applicable to all kinds of materials. Convective drying is able to provide dried product with satisfactory quality when operated under mild conditions. Using it under high temperature conditions can decrease the drying time but leads to lower quality. A combination of convective, microwave, and infrared drying techniques is considered an efficient way to improve the process (Hawlader et al., 2006; Kowalski and Pawlowski, 2011).

According to Qing-guo et al. (2006), the most widely used system for drying fruits and vegetables is the hot air drying method, but it still suffers from many drawbacks. It requires high temperature, which usually leads to many serious problems, such as reduction in the rehydration capacity and bulk density of the fruits and vegetables, reduction in nutritional value, and damage to the flavor and color of the products and biological properties of the fruits and vegetables (Baradey et al., 2015).

The convective drying method is still considered superior to other kinds of dehydration processes, which justifies its wide use in the food industry to date (Lewicki, 2006). It is considered one of the major food processing (drying) methods that helps in extending the shelf life of bio products, mainly fruits and vegetables. Compared to other methods it is still relatively inexpensive. Many factors have been given a lot of attention and considered significant in convective drying because they affect the quality of the final product and the drying kinetics. However, the most significant factors that impact the effectiveness of drying in convective drying are the product quality, drying, and energy consumption.

The hot air drying method is currently considered the most commonly used drying technique for dehydration of fruits and vegetables. The traditional convective drying process requires a long period of time (sometimes more than 3 h) to minimize or eliminate the moisture content of the product to the desired value. This is due to the value of operating temperature often used in the process, which plays an important role in affecting the rate of drying. Alibaş (2012) reported that the drying time reduced when drying temperature was increased. In his study, drying time was found to be 340, 270, and 220 min for operating drying temperatures of 55°C, 65°C, and 75°C, respectively. Therefore, it was increasing with increases in the thickness of the slices and the humidity level. However, compared to other dehydration techniques, such as freeze and vacuum drying, it provides the hardest texture appearance (Argyropoulos et al., 2008). Minimizing the negative effects of the hot air drying method on the final quality of the product has recently been intensively investigated by different researchers. One of the most attractive and effective methods to achieve that goal is intermittent (periodic) drying, which reduces the moisture gradient inside the dried materials as well.

2.3 INTERMITTENT DRYING

Intermittent drying is defined as a drying method where the conditions of drying such as amount of input of thermal energy, humidity, drying air temperature, and pressure can be changed and controlled with time (Kumar et al., 2014). Holowaty et al. (2012) defined it as "a drying method in which the heat is applied in a discourteous way." Changing the type of heat input or using more than one kind of heat input in drying (e.g., convection, conduction, radiation, or even microwave) can lead to intermittency in drying. From the definition, different intermittency techniques in drying can be applied and achieved, but the most common type that has been extensively studied by different researchers is changing drying air conditions. Choosing the proper type of intermittency in drying is a very important step and must rely on physics (heat and mass transfer) relevant to drying methods and properties of the product. Wrong intermittency will not lead to the expected minimization of energy or improvement in product quality. It has been reported that intermittency in drying is preferable for materials that are sensitive to crack formation such as wood and ceramics (Kowalski and Pawlowski, 2011). However, intermittency can be applied in different modes such as on/off mode, periodic variation mode, and ramp variation mode, as shown in Figure 2.1.

Intermittent drying technology differs from the conventional continuous drying methods where the product is placed inside the drying chamber and continuously exposed to heat until the desired amount of moisture is removed from the product. Applying the same amount of thermal energy to the product throughout the time interval of the drying process increases the cost and the energy consumption for drying. This procedure also negatively affects the quality of the product because the amount of moisture content of the product at the later stage of drying will be very small, which sometimes damages the surface and leads to degradation of quality. It has been reported that intermittent drying techniques improve the heat and mass transfer of the drying process, which leads to enhancement of the drying rate (Kowalski and Pawłowski, 2011). The philosophy behind using the intermittency technique is to increase the time

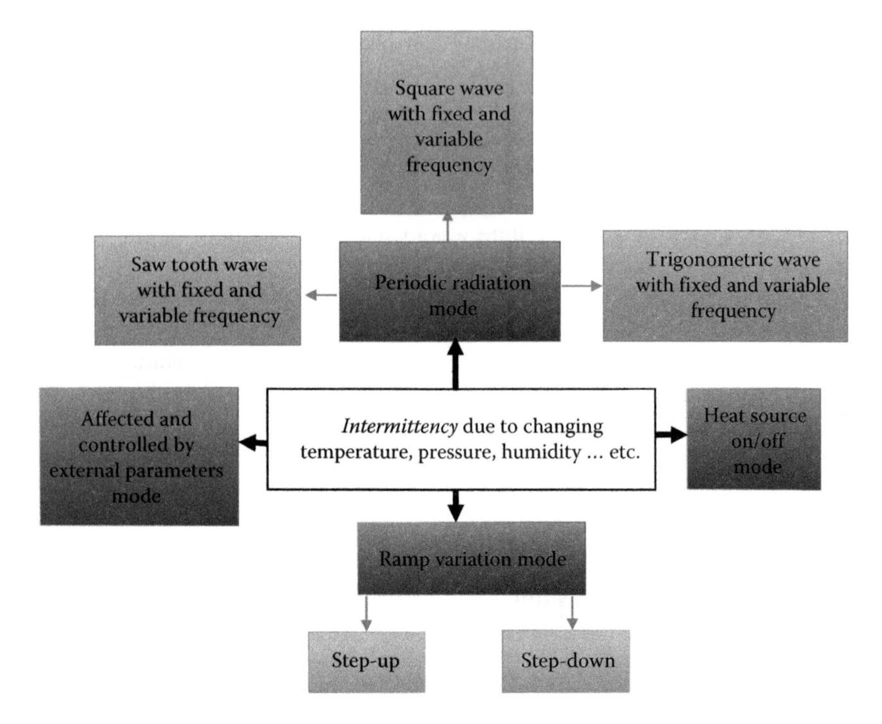

FIGURE 2.1 Different modes of intermittency.

interval required for the moisture content to be transferred and/or evaporated from the center to the surface of the product during the tempering period, which minimizes the degradation of quality and the damages that might occur during the drying process (Holowaty et al., 2012; Kumar et al., 2014; Tertoe 2010; Olalusi 2014).

2.4 QUALITY OF DRIED FOOD

The quality of dried food has been defined in different ways but the most popular one is "a combination of characteristics, attributes, or properties that give the commodity value as a human food" (Rahman, 1999). Judging the quality of dehydrated food could be totally different between researchers (including industry) and customers. From an economic point of view, customers are the only people who judge the quality of the final dried products. According to some researchers, customers include the whole industry series (e.g., farmers, packers, dispensers, sealers, retailers, managers, and shoppers). Instrumental measurements are preferred by researchers (e.g., electromagnetic and optical instruments for appearance, mechanical instruments for texture, chemical properties for flavor and aroma) because they eliminate the differences among customers, and are more precise and accurate.

The quality of food can be improved by taking some care and measurements in the preharvest stage, but can be improved to its highest level by drying in the postharvest stage. Preharvest factors (such as climatic factors), harvesting factors (such as maturity at harvest and harvesting methods factors), and postharvest factors (such as humidity, temperature, and postharvest diseases and infections factors) are very important issues influencing the level of food quality. Evaluation of the quality of dried food occurs based on studying the effect of the drying process on the physical, chemical, and biological parameters of the materials. Physical parameters are color, texture, shrinkage, porosity, rehydration ratio, and drying ratio. Chemical parameters are flavor, water activity, and shelf life. Biological parameters are antioxidants and nutritional values. Physical, chemical, and biological parameters are discussed in Section 2.4.

The physical properties of dried food play a crucial role in the acceptability of the product by customers. Customers prefer to see the original or almost original color of food. Unfavorable colors may lead to rejection of the product regardless of its taste and nutritional value (Hawlader et al., 2005). Changing of color occurs due to several factors, such as Maillard reactions, enzymatic browning, and color pigments. Detection and evaluation of the color of dried food is usually done by using destructive and nondestructive methods (Jangam et al., 2010). Level of browning is one aspect of color evaluation as it might affect acceptability as well. The browning issue negatively influences the physical and nutritional properties of the final product such as color, flavor, and softening. Browning in foods can occur through enzymic and nonenzymic (Maillard) reactions. The polyphenoloxidase group of enzymes catalyzes the oxidation of phenolic compounds in plants to o-quinones. Immediately, the quinines condense and react nonenzymatically with other phenolic compounds, amino acids, etc., to produce dark brown, black, or red pigments of indeterminate structure (Hawlader et al., 2004). Consequently, exclusion of oxygen and/or application of a low-pH environment can ease browning. So far, sulfating agents have been widely used in the drying industry, but these are considered harmful to health, which is unacceptable to some consumers. Therefore, lemon juice was used in this study as a natural inhibitor, as it contains plenty of citric acid (Hawlader et al., 2004).

One of the most important aspects of quality is texture, which is defined as the overall feeling or judgment that could be held about the dried food by taste or touch. Water or moisture removal from food during drying leads to changes in the overall texture and porosity of the product, which causes shrinkage. This occurs because the internal volume of the materials after drying decreases, which causes cracking in the structure of the final dried products and negatively affects the quality. Shrinkage level depends on the amount of water inside the product needing to be dried. The effect of drying on texture, porosity, and shrinkage and improving the textural properties of the materials have been studied by researchers. Rapid drying decreases the water content on the outer surface quickly to a very low level due to the fast evaporation process, which forms a rigid crust (also called a shell) on the outer surface of the final product. This happens because the shrinkage does not occur in a uniform way on all parts of the material in rapid dehydration conditions. Wang and Brennan (1995) confirmed the same results and observed formation of this shell on potato slices. They concluded that the level of shrinkage at high temperature drying is lower than at a low temperature. Level of shrinkage influenced the density and porosity of the potato. Surface cracking also occurred when the shrinkage of the outer surface of the dried product was not uniform along the surface. However, coupling the mass and heat transfer equations is the most common method used by different researchers for prediction of the cracking phenomenon (Baradey et al., 2015).

Chemical properties of food such as shelf life, water activity, and flavor have great effects on the quality of dried products. Changes that could occur, due to the drying process, to all or one of these issues might have a significant negative influence. Flavor is considered the most important attribute of chemical properties that has the greatest impact on quality. Sensory and chemical analysis are the two major methods used in evaluation of changes. Chemical analysis provides researchers with quantitative data, while sensory evaluation often depends on comparison between attributes of the dried product and the fresh one (Jangam et al., 2010). Storing the dried food products in a place where the level of oxygen is lower than 1% and CO_2 is higher than 8% is considered one of the most effective ways to maintain the quality of flavor at its highest value. This technique is usually used to avoid rancidity and other oxidation reactions (Perera, 2005). The percentage of water activity in dried food stuff must be less than 60% to prevent oxidation and enzymatic reactions. A higher percentage could increase the growth of microorganisms and bacteria, which is the main reason for spoilage. A place with relative humidity of 50%–70% is the best way to prevent bacteria from rapid growth and to increase the shelf life of the product.

The safety and quality of food also depend on biological attributes. Biological aspects related to microbial growth can lead to mold and yeast in dried materials, which most probably cause diseases in humans. Water content must be less than 0.65, which is the level at which mold starts, to prevent such problems. The shelf life of products is also influenced by these attributes. However, the government health sector usually provides standards and specifications on the biological properties of food.

2.5 EFFECT OF INTERMITTENT DRYING ON QUALITY OF THE PRODUCT

Special care must be taken with the quality of the final product, particularly for some sensitive materials which have high nutritional value such as biological materials, fruits and vegetables, and pharmacological agents. The aim of intermittent drying is to maximize the drying rate when crack formation is not possible and minimize it when crack formation starts to occur (Kowalski and Pawłowski, 2011). There is no consensus in the literature on the ability of intermittent drying to improve the quality of all materials. Some researchers claim that the intermittency technique only improves particular aspects of quality, while others have obtained different results for other aspects of quality. This could be due to the different materials used in their studies because the textural structure and water content are distinct from one product to another. But positive influences on other aspects of the drying process such as drying kinetics and energy consumption were observed in almost all of the studies conducted on dehydration of different materials.

Kowalski and Pawłowski (2011) studied the effect of intermittent drying on the quality of kaolin samples. Stationary fixed drying temperature tests were first conducted on the samples. The results showed that a strong crack occurred in the sample, which led to the transfer of the elastic energy accumulated in the material from the body to the surface of the sample. Drying with changeable air humidity showed the best product quality compared to stationary drying or variable air temperature. Szadzinska (2014) analyzed the effect of stepwise changing air temperature of the single-layer convective drying process on selected quality factors of green pepper such as color, Vitamin C, and water activity. Green pepper was cut into slices 8 cm long and 0.3 cm (please check dimension) thick with initial moisture content of 0.93 kg^{-1} wb. Excessive shrinkage and degradation in quality were observed due to the traditional drying process. The experimental results showed that about 88% of Vitamin C of dried green pepper was preserved, longer shelf life was achieved, water activity was decreased, color changing was between 32% and 48%, and the quality of the products was improved.

Another investigation on drying cherries, conducted by Kowalski and Szadzinska (2014), revealed that samples dried using stationary and intermittent drying at 60°C air temperature suffered from significant shrinkage and hardening. The study also showed that the effect of intermittent drying only led to poor quality of dried cherries. For that reason they proposed and experimentally tested another drying method prior to intermittent drying to improve the quality. The results of the experiments showed improvement in color, texture, and aroma for both intermittent and stationary methods preceded by osmotic and ultrasonic and osmotic drying. A remarkable reduction in water activity was also observed. However, moisture uniformity (average moisture content between internal and outer surface of dried product) plays an important role in affecting the enzyme activity and vitality of different dried foods. The lower the uniformity, the greater the negative influences. Figure 2.2 shows a comparison between the moisture uniformity of Chinese cabbage seeds dried using intermittent and stationary constant dryers conducted by Yang et al. (2013). It is clear from the figure that intermittent drying has a greater impact than constant drying.

The quality of dried food stuff in intermittent drying depends on the technique of using tempering or resting periods. It has been reported that implementing the proper tempering periods in drying has different merits such as reducing energy consumption,

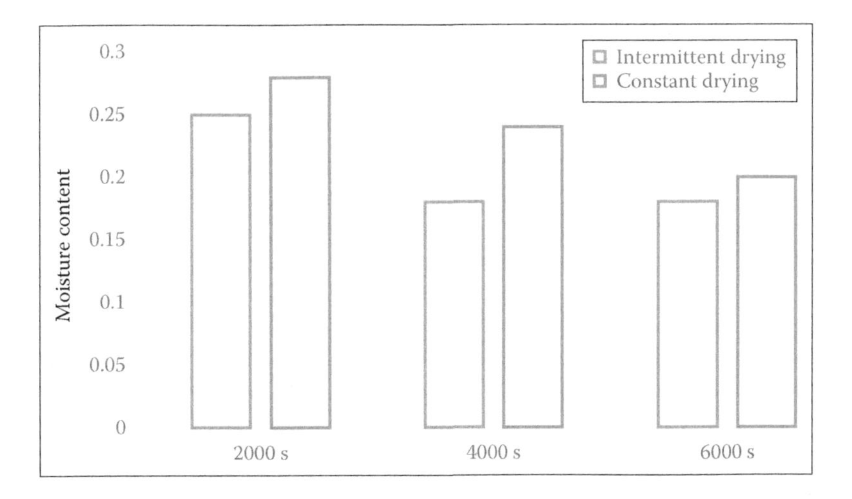

FIGURE 2.2 Moisture content difference between outer surface and inside of Chinese cabbage seed, with different time periods, resulting from using intermittent and constant drying methods. (Data from Yang, Z. et al., *Food Bioprod. Process.*, 9(1), 381, 2013.)

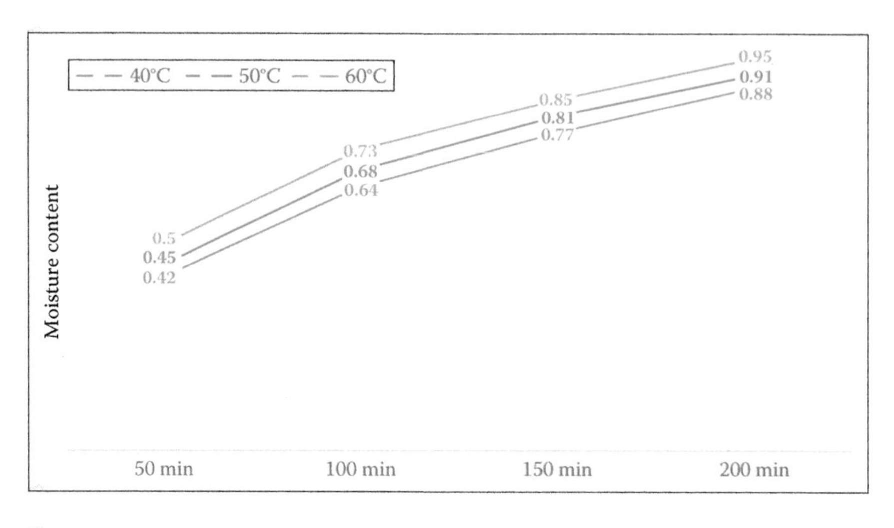

FIGURE 2.3 Effect of tempering period and air temperature on moisture removal of paddy rice. (Data from Golmohammadi, M. et al., *Food Bioprod. Process.*, 9(4), 275, 2015.)

increasing the drying rate, and improving the product quality (Yang et al., 2002; Baini and Langrish, 2007; Golmohammadi et al., 2015). The effect of tempering periods and air temperature moisture removal of paddy rice is shown in Figure 2.3, which emphasizes that drying rate (moisture removal) can be largely affected by tempering temperature, while the impact of air temperature is negligible. In the same area of research, the effect of resting period on drying rate was presented by Cihan and Ece (2001). They observed that moisture removal is considerably increased by increasing the tempering period of intermittent drying for rough rice. The impact of tempering temperature on tempering period for wheat drying was investigated by Nishiyama et al. (2006). They found that tempering temperature had a significant influence on tempering period.

2.6 EFFECT OF INTERMITTENT DRYING ON DRYING TIME

Shortening the drying time is considered a significant issue that has recently received great attention by researchers and industries. In response, different modified drying methods have been proposed in order to reduce the drying time as well as to improve the quality of the final dried products (Kowalski and Pawłowski, 2011). Intermittent drying is considered one of the most significant modified drying methods that helps in solving this problem. It can decrease the time by about 1 h compared to the traditional dehydration process, as shown in Figure 2.4, which represents the time required to reach the desired moisture value for stationary and intermittent drying methods at 70°C and 50°C. From the figure, it is clear that the time difference between stationary and intermittent drying processes is noticeable (e.g., time difference between stationary 70°C and 70°C [5 min on, 30 off] is 54 min).

Nishiyama et al. (2006) concluded that tempering temperature and period have positive effects on drying time. According to Kumar et al. (2014), intermittent drying can decrease the consumed drying air and drying time. Silva et al. (2016) predicted the

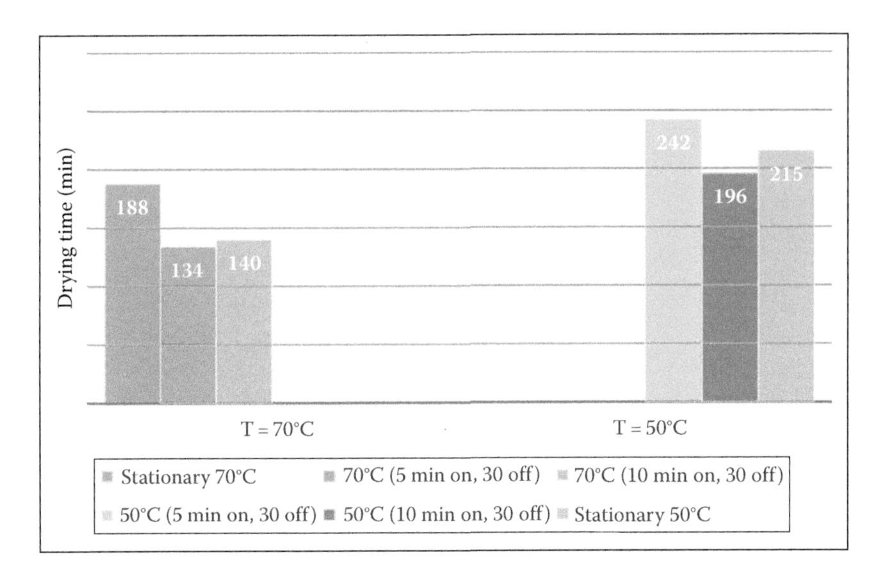

FIGURE 2.4 Comparison between drying time (in minutes) for traditional and intermittent drying methods at 70°C and 50°C. (Data from Szadzinska, J., *PhD Interdisciplinary J.*, 2014, Retrieved on September 2016. From: http://sdpg.pg.gda.pl/pij/wp-content/blogs.dir/133/files/2014/12/01_2014_13-szadzinska.pdf.)

impact on drying time of three-stage intermittent drying at 50°C for pears. A 26.2% reduction in drying time was obtained with 2.7 m/s air flow velocity compared to convective sun dryers, which are the conventional dryers usually used for dehydration of pears. The techniques used in this study is the industrialization process for drying pears in Portugal as the traditional dryers being used suffer from many drawbacks such as long drying time (more than 10 days required to complete the process), insects, meteorological conditions, and possible contamination from dust or mold. Yang et al. (2013) achieved a 23.7% reduction of total drying time at an air drying temperature of 40°C when the intermittent ratio changed from 1/3 to 2/3 in order to reach 5% moisture content for Chinese cabbage (*Brassica campestris* L. ssp) seeds using a heat pump dryer.

Other dehydration methods can be applied as pretreatment processes of the product (prior to intermittent drying) in order to maximize the efficiency of the whole drying process, reduce the drying time, and improve the quality of the product. Kowalski and Szadzinska (2014) studied the influence of using ultrasonic and osmotic dehydration methods, prior to stationary and intermittent convective drying, on the kinetics of the whole drying process. The intermittency technique was used in a mode of 5 min for heat and 30 min for tempering period. Figure 2.5 presents a comparison between the drying time for stationary drying and intermittent drying preceded by another method. A remarkable reduction in the time of the whole drying process was achieved, as shown in the same figure. Cihan and Ece (2001) also studied the effect on the drying time of the resting (no heat supply) period of intermittent drying for rough rice. It was concluded that a remarkable reduction was obtained in drying time by increasing the tempering period. In addition, Aquerreta et al. (2007) concluded that 38% reduction in drying time was achieved due to high tempering temperature (60°C) for post-drying of intermittent drying of rough rice.

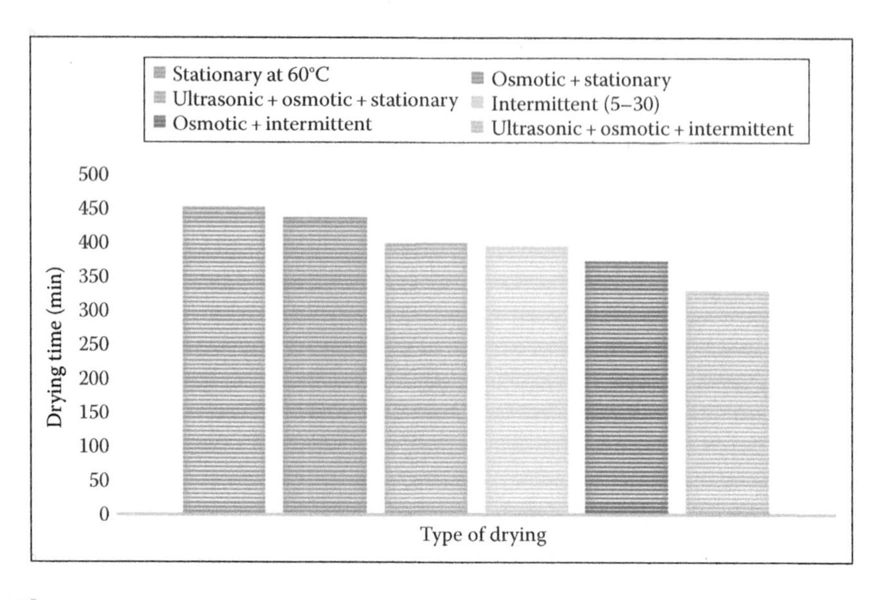

FIGURE 2.5 Drying time (in minutes) for different stationary and intermittent drying techniques at air temperature of 60°C. (Data from Kowalski, S. and Szadzinska, J., *Chem. Eng. Process.*, 82, 65, 2014.)

2.7 EFFECT OF INTERMITTENT DRYING ON ENERGY UTILIZATION

Drying technologies usually require a large amount of energy that sometimes accounts for up to 15% of all industrial energy usage, often with relatively low thermal efficiency in the range of 20%–25% (Chua et al., 2001). Different techniques have been used by researchers to minimize the energy required for drying in industries. The most common method used to reduce the energy is integration of drying technology with heat pump systems, utilizing waste heat for drying purposes, and intermittent drying. For instance, it has been reported that heat pump systems for clothes drying reduced energy consumption by 55% when compared to conventional electric dryers. It has also been reported that energy consumption per unit mass of dissipated water can be estimated if the average decrease of moisture with time and the increase in temperature of dried products during the drying process are known. The temperature of materials during drying can be measured and/or estimated mathematically or experimentally. Mathematical modeling is preferred for rapid estimation in case the design of the dryer is complicated (Silva et al., 2016).

Intermittent drying reduces the required energy of the process because it decreases the effective drying time and utilization of drying air. Numerous studies have been conducted to compare the energy saving between conventional continuous and intermittent drying methods. Kumar et al. (2014) reported that the most common type of intermittent drying that has been investigated and showed significant reduction in consumed energy is the on/off mode. Figure 2.6 shows energy saving achieved by using on/off intermittent drying for different products. Kowalski and Pawłowski (2011) concluded that intermittent drying (by periodically changing the temperature

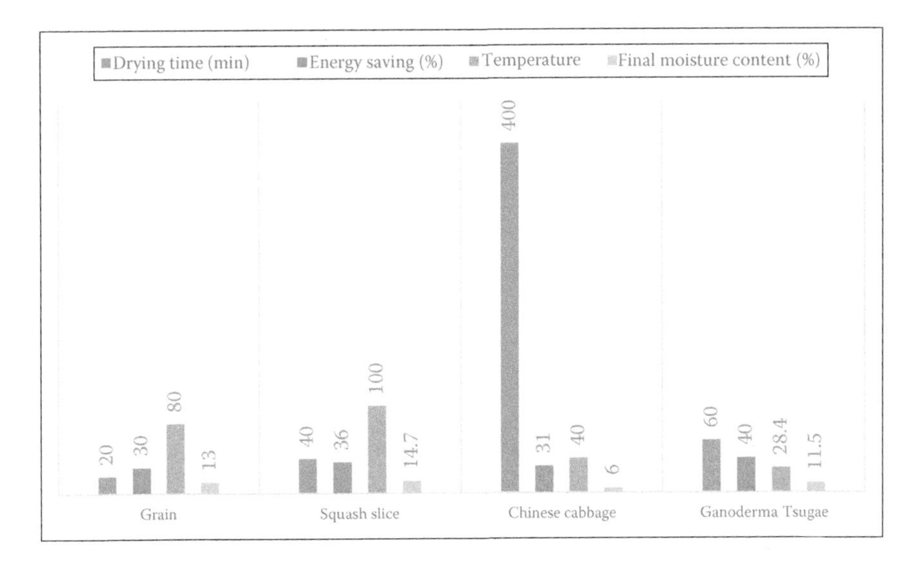

FIGURE 2.6 Effect of on/off intermittent drying mode on energy for different products. (Data from Kumar, C. et al., *J. Food Eng.*, 121, 48, 2014.)

of air drying between 100°C and 40°C on cylindrically shaped kaolin samples) decreased the energy consumption by about one third compared to constant and fixed drying conditions, as shown in Figure 2.7. The consumed electric energy for changeable air humidity was higher than for other drying modes because of the simultaneous operation of drier and humidifier, as shown in the same figure. This study also shows that the energy effectiveness (which is defined as the ratio of net energy consumed

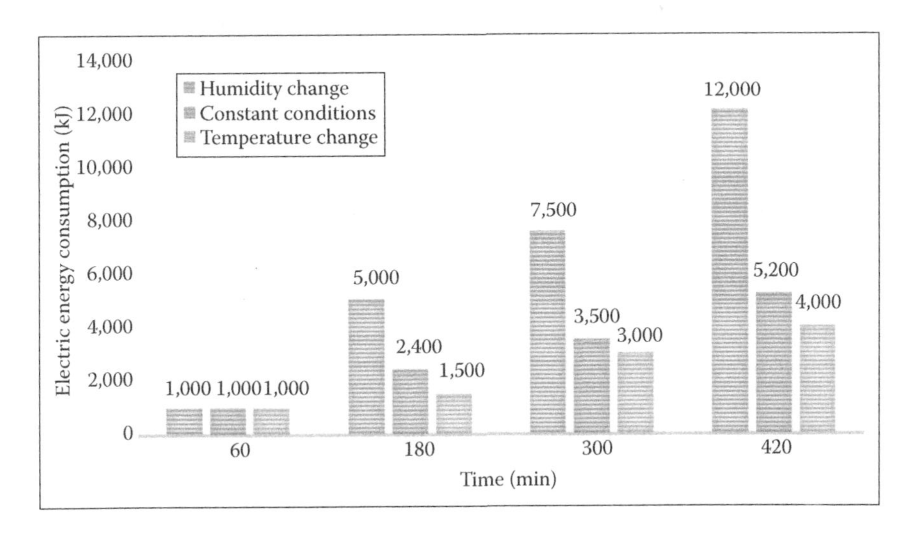

FIGURE 2.7 Comparison of electric energy consumption between intermittent drying and stationary drying of cylindrically shaped kaolin. (Data from Kowalski, S.J. and Pawłowski, A., *Chem. Eng. Process.*, 50, 384, 2011.)

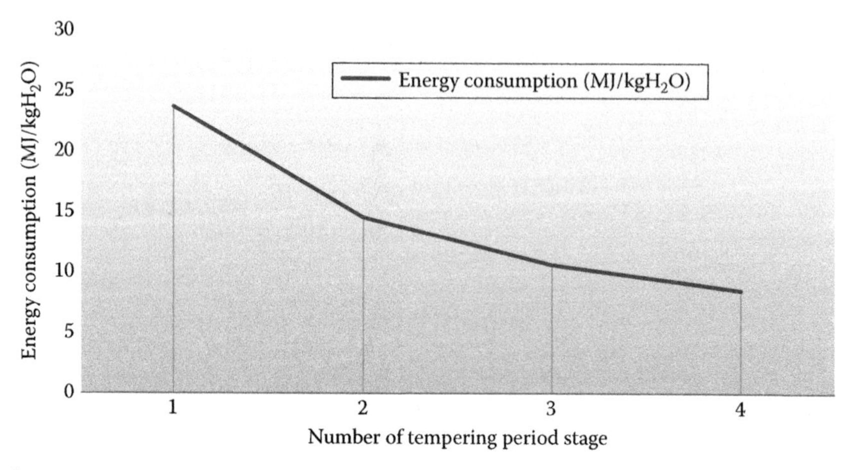

FIGURE 2.8 Impact of implementing rest period on energy consumption during intermittent drying for paddy rice. (Data from Golmohammadi, M. et al., *Food Bioprod. Process.*, 9(4), 275, 2015.)

for evaporating a unit of moisture mass to the total energy consumed for the process) was 1.92% for changeable air temperature intermittent drying and 0.7% for variable humidity. One of the drawbacks of this study is that a uniform temperature profile inside the product was not taken into account in modeling.

In response of that Putranto et al. (2011) considered a "good" drying model issue affects the performance of the dryer, product quality, and the energy consumption. For that, they investigated the impact of the reaction engineering approach used in intermittent drying for KOC kaolin clay by considering a uniform temperature profile inside the product. More accurate results were observed by implementing this approach under time-varying air temperature and humidity, and energy saving was noticeably achieved due to a decrease in the drying time of 48% and 61% for the intermittent dryer compared to 50% and 67% for conventional stationary dryers. Yang et al. (2013) investigated the effectiveness of a heat pump intermittent dryer on energy consumption, drying time, and moisture diffusion of Chinese cabbage seeds. About 48% less energy in the tempering period of 800 s over conventional convective drying was achieved in this study. The heat pump dryer worked in four different modes and the energy saving for these modes was 1,184.5, 1,572.3, 17,257.3, and 22,815.73 kJ, respectively. The effect of tempering period (also called resting period) on the energy consumption of the intermittent dryer for paddy rice was studied by Golmohammadi et al. (2015). The results of this investigation are shown in Figure 2.8, which indicates that tempering period strategy has a great impact on energy consumption, but the greater effect is for the first tempering period.

2.8 CONCLUSION

The effects of intermittent convective drying on drying time, energy consumption, and product quality have been presented. A comparison between stationary and intermittent drying and their effects on the mentioned factors is also presented.

Implementing the proper tempering temperature and period has a significant positive impact on drying time. In addition, it improves the quality of the dried products by reducing darkening and hardening and increasing shelf life. Therefore, a remarkable reduction in energy consumption has been observed and reported. This makes the intermittent convective drying technology competitive and more efficient when com pared with other options.

REFERENCES

Alibaş, I. 2012. Determination of vacuum and air drying characteristics of celeriac slices. *Journal of Biodiversity and Enviromental Sciences* 6(16): 1–13.

Argyropoulos, D., Heindl A., Müller J. 2008. Evaluation of processing parameters for hot-air drying to obtain high quality dried mushrooms in the Mediterranean region. *Conference on International Research on Food Security, Natural Resource Management and Rural Development. Tropentag 2008*, University of Hohenheim, Stuttgart, Germany, October 7–9.

Aquerreta, J., Iguaz, A., Arroqui, C., Virseda, P. 2007. Effect of high temperature intermittent drying and tempering on rough rice quality. *Journal of Food Engineering* 80: 611–618.

Baini, R., Langrish, T.A.G. 2007. Choosing an appropriate drying model for intermittent and continuous drying of bananas. *Journal of Food Engineering* 79: 330–343.

Baradey, Y., Hawlader, M.N.A., Ismail, A.F., Hrairi, M. 2015. Drying of fruits and vegetables: The impact of different drying methods on product quality. In Minea, V., ed., *Advances in Heat Pump-Assisted Drying Technology*. CRC Press, Taylor & Francis Group, Boca Raton, FL. 9781498734998.

Chua, K.J., Mujumdar, A.S., Hawlader, M.N.A., Chou, S.K., Ho, J.C. 2001. Batch drying of banana pieces—Effect of stepwise change in drying air temperature on drying kinetics and product colour. *Food Research International* 34: 721–731.

Cihan, A., Ece, M.C. 2001. Liquid diffusion model for intermittent drying of rough rice. *Journal of Food Engineering* 49: 327–331.

Golmohammadi, M., Assar, M., Rajabi-Hamaneh, M., Hashemi, S.J. 2015. Energy efficiency investigation of intermittent paddy rice dryer: Modeling and experimental study. *Food and Bioproducts Processing* 9(4): 275–283.

Hawlader, M.N.A., Perera, C.O., Tian, M. 2004. Heat pump drying under inert atmosphere. *Proceedings of the 14th International Drying Symposium (IDS2004)*, Vol. A, pp. 309–316, Sao Paulo, Brazil, August 22–25.

Hawlader, M.N.A., Perera, C.O., Tian, M. 2005. Influence of different drying methods on fruits' quality. *Eighth Annual IEA Heat Pump Conference*, Las Vegas, NV, May 30–June 2.

Hawlader, M.N.A., Perera, C.O., Tian, M. 2006. Properties of modified atmosphere heat pump dried foods. *Journal of Food Engineering* 74: 392–401.

Holowaty, S.A., Ramallo, L.A., Schmalko, M.E. 2012. Intermittent drying simulation in a deep bed dryer of yerba maté. *Journal of Food Engineering* 111(1): 110–114.

Jangam, S.V., Law, C.L., Mujumdar, A.S. 2010. *Drying of Foods, Vegetables and Fruits*. National University of Singapore, Singapore, 978-981-08-7985-3.

Karim, M.A., Hawlader, M.N.A. 2005. Drying characteristics of banana: Theoretical modelling and experimental validation. *Journal of Food Engineering* 70(1): 35–45.

Kowalski, S.J., Pawłowski, A. 2011. Energy consumption and quality aspect by intermittent drying. *Chemical Engineering and Processing* 50: 384–390.

Kowalski, S.J., Szadzinska, J. 2014. Convective-intermittent drying of cherries preceded by ultrasonic assisted osmotic dehydration. *Chemical Engineering and Processing* 82: 65–70.

Kumar, C., Karim, M.A., Joardder, M.U.H. 2014. Intermittent drying of food products: A critical review. *Journal of Food Engineering* 121: 48–57.

Lewicki, P.P. 2006. Design of hot air drying for better foods. *Trends in Food Science & Technology* 17: 153–163.

Nishiyama, Y., Cao, W., Li, B. 2006. Grain intermittent drying characteristics analyzed by a simplified model. *Journal of Food Engineering* 76: 272–279.

Olalusi, A. 2014. Hot air drying and quality of red and white varieties of onion (*Allium cepa*). *Journal of Agricultural Chemistry and Environment* 3: 13–19.

Perera, C.O. 2005. Selected quality attributes of dried foods. *Drying Technology* 23(4): 717–730.

Putranto, A., Chen, X.D., Devahastin, S., Xiao, Z., Webley, P.A. 2011. Application of the reaction engineering approach (REA) for modeling intermittent drying under time-varying humidity and temperature. *Chemical Engineering Science* 66: 2149–2156.

Qing-guo, H., Min, Z., Mujumdar, A.S., Wei-hua, D., Jin-cai, S. 2006. Effects of different drying methods on the quality changes of granular edamame. *Drying Technology* 24: 1025–1032.

Rahman, M.S. 1999. *Handbook of Food Preservation*. Marcel Dekker, Inc., New York.

Silva, V., Costa, J.J., Figueiredo, A.R., Nunes, A.J., Nunes, A.C., Ribeiro, T.I.B., Pereira, B. 2016. Study of three-stage intermittent drying of pears considering shrinkage and variable diffusion coefficient. *Journal of Food Engineering* 180(2016): 77–86.

Szadzinska, J. 2014. Influence of convective-intermittent drying on the kinetics, energy consumption and quality of green pepper. *PhD Interdisciplinary Journal*. Retrieved on September 2016. From: http://sdpg.pg.gda.pl/pij/wp-content/blogs.dir/133/files/2014/12/01_2014_13-szadzinska.pdf.

Tertoe, C. 2010. A review of osmodehydration for food industry. *African Journal of Food Science* 4(6): 303–324.

Wang, N., Brennan, J.G. 1995. Changes in structure, density and porosity of potato during dehydration. *Journal of Food Engineering* 24(1): 61–76.

Yang, W., Jia, C., Siebenmorgen, T.J., Howell, T.A., Cnossen, A.G. 2002. Intra-kernel moisture responses of rice to drying and tempering treatments by finite element simulation. *Transactions of the ASAE* 45: 1037–1044.

Yang, Z., Zhu, E., Zhu, Z., Wang, J., Li, S. 2013. A comparative study on intermittent heat pump drying process of Chinese cabbage (*Brassica campestris* L. ssp) seeds. *Food and Bioproducts Processing* 9(1): 381–388.

3 Conventional and Intermittent Food Drying Processes and the Effect on Food Quality

Shu-Hui Gan and Chung-Lim Law

CONTENTS

3.1 INTRODUCTION: BACKGROUND AND INTERMITTENCY

Drying is an important unit operation in process industry. It is often used to remove moisture to extend the product shelf life, and provide dry product either in solid or powder form, or even intermediate powders prior to extraction. As conventional drying involves extensive energy usage, drying is regarded as one of the most energy-intensive unit operations in process industry ranked after distillation. In addition, drying is the only unit operation that is used to remove bound moisture, whose removal is controlled by internal diffusion. Therefore, drying is a lengthy unit operation that takes time, especially for big-sized materials such as woods, bricks, and grains, and it may require days or even months to complete a batch.

Drying is a process that involves heat and mass transfer between the product and the drying medium. Therefore, it is a complex unit operation that involves different transport mechanisms, such as evaporation and diffusion. The removal of moisture affects the product quality, namely colour, appearance and texture, as well as the retention of bioactive ingredients (nutrients, flavour) and physical properties, such as bulk density, viscosity, flow ability, deformation and size distribution. Chemical properties such as bioactive contents as well as loss of nutrients are also affected. Change in colour of a product is often caused by non-enzymatic browning, while texture changes are attributed to shrinkage due to collapsing of the product's surface layer. Baker (1997) reported that 10% of the total energy usage in the food processing industries is consumed by the drying stage alone. Significant reduction in energy consumption in the drying processes would contribute to the bottom line of the whole operation.

Rough rice fissuring often occurs due to stresses experienced by the rice kernel during the drying process (Ondier et al. 2012). Severe colour change caused by browning reaction, especially non-enzymatic browning, has been observed in bananas (Pekke et al. 2013), potatoes (Mcminn and Magee 1997), papayas (Udomkun et al. 2015) and many other food products. Cernîşev (2010) reported that multistage drying may decrease non-enzymatic browning of dried tomatoes. Loss of nutrients is often reported in products that were subjected to high temperature drying, that is, higher than 60°C (Mujumdar and Devahastin 2008).

A balance between energy saving and quality retention, especially in terms of minimal colour change and deformation, high retention of bioactive ingredients as well as safety and environmental impact renders the dryer selection a challenging task for a design engineer. Among all the aforementioned aspects, product quality is the most important and it is the top priority in process industries, especially for food, pharmaceutical and healthcare industries.

The intermittent drying mode has been reported to be able to minimize energy usage while maintaining good product quality. Intermittent drying involves alternating drying periods with tempering periods sandwiched between active drying periods. In this regard, operating parameters such as temperature, flow rate and pressure may be manipulated. The most distinctive feature of intermittent drying is the allocation of a tempering period or 'resting time'. The moisture gradient causes moisture to migrate from the interior to the surface even during the tempering period. Hence, moisture distribution within the product becomes more uniform.

As such, the surface is sufficiently rewetted for drying in the subsequent active drying period. In this regard, thermal energy supplied by the drying medium is utilized to remove the surface moisture rather than raising the temperature of the drying material. Since the product surface's temperature is lower than the drying temperature, thermal degradation of heat-sensitive products can be avoided. The long total processing duration is offset by a higher thermal efficiency and shorter effective drying time.

In addition, intermittent drying offers advantages in drying either heat-sensitive or crack-sensitive materials. Physical appearance and food texture are preserved while nutrients are retained. Stresses that typically occur in conventional drying are reduced. Surface cracking intensity and its magnitude are minimized noticeably (Kowalski and Pawlowski 2011). Intermittent drying is beneficial for materials which dry primarily in the falling rate period, where internal diffusion controls the overall drying rate (Kumar et al. 2014).

The operating variables that are manipulated in intermittent drying are mostly energy-consuming components such as heater (air temperature), blower (air velocity) and input of other energy sources (e.g. infrared, microwave and radio-frequency). Among all, regulation of air temperature is considered to have the most substantial influence on product kinetics and various quality parameters (Chua et al. 2002). Generally, intermittency profile can be categorized into four types: pulsed energy input, cyclic variation, ramp variation and arbitrary variation.

Pulsed or periodic heat supply is the most frequently used profile. A simple and classic mode of pulsed intermittent drying is a series of continuous drying operations in initial stage (usually down to the critical moisture content), tempering period over intermediate stage (for redistribution of temperature and moisture content) and continuous drying in final stage (down to desired final moisture content) (Pan et al. 1998, 1999). Nevertheless, in recent applications the pulsed intermittent is regarded as a sequence of elementary cycles that consists of two heating periods separated by a tempering period. It should be noted that temporal (time-dependent) variation of the drying conditions is applicable to batch dryers as spatial (location-dependent) variation is to continuous dryers.

Cyclic and ramp profiles are another sub-category under intermittent drying technique. Cyclic variation applies a specified cyclic pattern of energy input such as sinusoidal, square-wave or saw-tooth pattern while ramp variation employs a step-up profile, step-down profile or combinations of them (Chua et al. 2001a, 2002; Ho et al. 2002). As for the case of arbitrary variation, intermittent occurs mainly due to fluctuation of energy supply that is caused by inherent features of the drying equipment design itself. For example, rotary drum drying, on/off microwave drying, rotating spouted bed drying and pulsed fluid bed drying (Kudra 2008).

Intermittent drying has been developed as a cost-effective unit operation for processing food products. Judicious selection of intermittent dryers is known to increase shelf life, reduce food wastage and preserve quality attributes such as colour, flavour, texture, appearance and nutrient retention (Chua et al. 2002).

3.2 INTERMITTENT DRYERS

3.2.1 Conventional Intermittent Dryers

Conventional intermittent dryers involve changing drying conditions to improve the drying performance. Drying conditions can be manipulated by periodically introducing drying medium into the drying chamber, or cyclic step-up of drying temperatures to dehydrate product samples, or varying the humidity in the drying stage. The choice depends on the optimized drying achievable.

Chong and Law (2011) studied intermittent drying with varying periods of hot air and cold air in order to dehydrate *Manilkara zapota* (ciku) samples. The higher the sample moisture content, the higher is the drying rate achievable. Intermittent hot air drying also produces a higher drying rate in high moisture content when compared to that of dehumidified-air drying. The hardness and chewiness of the dried product is found to increase with progression of dehydration. Li et al. (1998) studied the effect of intermittent drying on rough rice. Kowalski and Szadzińska (2014) studied the influence of intermittent drying on beetroot quality. Both of the intermittent heat-drying processes preserved better product quality compared to other dehydration methods. However, conventional intermittent dryers seem to be beneficial for thin-layer drying of foods.

3.2.2 Fluidized Bed Dryers

Fluidized bed drying involves suspending a column of particulate solids in a flowing gas stream, typically heated air. With increasing gas flow, a point is reached where the drag force imparted by the upward moving gas equals the weight of the particles. Generally, the advantages of using fluidized bed dryer with regard to improvement of drying rates are ascribed to excellent gas–particle contact, uniform drying with homogeneous temperature field and low operating temperatures resulting in a minimum thermal degradation. However, the fluidized bed drying technology is also limited by the high power consumption due to fluidization of entire bed—especially for fluidizing bigger particle size, the high potential of attrition, granulation and agglomeration depending on the properties of the fluidizing materials and also the narrow particle size distribution to avoid excessive entrainment.

Fluidized bed dryer has many variants after modifying the conventional fluidized bed dryer and some of the variants can be classified as intermittent dryer. The modified fluidized bed dryers such as pulsed fluidized bed dryers, spouting fluidized bed dryers and jetting fluidized bed dryers have been developed to address the problems encountered in the conventional fluidized bed dryer. The modified fluidized bed dryers that exhibit the characteristics of intermittent drying are briefly explained in Sections 3.2.2.1 through 3.2.2.3.

3.2.2.1 Pulsed Fluidized Bed Dryers

Intermittent or pulse fluidized bed drying either charge in drying medium into the bed periodically, or manipulate the particles in the fluidized bed cyclically by rotation. These actions introduce rest period (tempering) for the section of fluidized bed that does not receive charging of drying medium and this exhibits the behavior of an intermittent drying. Intermittent fluidized bed drying ensures a uniform drying of the product due to the high contact area between the drying medium and the product. The flow of hot drying air through the bed removes the moisture content away from the product, increasing the drying rate of the overall dehydration process. A moisture gradient diffusion through the product layers is achievable during the resting period to remove bound moisture content in products.

Burande et al. (2008) studied the effect of fluidized bed drying of green peas with periodic tempering time. The rehydration ratio of the dried green peas favoured higher temperatures and lower tempering time. The high contact area between the drying medium and the product ensured a uniform drying, increasing the product's rehydration capacity. Furthermore, Vega-Valencia et al. (2014) reported that high retention of the oligosaccharide content and high rehydration ability of dried cactus (Nopal) are achieved by intermittent drying using a fluidized bed with revolving chambers.

Li et al. (2006) studied the drying of green peas to assess the advantages of a pulsating fluidized bed over a conventional fluidized bed. The experiment was conducted with an initial static bed height of 0.05 mm at a pulsating frequency of 0.5 Hz. The fluidizing gas stream with superficial velocities ranging from 120 to 320 m^3/h was heated to the stipulated temperatures. To evaluate the respective product qualities, the green peas were rehydrated in distilled water at 25°C after drying for 100 min. The water uptake was observed to be discernibly higher for green peas dried in the pulsating fluidized bed. The greenish tint of the rehydrated peas was also maintained better through drying in the pulsating fluidized bed.

3.2.2.2 Spouting Fluidized Bed Dryers

Spouted fluidized bed dryers are often used for drying granular products that are either too coarse or too dense to be readily fluidized. The drying medium, introduced through a centrally located nozzle at the conical base, transports the particles to a certain height above the bed surface. After losing momentum, these particles fall back onto the bed surface. Through this spouting motion, good solid mixing is induced and accordingly a cyclical flow of particles is created. The propulsion only occurs in the centre of the bed, leading to down-flow of fresh product towards the nozzle area in a cyclic sequence. The products are in contact with the hot drying medium intermittently.

Sahin et al. (2013) observed better results in terms of colour, shrinkage, bulk and apparent densities, internal and bulk porosities, rehydration capacity and microstructure when the drying process of peas was conducted using an intermittent spouted fluidized bed. Balakrishnan et al. (2011) examined the effect of intermittent drying on the retention of volatile oil and oleoresin in dried cardamom capsules. The samples were fluidized at air temperatures of 40°C and 50°C. For a similar final moisture content (10% w.b.), intermittent drying at a lower temperature resulted in better yields of volatile oil and oleoresin as compared to continuous drying at the same superficial velocity. The volatile oil content was observed to be at a maximum of 9.7%, while there was a 10%–15% increase in oleoresin retention.

However, the high kinetic energy tossing the products into the air for drying can affect the product shaping. The risks for surface destruction of the dried products are higher. The surface of large particles can be easily damaged by high velocities of the drying medium, while fine particles can be easily lost during entrainment. Spouted fluidized bed is thus not suitable for the drying of surface-sensitive materials.

3.2.2.3 Jetting Fluidized Bed Dryers

Jetting fluidized bed dryers are also another type of intermittent dryer. One distinctive feature of jetting fluidized beds is that bubbles are formed and a fairly large jet is introduced through a centrally located nozzle at the conical base and transports the particles to a certain height above the bed surface, instead of the dilute-phase spout in the spouting fluidized bed dryers. Jumah et al. (1996) implemented the principle of intermittent drying in a novel rotating jet spouted bed for corn. The rotating jets introduced pulses of hot air into the bed and prevented the development of adverse temperature gradients. This feature enhanced air utilization and solids mixing when compared with conventional spouted fluidized beds. It was found that alternating short drying periods with equal or longer tempering periods yielded better product quality than prolonged continuous drying. Besides a reduction in drying-induced stresses, minimal mechanical damage to the kernels was observed due to reduced attrition caused by inter-particle collision.

3.2.3 Pulse Combustion Dryers

The pulse combustion drying technology is an intermittent dehydration technology. The process involves periodic ignition and extinguishing of combustibles (Wu and Mujumdar 2004). The common combustibles are ratios of fuel and air, which are mixed in the combustion chamber during the process. Heat energy is released upon instant combustion of the fuel–air blend. Pressure increment in the combustion chamber forces the high-energy flue gases to flow out, followed by a pressure drop, drawing fresh fuel and air into the chamber. The next cycle begins in a self-sustained ignition due to the remnant's hot flue gas contact. The acoustic temperature waves propagated from the combustion chamber supplies heat for drying purposes.

The benefits of pulse combustion drying method include short drying time, high evaporation rate, high energy efficiencies, environmental friendly operation, as well as the ability to process high-viscosity materials without agitation. Yanniotis et al. (2013) reported improved product quality from moisture removal utilizing pulse combustion technology. Hofsetz et al. (2007) studied the effect of high-temperature pulse combustion dehydration on the quality of banana slices. A uniform shrinkage was observed for low temperature drying, while puffing occurred at higher temperatures—short time dehydration.

3.2.3.1 Pulse Combustion Spray Dryers

Pulse combustion spray drying is a suspended particle processing technique that utilizes liquid atomization to create droplets and intermittent high-temperature waves periodically to dry product (Masters 2002). This method of moisture removal is suitable for heat-sensitive materials due to the low process temperature. Spray dryers utilizing pulse combustion have flexibility, as a wide range of fuels can be utilized. Other advantages of pulse combustion spray drying include, operating at atmospheric pressure and low temperatures; large-scale drying and can produce relatively uniform and spherical particles.

The drying process is also known to be rapid due to the large heat and mass transfer surface area available. The final product properties such as particle size and bulk density are dependent upon the choice of nozzle used for the pulse combustion spray dryers. Agglomeration in a multistage pulse combustion spray drying is thus nozzle-controlled. The centrifugal atomizer is more flexible in terms of processing capacity, although utilizing higher energy than pressure nozzles.

Huang and Mujumdar (2004) reported an improved quality of skim milk utilizing multistage pulse combustion spray dryers. The technology used was a first-stage pulse combustion spray dryer, a second stage with inner static fluid bed and a third-stage external vibrated fluid bed. The function of each stage is drying, post-dehydration, and cooling, respectively. Results showed a better solubility as well as flow ability, and higher bulk density of the dehydrated product. It was claimed by Huang and Mujumdar (2004) that pneumatic nozzles have high tendency to clog.

Besides, the pulse combustion spray dryers are highly favoured in the pharmaceutical industries. A study by Wang et al. (2007) prepared Nitrendipine–Aerosil–Tween 80, a poor solubility drug, using pulse combustion spray dryer and a conventional spray dryer. A better quality amorphous-state drug was produced using pulsed combustion spray dryer with no agglomeration, smaller particle size with narrower size distribution. A study by Xu et al. (2007) was carried out to investigate ibuprofen dispersity in dehydrated carriers, pulse-combustion-treated and non-treated ones. The modulated carriers showed a better dispersion of the ibuprofen. This is a significant breakthrough in the drug processing sector. Carriers with moisture removed using pulse combustion spray dryers may develop solubility strengths in dissolving highly insoluble drugs.

3.2.4 Heat Pump Dryers

A heat pump is used to perform dehumidification and heating of the convective drying medium, although auxiliary heaters are generally employed for better temperature control at the dryer inlet. The heat pump system consists of an evaporator where the refrigerant recovers both sensible and latent heats, a compressor where the vaporized refrigerant is compressed, a condenser where the refrigerant is cooled and the drying medium heated, and expansion valve where the condensed refrigerant is further cooled. Heat pump dryers provide excellent control of drying environment to meet specific production requirements. Among the benefits are higher energy efficiency, lower drying temperatures and better retention of volatile components.

However, this drying technology suffers from several drawbacks, such as the use of non-environmental-friendly refrigerants and limited range of drying temperature. Alternatively, intermittent drying in heat pump systems can reduce both capital and operating costs since it allows the use of a lower capacity heat pump or a single heat pump to service several drying chambers.

3.2.4.1 Intermittent Heat Pump Dryers

Intermittent drying technique can be applied on heat-pump-assisted dryer as well. Operating variables that are often manipulated in batch heat-pump-assisted dryers are: intermittent variation of air temperature, intermittent supply of airflow, intermittent regulation of air humidity and intermittent addition of other energy sources such as infrared, microwave and radio-frequency (Chua et al. 2002). Basically, reduced drying time and improved product quality were observed in most cases. Several studies done in the past include water-soluble polysaccharides retention of *Ganoderma tsugae* Murrill (Chin and Law 2010), colour change of dried salak fruit during storage (Ong et al. 2011), as well as ascorbic acid content and total phenolic content of dried salak fruit (Ong et al. 2012). All results of these studies showed that intermittent heat pump dehydration should be preferred when it comes to preservation of product quality.

The final quality of the dried products is affected by two major factors: the drying air temperatures and drying cycle times (Minea 2013). The intermittency in a heat pump dryer thus involves varying drying medium flow or drying air temperatures.

Several studies on the dried product characteristics obtained via intermittent heat pump drying are discussed. Chua et al. (2001b) studied the intermittent drying of banana slices using a two-stage heat pump dryer by changing drying medium temperature. The total colour difference was calculated for better comparison. The results showed that by applying the temperature step change, colour degradation as well as the drying time were reduced.

On the contrary, Ong and Law (2011b) maintained that the micro-porosity and optical properties of the salak samples were better when the product was dried using heat-pump-assisted drying in constant mode instead of intermittent mode. The intermittent drying produced dried fruits with high colour degradation rate during storage due to the high rehydration capacity. The microstructural properties of the dried fruit affect the water sorption characteristics during storage, which in turn shorten the storage life of the intermittent heat pump dried fruit.

Moreover, heat-pump-assisted dryer can be integrated with other heat input methods to enhance drying efficiency and also reduce thermal load on the heat pump system. During the last stage of drying, especially at about 10%–20% of moisture content, the change in humidity of air through heat-pump-assisted dryer is small due to reduction of driving force of mass transfer. Thus, energy may be supplied by different modes of heat transfer such as conduction, radiation and dielectric, simultaneously or in a pre-selected sequence in time-varying fashion to improve the drying kinetics. It was reported that combination of microwave and heat-pump-assisted drying in intermittent mode not only significantly shortened the drying time but also could reduce capital cost of a microwave-assisted dryer (Chua et al. 2002). According to Chua and Chou (2005), drying time was shortened by 42% and 31% for potato and carrot samples, respectively, when a suitable combination of heat pump/microwave drying time period was selected.

3.2.5 Dielectric Dryers

Radiation drying proceeds by means of non-ionizing electromagnetic waves which are manifested as heat through their interaction with food products. Radiation drying involves infrared drying and dielectric drying. Microwave drying and radio-frequency drying are two forms of dielectric drying. Based on its wavelength, the infrared spectrum can be divided into three regions: near (0.78–1.4 μm), mid- (1.4–3 μm) and far (3–1000 μm). The characteristic wavelengths of radio waves and microwaves are 10 and 0.1 m, respectively. Unlike infrared drying, which operates by surface impingement, dielectric drying has the unique ability to generate heat within the material. Radiation drying has gained popularity because of its superior thermal efficiency and high drying rate. Selective heating of food products means that the air and drying equipment are not heated, keeping the ambient temperature at normal levels. The wavelengths used in radiation drying also allow for instantaneous start-up and shut-down of the operation, resulting in accurate and more rapid process control.

3.2.5.1 Intermittent Microwave Dryers

During microwave drying, the microwave energy induces oscillation of the inner water molecules in a product and the heat is dissipated through the kinetic movements to vaporize the water content. The product is thus heated while dried with uniform heat distribution. Continuous microwave drying affects the quality attributes of the dried product. The overheating causes different pressures in the material layers, leading to puffing or shrinkage of the dried products. Surface of food materials are also depleted of aroma due to dehydration. Steed et al. (2008) studied a continuous microwave drying of purple-fleshed sweet potato puree based on different microwave energies. The results of increasing microwave energy used show an increase in the total phenolic content of sweet potatoes but a decrease in the dried product colour saturation. Both quality attributes are equally important in the competitive food market today.

In intermittent microwave drying, the microwave-powered heating is turned off from time to time, creating a periodic heating effect. This causes the moisture content in the product to be redistributed more uniformly during the resting period. The water content diffuses through the product layers, and is evaporated during the next heating cycle; in turn, more bound moisture is removed. This can also decrease the heating effect on the product quality and more bioactive ingredients can be retained.

Intermittent drying also precludes the case-hardening of food and the formation of large temperature gradients, both of which typically occur in continuous drying. Quality improvement when products are dried using intermittent microwave dryers was reported in several different studies using different products. These drying studies include Gong et al. (1998) on clay drying, Itaya et al. (2001) on ceramics dehydration, Gunasekaran (1999) on cranberry drying, Soysal et al. (2009) on red pepper drying, Xu et al. (2012) on litchi drying as well as Li et al. (2014) on wheat seeds drying.

However, there was a particular microwave intermittent drying of dill by Estürk and Soysal (2010) which concluded that continuous microwave drying resulted in better quality product than intermittent microwave drying. The assumption that oxidation occurred during resting periods of the intermittent drying caused the decrease in colour appearance of dried dill. This leads to the observation that intermittent microwave drying may not be suitable for all products' quality maintenance during or after dehydration.

3.2.5.2 Intermittent Infrared (IR) Dryers

Infrared (IR) drying is now commercially available for the purpose of rapid sample drying and has gained in popularity in heat-sensitive high-value foods that monitor sample moisture as a measure of quality control (Beary 1988). Ginzburg (1969) has suggested that IR radiation, operating under an intermittent mode, may be applied to dry heat-sensitive high-value biomaterials such as grains, flour, vegetables, pasta, meat and fish. To dry heat-sensitive materials, three options are available: to constantly monitor the product surface temperature, to regulate the intensity of the IR and to operate the IR in an intermittent mode. According to Ginzburg (1969), the effectiveness of drying is increased by the application of intermittent IR radiation and also by the combined radiant–convective method.

In many cases, intermittent radiation treatment is beneficial in decreasing duration of the drying process, as well as the colour change and browning index of the products, as shown in Table 3.2. Researchers such as Paakkonen et al. (1999) have shown that intermittent IR drying improves the quality of herbs, while Dontigny et al. (1992) have demonstrated that intermittent IR drying of graphite slurry significantly increases drying rate.

Besides, Zbicinski et al. (1992) investigated convective air drying and IR drying and have suggested that the use of an intermittent IR radiation drying mode coupled with convective air drying is best for heat-sensitive materials. Other researchers such as Dostie et al. (1989) and Carroll and Churchill (1986) have also reported shorter drying time with improved product quality for intermittent IR heating (Carroll and Churchill 1986; Dostie et al. 1989). Zhu and Pan (2009) also observed that intermittent heating mode can reduce colour degradation, whereas continuous heating causes severe colour degradation (Zhu and Pan 2009; Chandan et al. 2014).

Gan et al. (2015) also compared the effects of continuous and intermittent infrared drying on the quality of edible bird's nests. The drying chamber was operated at 40°C with $RH = 16.5\%$ when the heater was turned on, and at 25°C with $RH = 26.7\%$ when the heater was turned off. The intermittency ratios used were 0.20, 0.33, 0.67 and 1.00, respectively. An average radiation intensity of 0.23 W/m^2 was also employed. For a similar final moisture content of 10%–12% d.b., intermittent drying ($\alpha = 0.20$) at a lower temperature reduced the overall colour change of the dried product by 83.8% and improved the rehydration capacity. Higher rehydration rates imply shorter cooking times and hence, better product quality.

3.2.6 Multistage Intermittent Dryers

Multistage drying can be considered as an intermittent drying where comprising of two or more same or different types of dryers in the parallel process. The multistage drying aims to increase the energy efficiency of the drying process and at the same time increase the efficiency of bound moisture removal. One-stage dehydration process mostly only removes the free surface moisture of the product, leaving the trapped moisture in the inner layers of the dried product. This often decreases the product shelf life as the moisture content diffuses out of the product, accelerating microbial growth in the surroundings.

The multistage intermittent drying discussed in this text is made up of different types of intermittent dryers such as coupling of microwave drying with vacuum dryers, combined hot air with microwave drying, and also pulse-spouted microwave vacuum drying. This new emerging technology of coupling may have pros and cons than those of the single-stage drying processes. The research gap still exists on whether the impact of product quality is affect by different stages of intermittent drying.

Huang and Zhang (2015) studied the pulse-spouted microwave vacuum drying of okra. The results observed were those comparable to the industry's standard quality. Zhao et al. (2014) studied the two-stage intermittent microwave coupled with hot air drying of carrot slices. The two-stage intermittent drying is found to be promising for industrial applications. The shortest drying time and relatively low energy consumption is achieved in both cases with good quality of final dried products.

3.3 QUALITY ASPECTS IN INTERMITTENT DRYING

During drying, quality changes in food product are inevitable. Hence, many advances in drying technologies have been practised during the past decade with the main target of minimizing quality degradation of dried food products. Intermittent drying is recognized as one of the most effective methods to achieve that objective. Many theoretical and experimental investigations have been done to improve different quality attributes by applying intermittent drying. This section reviews studies that investigated the quality attributes in intermittent drying.

3.3.1 Retention of Bioactive Ingredients

The retention of bioactive ingredients (i.e. antioxidant phytochemicals, ascorbic acid, water soluble polysaccharides and beta-carotene) in the dried food products is one of the main quality attributes frequently investigated in intermittent drying. It is generally observed that if ascorbic acid is well retained, other components are also well retained. Hence, ascorbic acid can be taken as an index of the nutrient quality of foods (Marfil et al. 2008). Recent investigations depict that loss of ascorbic acid in food products could be attributed to a combination of the effects of thermal degradation and enzymatic oxidation due to high temperature and long drying time, respectively. Low temperature drying and intermittent drying may prevent the thermal degradation of ascorbic acid but excessive drying time may trigger enzymatic oxidation (Ong and Law 2011). Table 3.1 lists such studies and presents a summary of quality attributes on retention of bioactive ingredients by intermittent drying found in the literature.

3.3.2 Colour

Colour is one of the key factors that influence consumers' decision to buy a particular food. The effect of drying methods and types of intermittency, for instance, step-up and step-down temperatures, cosine and reverse cosine have been investigated on the colour changes for different food products. Commission Internationale de I'Eclairage (CIE) colour parameters have previously been demonstrated to be valuable in describing visual colour deterioration and providing important information for quality control in food products. The colour brightness coordinate CIE L^* measures the whiteness value of a colour and ranges from black at 0 to white at 100. CIE a^* measures red when positive and green when negative, whereas CIE b^* measures yellow when positive and blue when negative. Chroma value, for its turn, indicates the degree of colour saturation and is proportional to the strength of the colour; hue angle is used to characterize colour in food products and represents the purity of brown colour, which is an important parameter in drying processes where enzymatic and non-enzymatic browning takes place (Elcin and Belma 2009). The total colour change (ΔE) (Equation 3.1) is used to describe the colour change during drying:

$$\Delta E = \sqrt{\left(L_0^* - L_t^*\right)^2 + \left(a_0^* - a_t^*\right)^2 + \left(b_0^* - b_t^*\right)^2} \tag{3.1}$$

where

L_0^*, $a_0^* a^*$, and $b_0^* b^*$ are the initial colour measurements of raw bird's nest samples
L_t^* and, a_t^*, and b_t^* are the colour measurements at a pre-specified time

All the intermittent drying presented in Table 3.2 show improvement in colour change. However, it is noted that appropriate intermittency and drying conditions should be chosen based on energy efficiency of the process and expected quality of the dried food. There is a need to introduce an optimization scheme between reduction in colour change and achieving higher drying rate.

TABLE 3.1
Studies on the Retention of Bioactive Ingredients by Different Types of Intermittent Drying

Product	Intermittent Drying	Drying Conditions	Findings	Reference
Salak fruit	Intermittent heat pump drying	Periodic heat airflow supply Step-up air temperature (26°C–37°C)	Retained high concentration of ascorbic acid by intermittent mode (step-up air temperature) as compared to continuous drying	Ong and Law (2011a)
Carrot	Two-stage intermittent microwave (MW) coupled with hot air (HA) drying	HA (60°C) → MW (145 W) HA (60°C) → MW (175 W)	Highest retained alpha and beta-carotene contents by intermittent drying, with lowest drying time and energy consumption	Zhao et al. (2014)
Walnut	Intermittent oven drying	Step-down temperature (from 50°C to 30°C)	Highest retention of antioxidant activity by intermittent drying	Qu et al. (2016)
Ganoderma tsugae	Intermittent heat pump drying	40.6°C and $\alpha = 0.20–1.00$ 28.4°C and $\alpha = 0.20–1.00$	Retention of 52.37% and 53.52% of water-soluble polysaccharides at both drying temperatures at $\alpha = 0.20$	Chin and Law (2010)
Squash	Intermittent vibrating fluidized bed	Saw-tooth at 30°C Sinusoidal at 30°C Square wave at 25°C Square wave at 30°C	Retention of beta-carotene up to 87.2% with lowest drying time and energy consumption	Pan et al. (1998)

3.3.3 Physical Changes

The effect of intermittent drying on other physical changes such as microstructure modification, shrinkage and rehydration characteristics has been investigated experimentally. Ong and Law (2011a) studied the microstructure and optical properties of salak fruit under intermittent heat pump drying and pre-treatment conditions. They observed that the sample under periodic heat supply showed fewer wrinkled cells compared to step-up temperature and constant mode. Other physical characteristics, for example, rehydration ability of squash (Pan et al. 1998) and fissuring of rice kernel (Aquerreta et al. 2007), have been found to improve by incorporating intermittent drying. Pan et al. (1998) proved that the rehydration ability of the squash gained by intermittent drying is higher when compared to continuous drying.

TABLE 3.2
Comparison of Quality Attributes of Food Products by Different Types of Intermittent Drying

Product	Mode of Drying with Intermittency (∝)	Drying Conditions	Drying Time (min)	% Reduction in Drying Time	Browning Index (BI)	Colour Degradation (ΔE)	References
Potato	Hot air ∝ = 1	80°C	345	—	42.53 ± 1.11	—	
	IR ∝=1 = 1	1.4 m/s	208	39.71	38.77 ± 0.62	10.4 ± 0.83	Chua and Chou (2005);
	IR ∝=0.25 = 0.25		174	49.57		8.6 ± 0.68	Kalathur and
	IR ∝=0.50 = 0.50		119	65.51		11.8 ± 0.94	Kurumanchi (2010)
	IR ∝=0.60 = 0.60		115	66.67		12.1 ± 0.97	
	Combined hot air and IR		195	43.48	32.48 ± 0.44	—	
Carrot	Hot air ∝=1 = 1	80°C	345	—	107.49 ± 1.12	—	
	IR ∝=1 = 1	1.4 m/s	300	13.04	106.07 ± 0.85	10.0 ± 0.80	Chua and Chou (2005);
	IR ∝=0.25 = 0.25		220	36.23		3.2 ± 0.25	Kalathur and
	IR ∝=0.50 = 0.50		150	56.52		6.2 ± 0.49	Kurumanchi (2010)
	IR ∝=0.60 = 0.60		145	57.97		13.1 ± 1.00	
	Combined hot air and IR		195	43.48	100.53 ± 0.72	—	
Onion	Hot air	60°C	340	—	18.99	26.60	Praveen et al. (2005)
	IR	2 m/s	280	17.65	15.85	30.62	
	Combined hot air and IR		220	35.29	13.74	20.63	
Blueberry	Hot air	60°C 4 m/s	960	—	IR drying produced much firmer-texture product with much increased drying efficiency compared to conventional air drying		Shi et al. (2008)
	Infrared, IR = 4000 W/m^2	60°C	540	44.00			
		70°C	210	78.00	Effective moisture diffusivity and activation energy were higher at IR drying compared to conventional air drying		
		80°C	120	88.00			
		90°C	90	91.00			

(*Continued*)

TABLE 3.2 (*Continued*)
Comparison of Quality Attributes of Food Products by Different Types of Intermittent Drying

Product	Mode of Drying with Intermittency (∝)	Drying Conditions	Drying Time (min)	% Reduction in Drying Time	Browning Index (BI)	Colour Degradation (ΔE)	References
Ganoderma tusgae	Heat pump, ∝=1.00 = 1.00	28.4°C	1800	—	—	—	Chin and Law (2010)
	Heat pump, ∝=0.67= 0.67		2400	33.00	—	12.72	
	Heat pump, ∝=0.33 = 0.33		3120	47.69	—	9.30	
	Heat pump, ∝=0.20 = 0.20		4860	62.96	—	10.88	

Mean values ± standard deviation (n = 3 replications) within the same column are not significantly different ($p > 0.05$).

During continuous drying, same amount of energy was supplied throughout the drying process, resulting in quality degradation, heat damage to the surface and wastage of heat energy. On the contrary, these improvements by intermittent drying possibly can be attributed to the temperature and moisture redistribution during the tempering period. The strategy of using intermittency allows time to transfer moisture from the center to sample surface during the tempering period. Thus, quality degradation and heat damage can be minimized by applying intermittent drying.

3.4 CONCLUSION

Types of dryers used in intermittent dehydration discussed in this report include conventional intermittent dryers, pulse fluidized bed dryers, pulse combustion dryers, intermittent heat pump dryers and periodic microwave-powered dryers. The choice of dryers for dried product quality preservation depends on the products' sensitivity towards heat and its surroundings. In general, intermittent drying retains quality attributes of dried products better than that of conventional continuous drying. The multistage intermittent drying still has room for research gaps and further improvements.

Different intermittent drying methods and their impact on product quality are presented. This review paper conclusively demonstrates that intermittent drying is an effective method for enhancing food product quality. The choice of dryers is largely dependent on the sensitivity of dried products towards heat and their surroundings. However, a general comparison is not possible since previous studies are conducted under various drying conditions and intermittency strategies. Furthermore, the complexity of food materials as well as the heat and mass transfer involved during tempering preclude better understanding of the process. As such, the drying conditions and the equation for intermittency ratio should be standardized across the research community. It is proposed that the operating conditions should be manipulated according to the moisture transport mechanism and energy efficiency at different stages of drying in order to optimize the drying kinetics and improve the product quality. Further experimental and theoretical studies on intermittent drying are required to establish a better strategy.

REFERENCES

Aquerreta, J., Iquaz, A., Arroqui, C., and P. Virseda. 2007. Effect of high temperature intermittent drying and tempering on rough rice quality. *Journal of Food Engineering*, 80: 611–618.

Baker, C. 1997. *Industrial Drying of Foods*. London, U.K.: Blackie Academic & Professional.

Balakrishnan, M., Raghavan, G., Sreenarayanan, V., and R. Viswanathan. 2011. Batch drying kinetics of cardamom in a two-dimensional spouted bed. *Drying Technology*, 29(11): 1283–1290.

Beary, E.S. 1988. Comparison of microwave drying and conventional drying techniques for reference materials. *Analysis Chemistry*, 60: 742–746.

Burande, R., Kumbhar, B., Ghosh, P., and D. Jayas. 2008. Optimization of fluidized bed drying process of green peas using response surface methodology. *Drying Technology*, 26(7): 920–930.

Carroll, M.B. and S.W. Churchill. 1986. A numerical study of periodic on-off versus continuous heating by conduction. *Numerical Heat Transfer*, 10: 297–310.

Cernîşev, S. 2010. Effects of conventional and multistage drying processing on non-enzymatic browning in tomato. *Journal of Food Engineering*, 96(1): 114–118.

Chandan, K., Karim, M.A., and U.H.J. Mohammad. 2014. Intermittent drying of food products: A critical review. *Journal of Food Engineering*, 121: 48–57.

Chin, S. and C.L. Law. 2010. Product quality and drying characteristics of intermittent heat pump drying of *Ganoderma tsugae* Murrill. *Drying Technology*, 28(12): 1457–1465.

Chong, C. and C.L. Law. 2011. Application of intermittent drying of cyclic temperature and step-up temperature in enhancing textural attributes of dehydrated *Manilkara zapota*. *Drying Technology*, 29(2): 245–252.

Chua, K.J. and S.K. Chou. 2005. A comparative study between intermittent microwave and infrared drying of bioproducts. *International Journal of Food Science and Technology*, 40: 23–39.

Chua, K.J., Hawlader, M.N.A., Chou, S.K., and J.C. Ho. 2002. On the study of time-varying temperature drying-effect on drying kinetics and product quality. *Drying Technology*, 20(8): 1559–1577.

Chua, K.J., Mujumdar, A.S., Hawlader, M.N.A., Chou, S.K., and J.C. Ho. 2001a. Convective drying of agricultural products, effect of continuous and stepwise change in drying air temperature. *Drying Technology*, 19(8): 1949–1960.

Chua, K.J., Mujumdar, A.S., Hawlader, M.N.A., Chou, S.K., and J.C. Ho. 2001b. Batch drying of banana pieces—Effect of stepwise change in drying air temperature on drying kinetics and product colour. *Food Research International*, 34(8): 721–731.

Dontingy, P., Angers, P., and M. Supino. 1992. Graphite slurry dehydration by infrared radiation under vacuum conditions. *Drying*, 92: 669–678.

Dostie, M., Seguin, J.N., Maure, D., Ton-That, Q.A., and R. Chatingy. 1989. Preliminary measurements on the drying of thick porous materials by combinations of intermittent IR and continuous convection heating. *Drying*, 92: 513–520.

Elcin, D. and O. Belma. 2009. Colour change kinetics of microwave-dried basil. *Drying Technology*, 27: 156–166.

Estürk, O. and Y. Soysal. 2010. Drying properties and quality parameters of dill dried with intermittent and continuous microwave-convective air treatments. *Journal of Agricultural Sciences*, 16: 26–36.

Gan, S.H., Ong, S.P., Chin, N.L., and C.L. Law. 2015. Colour change, nitrite content and rehydration capacity of edible bird's nest by advanced drying method. *Drying Technology*, 34(11): 1330–1342.

Ginzburg, A.S. 1969. *Application of Infrared Radiation in Food Processing*. London, U.K.: Leonard Hill, pp. 174–254.

Gong, Z.X., Mujumdar, A.S., Itaya, Y., Mori, S., and Hasatani, M. 1998. Drying of clay and non-clay media: Heat and mass transfer and quality aspects. *Drying Technology* 16(6): 1119–1152.

Gunasekaran, S. 1999. Pulsed microwave-vacuum drying of food. *Drying Technology* 17(3): 395–412.

Ho, J.C., Chou, S.K., Chua, K.J., Mujumdar, A.S., and M.N.A. Hawlader. 2002. Analytical study of cyclic temperature drying: Effect on drying kinetics and product quality. *Journal of Food Engineering*, 51(1): 65–75.

Hofsetz, K., Lopes, C., Hubinger, M., Mayor, L., and A. Sereno. 2007. Changes in the physical properties of bananas on applying HTST pulse during air-drying. *Journal of Food Engineering*, 83(4): 531–540.

Huang, J. and M. Zhang. 2015. Effect of three drying methods on the drying characteristics and quality of okra. *Drying Technology*, 34(8): 900–911.

Huang, L. and A. Mujumdar. 2004. Spray drying technology—Principles and practice. In: A. Mujumdar, ed., *Guide to Industrial Drying: Principles, Equipment and New Developments*, 1st ed. Mumbai, India: Colours Publications, pp. 101–132.

Itaya, Y., Uchiyama, S., and S. Mori. 2001. Internal heating effect on ceramic drying by microwaves. In: A. Mujumdar, ed., *Proceedings of Asia-Australia Drying Conference Malaysia*. Bali, Indonesia: Institution of Chemical Engineers Publication.

Jumah, R.Y., Mujumdar, A.S., and G.S.V. Raghavan. 1996. A mathematical model for constant and intermittent batch drying of grains in a novel rotating jet spouted bed. In: I.W. Turner and A.S. Mujumdar, eds., *Mathematical Modelling and Numerical Techniques in Drying Technology*, Marcel Dekker, New York, pp. 339–380.

Kalathur, H.V., Hungalore, U.H., and S.M.S.R. Kurumanchi. 2010. Hot air assisted infrared drying of vegetables and its quality. *Food Science and Technology Research*, 16(5): 381–388.

Kowalski, S. and A. Pawlowski. 2011. Intermittent drying: Energy expenditure and product quality. *Chemical Engineering & Technology*, 34(7): 1123–1129.

Kowalski, S. and J. Szadzińska. 2014. Kinetics and quality aspects of beetroots dried in non-stationary conditions. *Drying Technology*, 32(11): 1310–1318.

Kudra, T. 2008. Pulse-combustion drying: Status and potentials. *Drying Technology*, 26(12): 1409–1420.

Kumar, D.G.P., Hebbar, H.U., Sukumar, D., and M.N. Ramesh. 2005. Infrared and hot-air drying of onions. *Journal of Food Processing and Preservation*, 29: 132–150.

Li, Y., Cao, C., Yu, Q., and Q. Zhong. 1998. Study on rough rice fissuring during intermittent drying. *Drying Technology*, 17(9): 1779–1793.

Li, Z.Y., Ye, J.S., Wang, H.T., and R.F. Wang. 2006. Drying characteristics of green peas in fluidized beds. *Transactions TSTU*, 12(3A): 668–675.

Li, Y., Zhang, T., Wu, C., and C. Zhang. 2014. Intermittent microwave drying of wheat (*Triticum aestivum* L.) seeds. *Journal of Experimental Biology and Agricultural Sciences*, 2(1): 32–36.

Marfil, P.H.M., Santos, E.M., and V.R.N. Telis. 2008. Ascorbic acid degradation kinetics in tomatoes at different drying conditions. *LWT—Food Science and Technology*, 41(9): 1642–1647.

Masters, K. 2002. *Spray Drying in Practice*. Denmark: Spray Dry Consult International ApS, Charlottenlund, Denmark, p. 464.

Mcminn, W. and T. Magee. 1997. Kinetics of ascorbic acid degradation and non-enzymatic browning in potatoes. *Food and Bioproducts Processing*, 75(4): 223–231.

Minea, V. 2013. Heat-pump–assisted drying: Recent technological advances and R&D Needs. *Drying Technology*, 31(10): 1177–1189.

Mujumdar, A.S. and S. Devahastin. 2008. Fundamental principles of drying. *ME5202 Industrial Transfer Processes*. Singapore: National University of Singapore. https://www.arunmujumdar.com/file/Publications/books/ME5202_2011_Mujumdar.pdf. Accessed December 12, 2015.

Ondier, G., Siebenmorgen, T., and A. Mauromoustakos. 2012. Drying characteristics and milling quality of rough rice dried in a single pass incorporating glass transition principles. *Drying Technology*, 30(16): 1821–1830.

Ong, S. and C.L. Law. 2011a. Drying kinetics and antioxidant phytochemicals retention of salak fruit under different drying and pre-treatment conditions. *Drying Technology*, 29(4): 429–441.

Ong, S. and C.L. Law. 2011b. Microstructure and optical properties of salak fruit under different drying and pre-treatment conditions. *Drying Technology*, 29(16): 1954–1962.

Ong, S., Law, C.L., and C.L. Hii. 2011. Effect of pre-treatment and drying method on colour degradation kinetics of dried salak fruit during storage. *Food Bioprocess Technology*, 5(6): 2331–2341.

Ong, S., Law, C.L., and C.L. Hii. 2012. Optimization of heat pump–assisted intermittent drying. *Drying Technology*, 30(15): 1676–1687.

Paakkonen, K., Havento, J., Galambosi, B., and M. Pyykkonen. 1999. Infrared drying of herb. *Agricultural and Food Science in Finland*, 8: 19–27.

Pan, Y., Zhao, L., Dong, Z., Mujumdar, A., and T. Kudra. 1999. Intermittent drying of carrot in a vibrated fluid bed: Effect on product quality. *Drying Technology*, 17(10): 2323–2340.

Pan, Y., Zhao, L., and W. Hu. 1998. The effect of tempering-intermittent drying on quality and energy of plant materials. *Drying Technology*, 17(9): 1795–1812.

Pekke, M., Pan, Z., Atungulu, G., Smith, G., and J. Thompson. 2013. Drying characteristics and quality of bananas under infrared radiation heating. *International Journal Agricultural & Biological Engineering*, 6(3): 58–70.

Prasertsan, S. and P. Saen-saby. 1998. Heat pump drying of agricultural materials. *Drying Technology*, 16(1): 235–250.

Praveen, K.D.G., Umesh, H.H., Sukumar, D., and M.N. Ramesh. 2005. Infrared and hot-air drying of onions. *Journal of Food Processing and Preservation*, 29(2): 132–150.

Qu, Q.L., Yang, X.Y., Fu, M., Chen, A.M., Zhang, X.H., He, Z.P., and A.G. Qiao. 2016. Effects of three conventional drying methods on the lipid oxidation, fatty acids composition, and antioxidant activities of walnut (*Juglans regia* L.). *Drying Technology*, 34(7): 822–829.

Sahin, S., Sumnu, G., and F. Tunaboyu. 2013. Usage of solar-assisted spouted bed drier in drying of pea. *Food and Bioproducts Processing*, 91(3): 271–278.

Shi, J., Liu, D.H., Ibarra, A.C., Kakuda, Y., and S.J. Xue. 2008. The scavenging capacity and synergistic effects of lycopene, vitamin E, vitamin C, and β-carotene mixtures on the DPPH free radical. *LWT-Food Science and Technology*, 41(7): 1344–1349.

Soysal, Y., Ayhan, Z., Eştürk, O., and M. Arıkan. 2009. Intermittent microwave–convective drying of red pepper: Drying kinetics, physical (colour and texture) and sensory quality. *Biosystems Engineering*, 103(4): 455–463.

Steed, L., Truong, V., Simunovic, J., Sandeep, K., Kumar, P., Cartwright, G., and K. Swartzel. 2008. Continuous flow microwave-assisted processing and aseptic packaging of purple-fleshed sweet potato purees. *Journal of Food Science*, 73(9): E455–E462.

Udomkun, P., Nagle, M., Mahayothee, B., Nohr, D., Koza, A., and J. Müller. 2015. Influence of air drying properties on non-enzymatic browning, major bio-active compounds and antioxidant capacity of osmotically pre-treated papaya. *LWT—Food Science and Technology*, 60(2): 914–922.

Vega-Valencia, Y., Cruz y Victoria, M., Vizcarra Mendoza, M., and I. Anaya Sosa. 2014. Intermittent drying of nopal (*Opuntia Ficus Indica*) in a fluidized bed pilot dryer adapted with revolving chambers. *Journal of Food Process Engineering*, 37(3): 211–219.

Wang, L., Cui, F., and H. Sunada. 2007. Improvement of the dissolution rate of nitrendipine using a new pulse combustion drying method. *Chemical & Pharmaceutical Bulletin*, 55(8): 1119–1125.

Wu, Z. and A. Mujumdar. 2004. Pulse combustion drying. In: A. Mujumdar, ed., *Guide to Industrial Drying: Principles, Equipment and New Developments*, 1st ed. Mumbai, India: Colours Publications, pp. 133–155.

Xu, F., Chen, Z., Li, C., Liao, J., Zhang, H., and C. Chen. 2012. Intermittent microwave drying of litchi: Drying kinetics and quality formation in colour difference. *Advanced Materials Research*, 482–484: 2090–2095.

Xu, L., Li, S., and H. Sunada. 2007. Preparation and evaluation of ibuprofen solid dispersion systems with kollidon particles using a pulse combustion dryer system. *Chemical & Pharmaceutical Bulletin*, 55(11): 1545–1550.

Yanniotis, S., Taoukis, P., Stoforos, N., and V. Karathanos. 2013. *Advances in Food Process Engineering Research and Applications*. New York: Springer.

Zbicinski, I., Jakobsen, A., and J.L. Driscoll. 1992. Application of infrared radiation for drying of particulate material. *Drying*, 92: 704–711.

Zhao, D., An, K., Ding, S., Liu, L., Xu, Z., and Z. Wang. 2014. Two-stage intermittent microwave coupled with hot-air drying of carrot slices: Drying kinetics and physical quality. *Food Bioprocess Technology*, 7(8): 2308–2318.

Zhu, Y. and Z. Pan. 2009. Processing and quality characteristics of apple slices processed under simultaneous infrared dry-blanching and dehydration with continuous heating. *Journal of Food Engineering*, 90(4): 441–452.

4 Influence of Process Parameters Variation on Hybrid Nonstationary Drying

Stefan Jan Kowalski, Justyna Szadzińska, and Andrzej Pawłowski

CONTENTS

4.1 INTRODUCTION

Drying is a fundamental unit operation in the production of many goods like ceramics, wood, and biological products like fruits and vegetables. Convective drying of engineering or biological materials is one of the oldest methods of their preservation, often realized in natural conditions such as solar drying. Various industrial sectors consume different fractions of energy for drying; for example, food and agriculture (15%), paper (33%), timber (15%), ceramic and building materials (18%), and textiles (5%) together amount to a significant value of 16%, on average. Hot air drying reveals several advantages that explain its wide use in industry such as in civil engineering and the food industry. It is a relatively inexpensive and easily operated drying technology. On the other hand, hot air drying is considered a highly destructive method, particularly for thermally sensitive materials like biological materials. Therefore, in the convective drying mode it is important to select the proper drying conditions and suitable drying medium parameters (air temperature, humidity, velocity, etc.), as well as a rational heat dosage. All these factors affect the final quality. Strong shrinkage and damage due to drying-induced stresses affect engineering materials like ceramics and wood, but color, flavor changes, and the reduction of nutritional value are the most important issues associated with the quality of dried biological products (Ho et al. 2002; Chua et al. 2003; Kumar et al. 2014).

One of the technical solutions to improve drying efficiency is so-called *intermittent drying*, which is based on controlled supply of thermal energy that varies periodically over time by the drying design or by its operation. Such an approach enables controlled supply of energy affecting the drying efficiency, that is, the time of drying as well as the quality of dried products. Time-varying drying conditions improve heat and mass transfer and reduce the time required to achieve a desired moisture content, and thereby minimize energy consumption (Pan et al. 1999; Thomkapanich et al. 2007; Ong et al. 2012). Several studies (Chua et al. 2001; Kowalski et al. 2013b) have reported that intermittent drying prevents overheating, which gives much better quality as well as higher retention of nutrition of biological products than drying under constant conditions.

In some cases, the acoustic emission (AE) method is used to control drying of engineering materials and to estimate the moment of change in drying parameters that should be initiated to avoid destruction of these products. It also helps to arrange suitable schedules of intermittent drying for the individual kinds of engineering products (Kowalski and Mielniczuk 2007; Kowalski and Szadzińska 2012).

On the other hand, it is possible to maintain good quality of dried fruits and vegetables not only through a suitable arrangement of drying conditions but also through their pretreatment, for example, osmotic dehydration (OD), which enables moisture removal up to 50% w.b. The experiences of many researchers (Lewicki and Lenart 2006; Konopacka et al. 2009; Kowalski and Mierzwa 2013) suggest that OD has a significant influence on the physical and chemical properties of the final product, that is, it extends the shelf life and improves its storage capacity due to lower water activity (a_w).

Intermittent drying offers better energy efficiency due to reduced heat input, shorter drying time, and lower energy consumption, as well as improved product quality as a result of reduction of drying-induced stresses and shrinkage (Zhang and Mujumdar 1992; Kowalski 2003; Bon and Kudra 2007). The essence of this kind of drying consists in slowing down the drying rate just before the material starts to crack and increasing the drying rate when there is no danger of crack formation (Herrithsch et al. 2008; Kowalski and Pawłowski 2015). The AE method is used to detect the incipient crack formation, and thus to identify the moment at which the change of drying conditions should be initiated (Kowalski 2010).

Due to the adverse effects of heat treatment, the combination of novel and traditional technologies has become very popular in recent times. The use of microwave and ultrasound in drying is of great interest because it influences the main characteristics like drying time and product quality. Application of alternative techniques in drying of foods enhanced by ultrasound accelerates moisture removal and improves the nutritional aspect of biological products because of lower drying temperatures than in conventional methods (Knorr et al. 2004; Kowalski et al. 2016). Ultrasound used in product pretreatment can also be considered as an alternative to blanching before the drying process. In turn, ultrasound-assisted OD is mainly used to increase effective water diffusivity. One can generally expect an increase in drying time but this is reflected in a compensating increase in water loss and solid gain and quality improvement, that is, greater nutrition content, as well as better flavor and color (Rawson et al. 2011; Nowacka et al. 2012).

In this chapter, the kinetics of intermittent drying was determined experimentally for kaolin clay, wood, and food products. The AE method in drying has been presented and described. The results of drying in intermittent conditions are compared with the data obtained by drying processes carried out in stationary conditions to demonstrate the advantages of the nonstationary drying method. Moreover, the mathematical model of drying kinetics was developed and used to calculate numerically the kinetic curves of intermittent drying and to assess the drying efficiency enhanced with microwaves and/or ultrasound.

4.2 EXPERIMENTAL

4.2.1 Equipment and Tested Materials

The drying experiments were carried out in two different dryers. The first one was the laboratory chamber dryer Zalmed SML 42/250/M modified for the purpose of nonstationary (intermittent) drying, realized by periodic inlet of vapor or cold air into the dryer. Additionally, the AE sensor was connected to the bottom of the scale pan and the AE descriptors like number of AE events and AE energy to monitor *online* the development of sample crack formation (Kowalski and Pawłowski 2011a,b; Kowalski et al. 2013a,b). The second dryer was a hybrid one which allows the simultaneous or separate utilization of different energy sources such as convection, microwaves, and ultrasound (Kowalski et al. 2016). The measurements of total color change ΔE were carried out using the colorimeter Konica Minolta (Japan) model CR-400 and indicated in CIELab color space. Water activity a_w was measured using the temperature and humidity converter with a function of water activity measurement with Testo (Germany) model 650/0628.0024. The absorbance measurements were performed with the spectrophotometer Shimadzu (Germany) model UV-2401PC. Strength tests, which allowed the evaluation of the mechanical cohesion of dried kaolin, were performed using the universal strength machine Cometech QC-508A1 with an appropriate head.

The materials used in experiments were industrial raw materials particularly sensitive to heat with different physicochemical properties and performance, that is, engineering (construction) materials such as ceramics (kaolin clay KOC type) and wood (walnut wood *Juglans regia* L.) and biological materials such as red beetroots (*Beta vulgaris* L.) and green pepper (*Capsicum annuum* L.).

4.2.2 Drying Programs

In the case of intermittent drying of kaolin and wood the periodic changes of air temperature (T_a) as well as air humidity were programmed. Additionally, nonstationary drying of the kaolin KOC clay was carried out in a laboratory hybrid dryer with periodic application of microwave and/or ultrasound. In the case of intermittent drying of biological materials there were periodic changes of air temperature. Moreover, an effort was made to find the best drying conditions from the kinetics point of view, combining pretreatment and convective drying in constant and intermittent conditions. The red beetroot was dehydrated in 5% aqueous solution of sodium chloride for 30 min at $T = 23°C$ or blanched in hot water at $T = 95°C$ for 5 min.

The main problem during investigation of engineering and biological materials is their different behavior during dehydration. Therefore, the materials were first dried convectively in constant conditions and were controlled and monitored continuously to determine the best moment for applying intermittent conditions (changes in air temperature/humidity). This moment can also be determined based on the calculations of strains and stresses generated during drying (Kowalski et al. 1997). However, it is not always accurate as the strains and stresses are strongly dependent on the internal structure of the material, being in some cases very complicated and not uniform.

TABLE 4.1
Drying Programs

Constant Conditions	Intermittent Conditions
	Engineering Materials
Kaolin KOC clay	
T_a = 100°C constant RH	T_a = 100°C during CDRP, periodic air temperature changes between 100°C and 50°C during FDRP, equilibrium RH
	T_a = 100°C, RH changes during FDRP between 4% and 60%–80%
T_a = 70°C constant RH	T_a = 70°C, periodic microwave application (100 W)
	T_a = 70°C, periodic microwave (100 W) and ultrasound application (200 W)
Walnut wood	
T_a = 100°C constant RH	Periodic air temperature changes between 100°C and 40°C during whole process, equilibrium RH
	T_a = 100°C with periodic air humidity changes between 5% and 55%–70% during whole process
	Biological Materials
Red beetroot	
T_a = 80°C constant RH	T_a = 80°C during CDRP, periodic air temperature changes between 50°C and 90°C during FDRP, equilibrium RH
Green pepper	
T_a = 70°C constant RH	T_a = 70°C during CDRP, periodic air temperature changes between 58°C and 72°C during FDRP, equilibrium RH

The information received from the convective drying processes carried out in constant conditions allows the creation of suitable and appropriate programs of intermittent drying, which are presented in Table 4.1.

4.3 DRYING CONTROL DUE TO ACOUSTIC EMISSION

Drying processes ought to be appropriately arranged and operated to obtain high-quality dried products, that is, products without excessive deformations, surface cracks, and above all crosswise fractures in case of engineering materials. Nonuniform moisture distribution in products arising during drying causes nonuniform material shrinkage and generates stresses, which are responsible for permanent deformations and material fracture. It is possible to analyze the risk of fracture in drying samples both theoretically and experimentally. A mechanistic drying model forms the basis for numerical simulations of drying kinetics and analysis of the drying-induced stresses (Kowalski 2003). In this way, it is possible to determine the spots where the drying-induced stresses reach their maximum level and where a crack may possibly occur (Kowalski and Rybicki 2007). The theoretical predictions are confronted with the experimental data obtained due to application of the AE, which enables monitoring *on line* the development of the drying-induced fractures caused by stresses (Kowalski et al. 2000). The exemplary AE equipment (AMSY5 by Vallen) is presented in Figure 4.1.

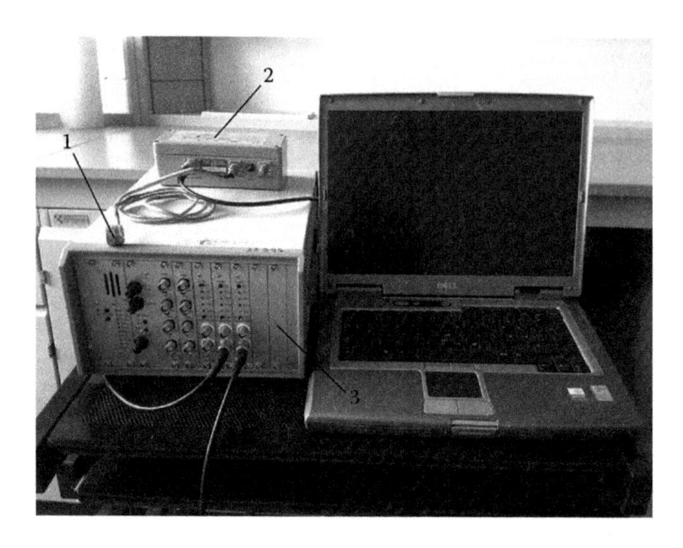

FIGURE 4.1 Vallen AMSY5 AE setup.

Acoustic emission is a nondestructive method allowing indirect control of microfracture and macrofracture development during drying and above all the identification of the period and also the place where the fractures start to develop. In this sense, AE is a method that enables control of the drying process and helps to protect the material against destruction (Kowalski 2010; Kowalski et al. 2013a,b).

Acoustic wave created during material fracture is collected through piezoelectric probe (1) attached directly or indirectly to the sample. The signal in the probe is transformed from the mechanical wave into an electric signal, which is first amplified in preamplifier AEP3 (2). Next, this signal can be transmitted to the main unit (3), which consists of an amplifier and a processing unit with the ability to connect a number of channels to allow fracture localization, for example. Finally, after appropriate transformation the signal is collected in the Vallen software through the data acquisition card. The AE signal is transformed in various ways so that one can identify received signals with the use of a number of different AE descriptors.

Based on the authors' experience and experiments performed to date, it can be stated that the AE descriptors best reflecting the character of mechanical phenomena occurring in drying materials are

- *Hits rate*: This descriptor shows the dynamics of the destruction development (e.g., a rise of drying temperature involves rapid growth of the AE hits rate). Moreover, this descriptor indicates the stages of drying, in which the reduction or increase of the AE activity takes place.
- *The hit of maximum energy*: This descriptor is more useful than "energy of hits" as it shows the single hit with maximum energy in a given time interval. The descriptor "energy of hits" presents the energy of all hits in a given time interval.
- *The total number of hits and the total energy of hits*: These parameters show some individual phenomena occurring during drying. Thanks to these

descriptors it is possible to distinguish stages of drying in which some irregular changes of the AE energy or the AE hits rate appear. These descriptors point out the critical moments of drying, in which the fracture of drying material may occur.

The realized tests allow the interpretation of the AE signals that may occur during drying of, for example, kaolin clay (Figure 4.2). The first (I) characteristic group of AE signals appears at the beginning of the drying process, the second (II) one appears in the period when the surface layer shrinks intensively, and the third (III) one is sometimes noticeable in the final stage of drying and is identified as being generated by the reversed stresses.

Figure 4.3 presents the rate of AE hits for the five different temperatures of drying. For the conditions of high drying rates created by high temperatures, the rate of AE hits achieves higher values than for lower temperatures. The high active emission of AE signals is reflected in the drying-induced stresses.

Note that the highest peak of the AE hits, which corresponds to a temperature of 120°C, appears earlier than the lower peaks corresponding to the lower drying temperatures.

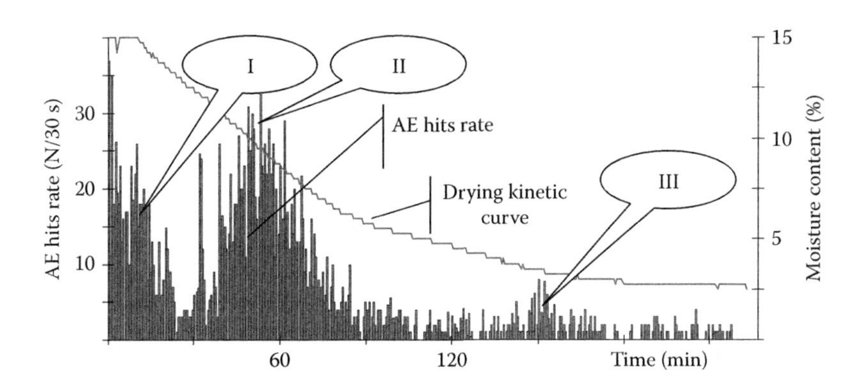

FIGURE 4.2 AE signals and the drying curve.

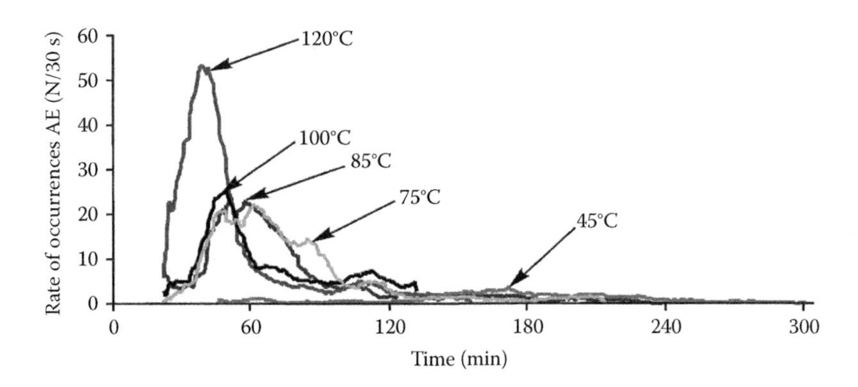

FIGURE 4.3 The rate of AE hits during drying at different temperatures.

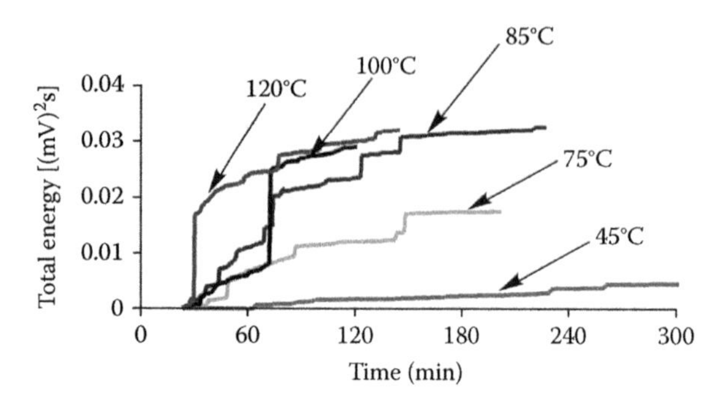

FIGURE 4.4 Total energy of hits during drying process for various conditions.

The primary peak of AE hits appeared in 35 min of drying time, that is, when the tensional stresses at the cylinder surface reached maximum. The secondary peak is visible at about 50 min drying time for a temperature of 100°C. At this time the core of the body starts to dry. The wet core wants to shrink but the surface layer is not able to deform itself because it is almost dry. So, in these circumstances the tensional stresses arise in the core. The secondary maximum is of course much lower than the first one.

Figure 4.4 shows several curves of total AE energy released from kaolin cylinders during drying at different temperatures. Each AE signal carries a certain portion of energy. The flat horizontal lines represent the low-energy signals. Hits of high energy create sudden vertical lines, for example, those visible on the energy curves obtained for drying at temperatures of 120°C and 100°C. These very energetic signals are generated by strong material cracks.

By analyzing Figure 4.4, one can see the differences in released energy for different drying conditions. The curve for 45°C, being almost horizontal, represents low-energy AE signals. This means that drying at this temperature is suitable for a given type of material due to low fracture potential. Thus, the manufactured products are of good quality and without residual stresses. Unfortunately, drying at such a low temperature takes a long time and is unsatisfactory from an economic point of view.

4.4 INTERMITTENT DRYING

4.4.1 Intermittent Drying of Engineering Materials

Drying of engineering materials such as ceramics and wood is often accompanied by their deformation and, more importantly, cracking. Fractured product is usually useless and cannot be processed further or used for any engineering purpose. This is why great attention is paid to reduction of drying-induced stresses, which are the main reason for the abovementioned negative effects (Sherer 1990; Kowalski and Rybicki 1994; Katekawa and Silva 2006). Typical convective drying allows the achievement of desired, good-quality dried engineering materials if the processes are carried out very carefully. However, it comes at the price of a very long-lasting process and high energy consumption, because of their low efficiency. During past decades global legislation related to reduction of energy consumption in industry led to the improvement of commonly used inefficient drying techniques and the promotion of safety processes. Apart from hybrid drying techniques, which need high financial outlays on the apparatus, other methods to improve heat and mass transfer were proposed. One group of such methods is the nonstationary drying technique, also called *intermittent* drying. This method is based on periodic changes of drying conditions/parameters, as shown in Figure 4.5, and enables a large variety of process parameters.

The main advantage of intermittent drying is the necessity for only small changes in the drying apparatus or in the drying program. Application of intensive and so-called relaxation or tempering periods during drying will affect reduction of moisture and temperature gradients throughout the processed material, which in fact will lead to decrease of drying-induced stresses.

The described effects occurring during intermittent drying lead to the conclusion that the essence of this kind of drying consists in slowing down the drying process before the moments when the material is the most prone to crack and maximizing the drying rate, when no danger exists of material fracturing. This is why nonstationary drying techniques are highly valuable for constructive material in civil industry, where large batches are dried and easy process control needs to be available. The following section presents the selected results of intermittent drying processes realized by periodic changes of temperature and humidity of drying medium.

4.4.1.1 Intermittent Drying of Walnut Wood

The influence of periodic air temperature and humidity changes on drying kinetics was considered, and the obtained results were compared to pure convective stationary drying. The solid state of wood allows the measurement of only its surface temperature and the control of the behavior of the sample during drying using the AE method. The effect of such a process is that just at the beginning of drying some changes inside the material occur as AE descriptors rise quickly in the heating period.

Intermittent processes were designed on the basis of information received from drying at constant conditions. Figure 4.6 presents the kinetics of this process with 1 h periods of temperature changes. It can be seen that high reduction of drying medium temperature entails significant reduction in the drying rate and below critical moisture content can even lead to cessation of the process (period from 510 to 570 min). This effect unnecessarily lengthens the drying time, increases the drying costs, and causes reduction in productivity. By comparing stationary and nonstationary drying curves, one can draw the conclusion that the best way to carry out intermittent drying with temperature changes for wood would be constant decrease of the relaxation period time in reference to constant intensive drying periods.

Figure 4.7 presents the drying kinetics carried out with periodic changes of air humidity. Drying with very high air humidity during relaxation periods leads to an increase in the moisture content of the material. In fact, these changes follow from vapor condensation on the sample surface, which in the saturated state during drying at air temperature of 100°C will always be colder than the introduced vapor. The vapor condensation on the material surface, in contrast with the former temperature changes, will lead to the drying process stopping completely. In such a process, when condensation appears it is necessary to extend the heating periods in relation to relaxation periods in order to achieve a desired final moisture content and reduce drying time.

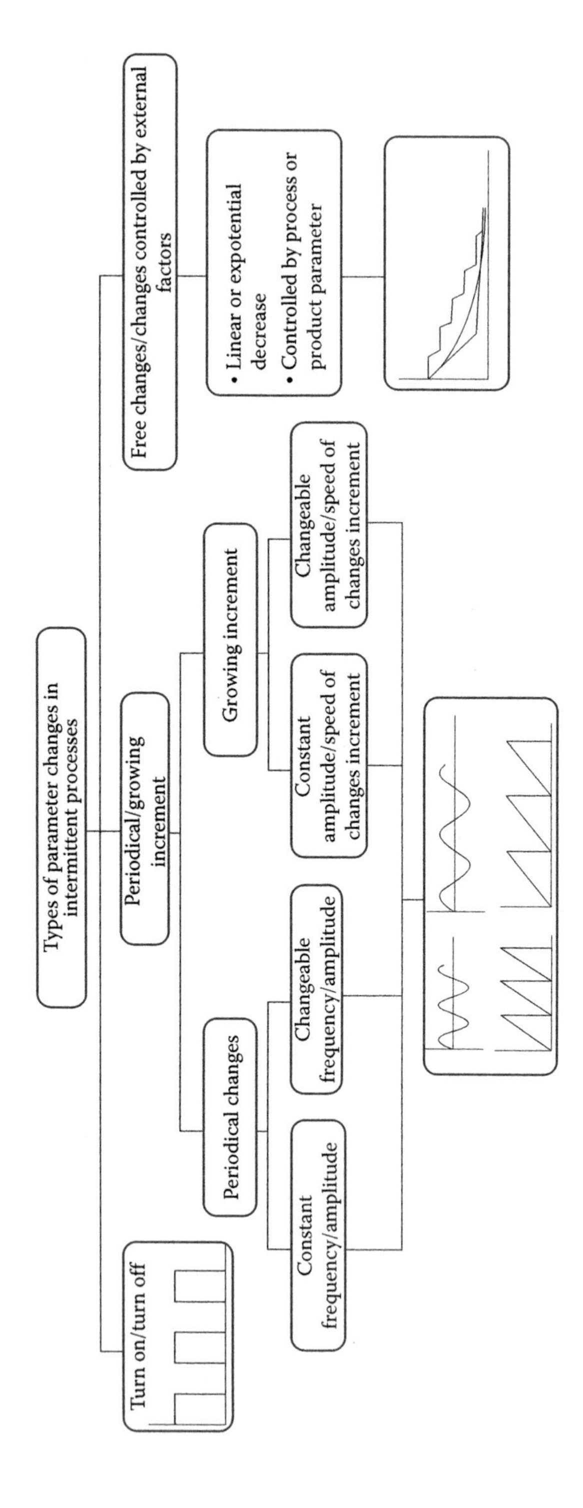

FIGURE 4.5 Parameter changes in intermittent processes.

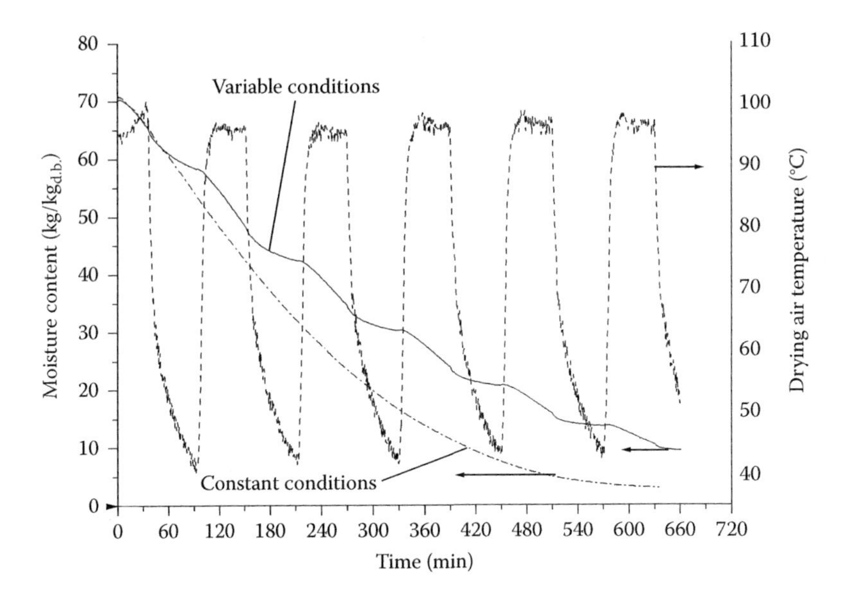

FIGURE 4.6 Drying curve and air temperature during intermittent drying of walnut wood with temperature changes.

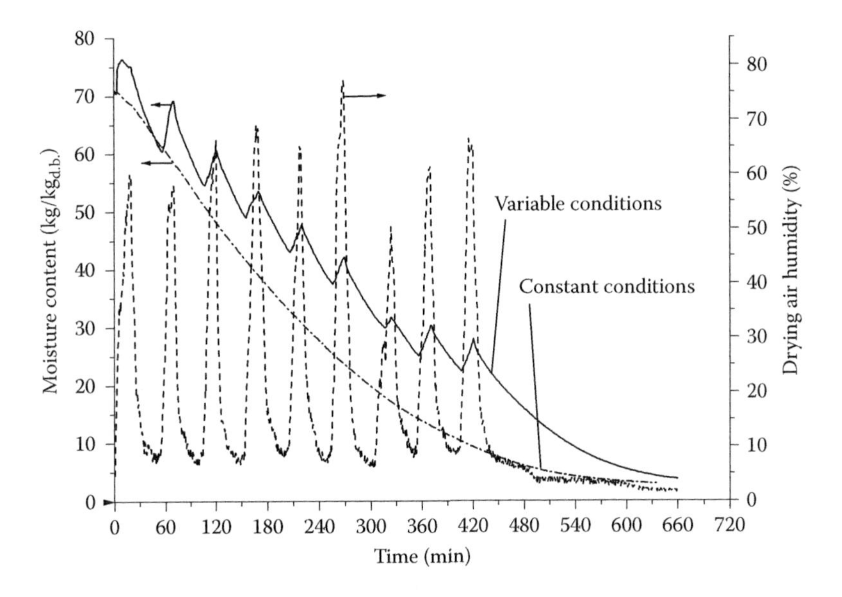

FIGURE 4.7 Drying curve and air humidity during intermittent drying of walnut wood with humidity changes.

The other possibility presented in Figure 4.7 utilizes the stationary process after intermittent drying, when there is no danger of material fracture. This option is quite promising as the experimental results show very intensive drying when the material reveals low moisture content. Such drying behavior exists not only at the end of drying but also after each relaxation period. This can be explained by an increase of material temperature during vapor condensation and more uniform moisture distribution throughout the material. The appropriate choice of intermittent drying parameters can finally lead to only a very small increase in drying time even when some increase in moisture content appears during relaxation periods.

4.4.1.2 Intermittent Drying of Kaolin KOC Clay

The properties and structure of kaolin clay enable the measurement of the internal and external temperature of the material during drying. This is why kinetics investigation for ceramic material can be done through analysis of moisture content and sample temperature curves vs. drying time.

In order to investigate the intermittent drying, first the reference values from the continuous drying process have to be determined. They show that fractures for clay-like materials appear mainly in the falling drying rate period (FDRP) when drying-induced stresses are high enough to destroy material. This is why first only pure convective processes are considered. In such conditions the best way is to dry material intensively in the constant drying rate period (CDRP) and after this stage apply periodic relaxation periods.

In the first intermittent drying tests the temperature variations were applied, where consecutive periods lasted 1 h. The kinetics curves for this process (Figure 4.8) show slight reduction in drying rate during tempering periods, and rise of drying rate

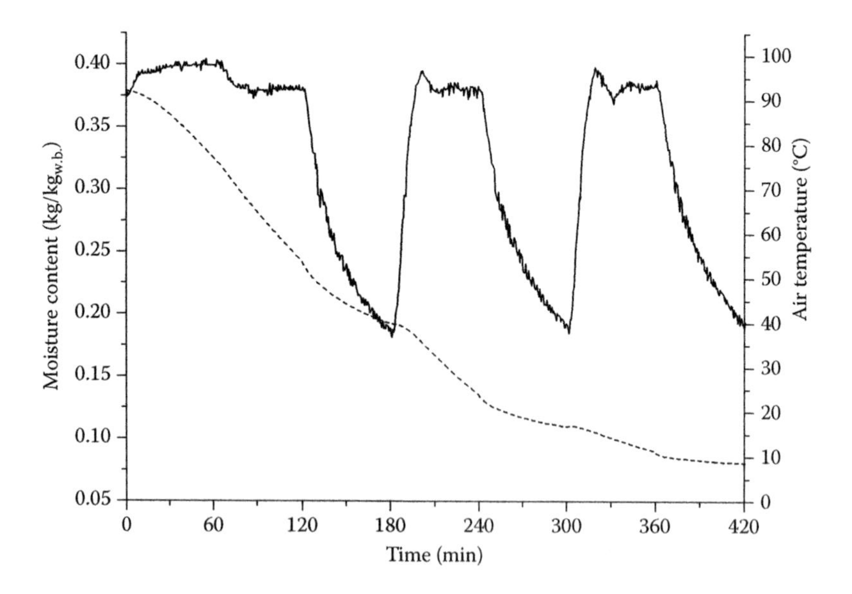

FIGURE 4.8 Drying curve and air temperature for kaolin clay dried with periodic air temperature changes.

during consecutive intensive drying periods. It is worth noting that during intensive drying periods the process is faster than during stationary drying for corresponding moisture content in material. This only slightly lengthens the drying time, which is a very important issue from the industrial application point of view.

The other convective process with humidity changes of drying medium can be considered, and the example of drying kinetics for this process is shown in Figure 4.9.

Similarly to the abovementioned case, here also the parameter changes were made just before and during FDRP. Based on the drying curve, it can be stated that drying is stopped during the humidification period due to change in the driving force, but also by the vapor condensation on a cooler sample, the temperature of which was about 50°C. Analysis of material temperature also confirms this phenomenon as the material temperature increased immediately as an effect of condensation and thus transferred heat into the material.

Apart from intermittent convective drying, such processes can be carried out using hybrid drying techniques where pure convective drying is enhanced with periodic application of an additional energy source. Hybrid intermittent drying was carried out in the second drier and compared with a representative convective process, which presents slightly different drying kinetics than in the earlier case. Considering intermittent hybrid drying where other energy sources are used, the periods of their application have to be considered once again and will be different than in pure convective intermittent drying. In hybrid processes, their intensification should take place during CDRP when material does not reveal cracks, and partially in FDRP to intensify the process. In such processes, pure convection will play the role of relaxation periods, in contrast to nonhybrid processes.

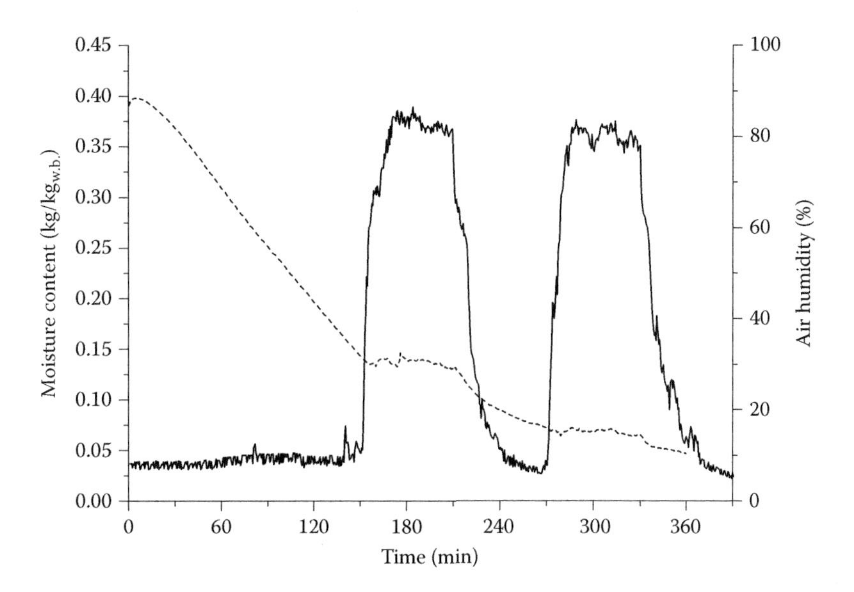

FIGURE 4.9 Drying curve and air humidity for kaolin clay dried with periodic air humidity changes.

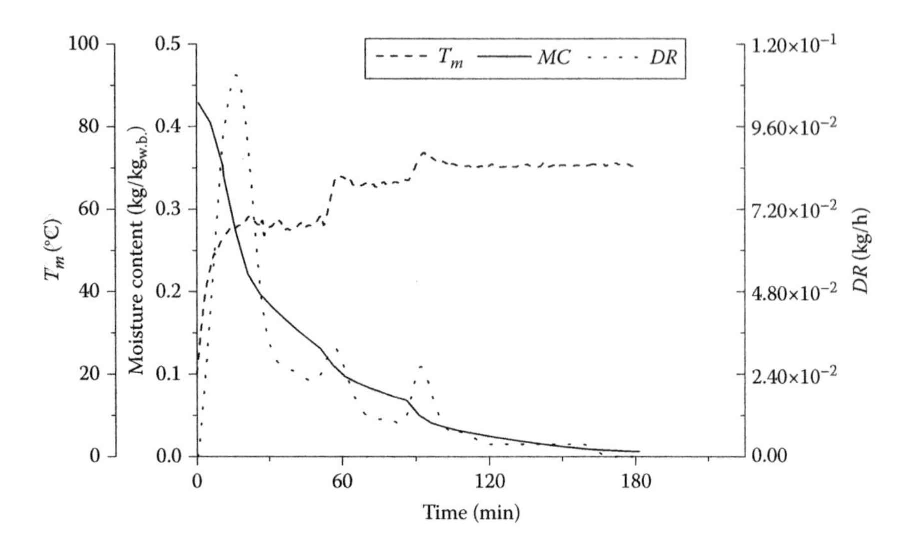

FIGURE 4.10 Drying curve, material temperature, and drying rate for kaolin clay dried at 70°C with periodic microwave application.

The first designed convective-microwave intermittent drying process was carried out with only three microwave periods of 20, 5, and 5 min, because microwaves are very effective and their constant or even longer application could lead to unwanted effects such as material overheating or destruction. Figure 4.10 shows that even short periodic microwave radiation allows significant increase of the drying rate, and shortens the process by almost half in comparison with the analogical convective process. This positive effect is very important in industry, where productivity is one of the main economic factors.

Further improvement of intermittent drying can be done by utilization of additional energy sources; however, it is very important to choose different combined drying techniques wisely with respect to their energy transfer mechanism.

Figure 4.11 presents drying kinetics for the convective process enhanced periodically by microwaves and ultrasound. High-power acoustic energy was applied with the purpose of intensification of heat and mass transfer near the sample surface as well as inside it. Ultrasound power of 200 W together with microwaves appears to be insufficient to visibly improve the drying kinetics. The overall drying time was not noticeably reduced with respect to nonstationary drying without ultrasound where a slight increase in drying rate is observed. However, it can be observed that the temperature of dried material was slightly lower in the intensive drying periods, which may result from faster evaporation of atomized water near the sample surface. The possibility of the sample temperature decreasing is a great advantage for drying of any heat-sensitive materials, not only constructive materials.

4.4.1.3 Summary

As productivity is one of the most important parameters in industry, the intermittent drying technique can be applied to meet the relevant expectations. As not all

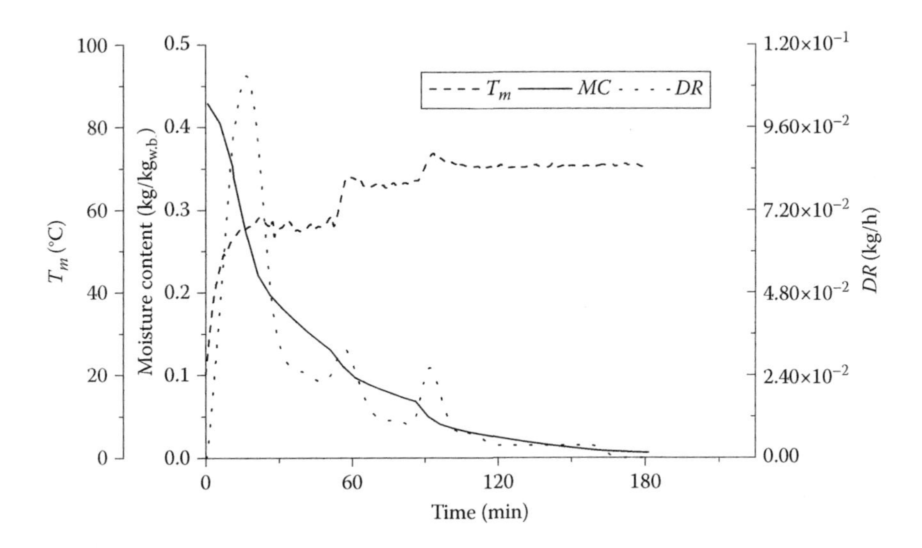

FIGURE 4.11 Drying curve, material temperature, and drying rate for kaolin clay dried at 70°C with periodic microwave and ultrasound application.

of the techniques presented above shorten the overall drying time, their advantage is homogenization of temperature and moisture distribution and thus reduction of drying-induced stresses generated in material. Typical convective intermittent drying is a promising alternative to stationary drying if only the moments where drying-induced stresses rise above critical values are taken into account. However, the better, more expensive form from the investment point of view are the hybrid intermittent drying techniques. Here, even a very brief application of an additional energy source leads to significant shortening of drying time in combination with material moisture and temperature homogenization, preventing stresses from arising.

4.4.2 Energy and Quality Aspect in Intermittent Drying of Engineering Materials

Drying of engineering materials is focused mainly on product performance but also process productivity and its costs. The selection of an appropriate nonstationary convective or hybrid drying program and determination of its optimal parameters is essential to achieve high-quality products at a potentially minimal cost.

From the energetic point of view, intermittent drying, which is based on controlled supply of energy (thermal or electrical) that varies periodically, can be a solution to decrease its consumption. This follows from the fact that nonstationary drying may offer higher energy efficiency as a result of shorter drying time period, reduced total heat input, and lower air consumption (Jumah et al. 2007; Thomkapanich et al. 2007; Kowalski and Pawłowski 2011a,b).

Apart from the economic aspect, processing of engineering material needs to be considered from the quality and property point of view as the main material determinant for further usefulness. This is why such materials often undergo strength tests, which allow the determination of their required durability. On the other hand, some of the engineering materials also play a decorative role, for example, wood elements where strength is not as important as material appearance. Drying with periodic changes of process parameters should meet the expectations of both the abovementioned aspects that are revealed by new and modified drying processes.

4.4.2.1 Quality of Intermittent Dried Walnut Wood

Intensification of wood drying, which reveals anisotropic properties, needs to be conducted with great care as it can promote higher moisture and temperature gradients throughout the material and generate higher drying-induced stresses. Figure 4.12 shows the propagation of crack formation in stationary convective drying.

The wood samples presented in Figure 4.12 were located in such a way to maximize the surface of heat and mass transfer, which is not applicable to industry.

Analysis of these photos gives the background for designing intermittent drying processes. It can be observed that significant material cracks develop just at the beginning of the drying process and the cracks increase until 240 min. After this time, stress reversion is observed as the cracks became smaller, and finally the material becomes stable until the end of drying. The reversion of tension decreases the size of cracks; however, it does not "close" them totally. Even small fractures influence the material properties as the wood tissue is ripped and less prone to further processing.

In Figure 4.13 the samples with insulated and uninsulated surfaces for mass exchange are presented and show different fractures from each side. The intermittent dried samples show reduction in the number and size of cracks on both insulated and uninsulated surfaces; however, none of the applied temperature changes proved to be sufficient to prevent material fracture.

The intermittent process carried out with air humidity changes enables significant reduction of drying stresses and avoids material fracture. It is visible on both sample surfaces which have insignificant cracks not disqualifying the wood from further processing. During relaxation periods, the process was stopped due to vapor condensation on the sample surface. This phenomenon reduces the moisture gradient inside the material and as a result drying-induced stresses. This can explain the difference in quality between both samples dried in intermittent conditions with temperature and air humidity changes.

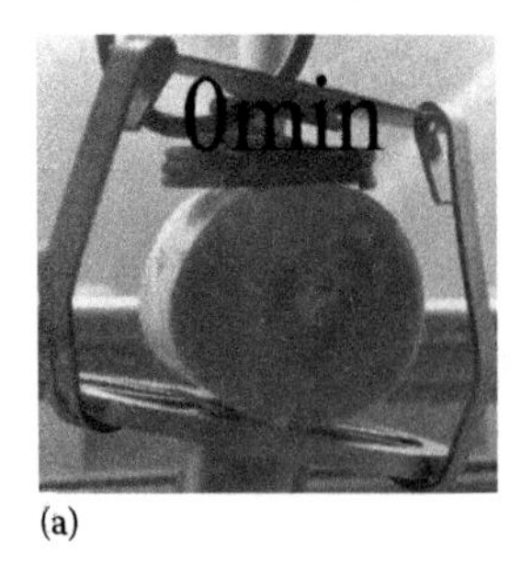

(a)

(b)

(c)

FIGURE 4.12 Photos of walnut wood fracture propagation over time during drying at 100°C. (a) 0 min, (b) 360 min, and (c) 600 min.

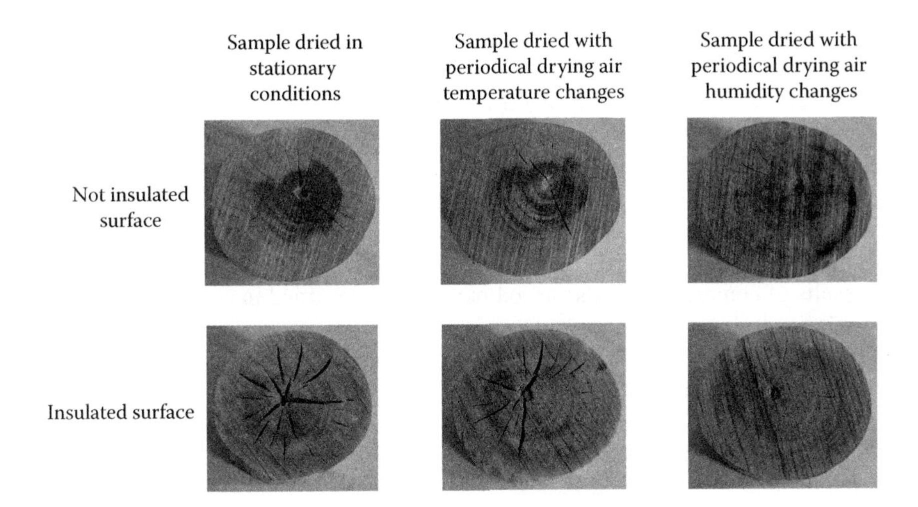

FIGURE 4.13 Photos of walnut wood fracture after drying for processes carried out in stationary and intermittent conditions.

4.4.2.2 Quality of Intermittent Dried Kaolin KOC Clay

Ceramic materials belong to a group of engineering materials and reveal a number of interesting properties such as high temperature resistance, good mechanical properties, good insulation, and dielectric properties. As the quality of ceramic products is determined by a number of parameters dependent on their intentional application, two of them will be shown as the sample appearance and its strength.

Kaolin samples dried stationary are disqualified from further use as after firing they would not achieve appropriate strength and appearance due to many cracks appearing during drying. Application of the drying parameter variation over time enables reduction of the abovementioned negative drying effects. Figure 4.14 shows a comparison of the final sample appearance achieved after stationary and nonstationary drying processes. The temperature oscillations result in only a few small

FIGURE 4.14 Quality of kaolin clay samples drier: (a) in constant conditions, (b) with periodic air temperature changes, (c) with periodic air humidity changes.

cracks which appear near the upper surface, where the drying-induced stresses arise first. The second nonstationary drying process with air humidity changes prevents fracture as hot vapor applied in the drier results in almost immediate cessation of the process and increases drying stresses. This shows that small modification in pure convective drying can improve the material quality significantly (Figure 4.14c).

Another very important quality parameter is the material strength when the interior of the product can undergo fractures, even when its surface does not show any negative changes, for example, during hybrid drying techniques. Figure 4.15 presents the results of compression tests carried out for samples dried in constant and hybrid intermittent drying conditions.

The compression force for materials dried in both hybrid intermittent drying programs was higher than that achieved in the pure convective process. Microwave periodic application enables the improvement of material resistance at about 20%, while additional ultrasound application increases the resistance by about 35% in comparison to pure convectively dried samples.

The increase of material strength is a big advantage of intermittent hybrid drying; however, it is not always connected to material appearance, as confirmed by the sample photos presented in Figure 4.16.

The relatively low temperature of convective drying allows the slowing of the process and saving of the material properties, reflected in good appearance without visible fractures. The improvement of this drying process through periodic application of microwaves and/or ultrasound enables an increase in product strength; however, the surface of such samples has a visibly worse appearance. The most interesting fact is that the quality of the samples determined on the basis of their appearance cannot be correlated in any way with their mechanical properties for samples dried with the intermittent hybrid technique.

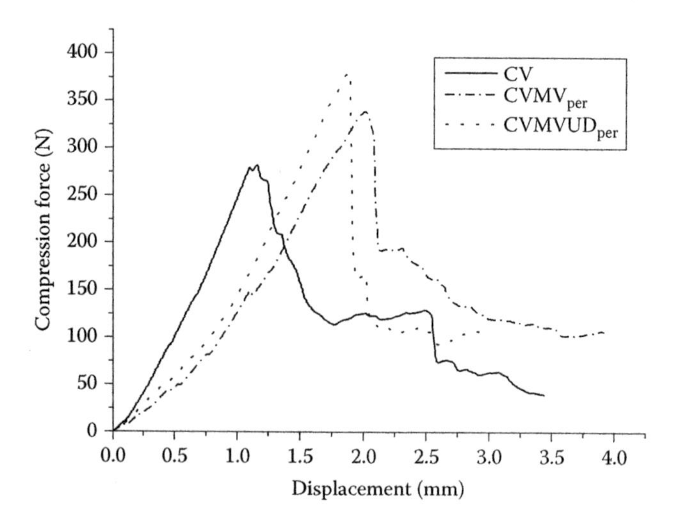

FIGURE 4.15 The curves of material strength for kaolin clay dried in constant and intermittent processes (with periodic microwave and ultrasound application).

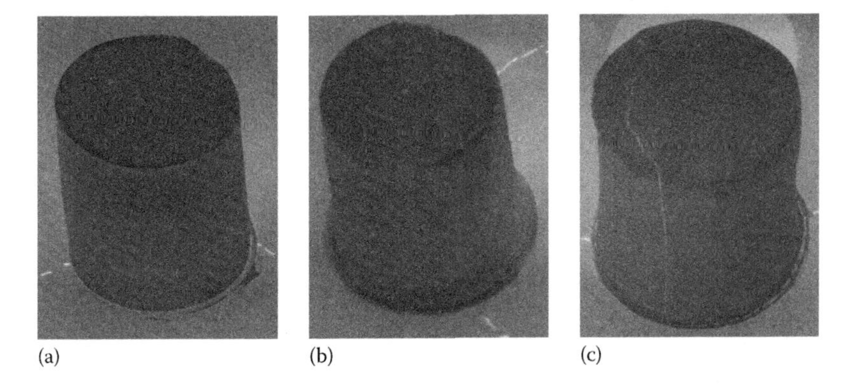

(a) (b) (c)

FIGURE 4.16 Quality of samples dried in hybrid drier: (a) in constant conditions, (b) with periodic microwave application, (c) with periodic microwave and ultrasound application.

4.4.2.3 Energy Aspect in Intermittent Drying of Engineering Materials

The main problem often considered during the evaluation of energy consumption is how to compare different processes carried out in different driers as they each have their own specific energy consumption. To evaluate the reliability of the energetic aspect for different drying processes, they have to be carried out in the same drier where different process conditions are examined. Additionally, the material undergoing drying should, if possible, be of the same size, weight, and initial moisture content.

Analyzing the energetic aspect for drying of kaolin clay, one can state that under the constant conditions described in the previous section, the energy consumption increases proportionally to drying time, and total energy used for the process amounts to 1.5 kWh. The application of process variation will affect energy consumption; for example, during the process carried out with temperature changes in relaxation periods the energy can be significantly reduced. Unfortunately, in the following intensive drying period the energy consumption rises significantly to heat up the drying medium once again. Nevertheless, the overall energy consumption for this process amounts to only 1.15 kWh. In other processes with variable air humidity the total consumed energy amounted to 3.3 kWh, which results from the large amount of energy used for vapor production in a humidifier. Figure 4.17 presents the energy consumption over time, where the characteristic intermittent periods are highly visible.

As shown in Figure 4.17, the final energy efficiency mainly affects the periodic changes of process parameters in the second stage of drying. The situation looks different in hybrid intermittent processes where the added energy usually results in significant reduction of total drying time. This in turn leads to reduction of the overall energy utilized more effectively, especially in the intensive drying periods. The pure convective process carried out in the hybrid drier consumed a total of 4.13 kWh. Unfortunately, this energy cannot be compared to the examples presented above as two different driers were used for those experiments. In the case of hybrid intermittent

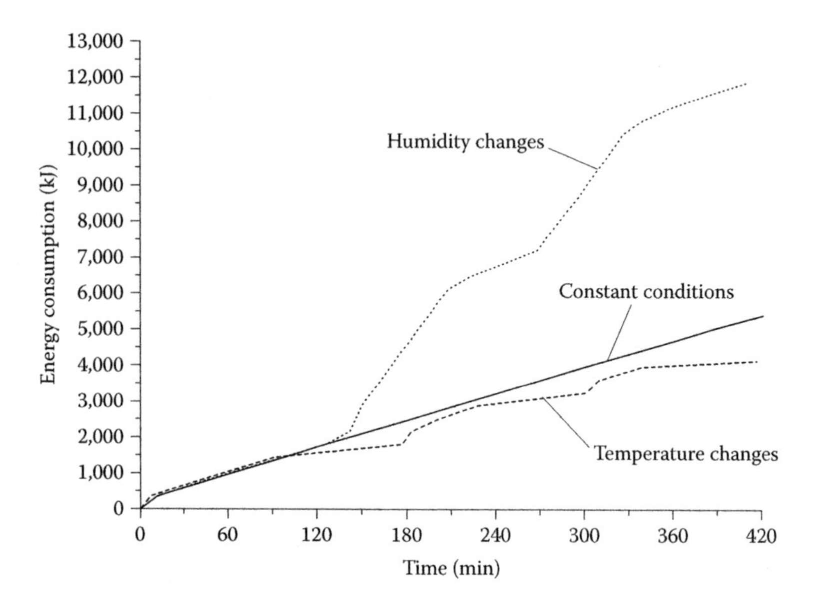

FIGURE 4.17 Energy consumption during constant and intermittent drying processes.

drying processes, due to the application of additional energy, its consumption in a time unit was higher. Nevertheless, because the overall drying time reduction was almost half, the total energy consumption (EC) for each of them was lower than for the purely convective process. This is why in the process assisted periodically with microwaves the overall energy was reduced to 2.88 kWh and the additional application of ultrasound allowed the reduction of this value even further to 2.72 kWh.

4.4.2.4 Summary

Depending on the purpose of the considered dried material, as well as its structure and properties, different drying schemes and techniques can be used. The energetic and quality aspects are not always coherent, and thus the appropriate design of intermittent drying processes is very important. The examples presented in this section can have a direct insight into the most important parameters such as productivity, energy consumption, or the best material quality. Nonetheless, using each of the intermittent drying processes will lead to improvement of the process parameters in comparison with traditional stationary convective drying.

4.4.3 Intermittent Drying of Biological Materials

Different physical, chemical, mechanical, and biochemical properties of biological materials, as well as quality requirements for dried food products, caused intensive development of various drying technologies and drying equipment. In the case of conventional drying methods for agricultural products (e.g., convective drying), there are many problems associated with their effective usage, that is, long drying time, low efficiency of the process, nonuniform product quality, and relatively high operating costs. Striving to overcome the above mentioned limitations typical for the convective drying method led to the creation of numerous new drying technologies (Mujumdar 2007; Mujumdar and Law 2010). The development of alternative drying techniques is dictated by the requirements of the market and society's increasing demands for better-quality products, as well as by energy limitations. One of the most interesting and recommended solutions in drying technology is nonstationary drying, also called intermittent drying, based on the changes in drying conditions. The first mention of intermittent drying appeared in the 1950s. It was described by Łykow in his fundamental work "Drying Theory" (Łykow 1968). For some time this method has attracted the interest of researchers, but its potential has not been sufficiently used in industry. In this method, at given time intervals there is a slowing down of the drying process (relaxation/tempering) when there is a danger of material destruction and an acceleration when the danger is over. Such relaxation plays a key role as the moisture and temperature profiles in the dried material during intensive drying are leveling. In general, the mechanism of intermittent drying is based on a controlled supply of heat energy that changes at regular or irregular time intervals (Chua et al. 2003). Programming such changes in operating parameters is related to their frequencies and amplitudes, which after application remain constant or periodically increase and decrease over time. Modification of the commonly used convective drying method proved to be highly valuable when applied to drying of biological materials, such as fruits and vegetables, which are extremely sensitive to thermal conditions (Kowalski et al. 2013a,b; Kowalski and Szadzińska 2014). The following section describes the selected results of convective-intermittent drying realized in periodic air temperature changes. The analysis of nonstationary drying kinetics, including comparison between intermittent and continuous convective drying, is presented below.

4.4.3.1 Intermittent Drying of Red Beetroot

At the beginning of the studies, the convective drying test at a constant air temperature of 80°C was investigated. Furthermore, the effect of stepwise changed air temperature and pretreatment methods such as OD and blanching (BL) on the drying kinetics of red beetroot was investigated. Figure 4.18 presents the drying kinetics for constant and intermittent conditions and the stages of drying: heating (H), constant drying rate period (CDRP), and falling drying rate period (FDRP). A relatively high drying rate and a high moisture capacity between 15 and 120 min of drying and stable material temperature of approximately 43°C were observed in the CDRP. Under constant conditions the longest drying time obtained amounted to about 305 min. For intermittent conditions the total drying time amounted to 248 min.

Periodic changes in air temperature in the range of 50°C–90°C, on average, caused a temporary decrease and increase of the material temperature, which preserved the beets against overheating. The final effect of cooling-heating cycles in the FDRP was an 18% reduction in total drying time, as compared to constant conditions. Figure 4.19 shows intermittent drying processes followed by BL and OD.

The total drying time of red beetroot in the process preceded by BL amounted to 280 min, on average. Due to higher initial moisture content of approximately 9.07 $kg/kg_{d.b.}$, the overall drying time lengthened by about 30 min, as compared to intermittent drying without pretreatment. In turn, the intermittent drying

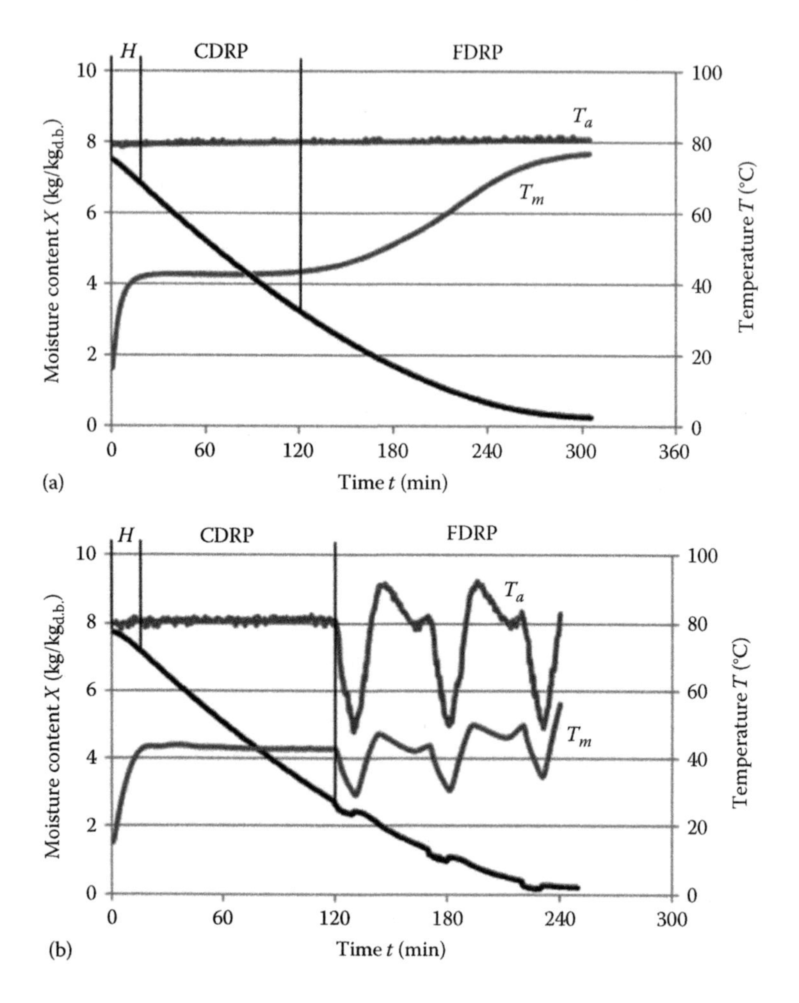

FIGURE 4.18 Drying of red beetroot: (a) constant conditions, (b) intermittent conditions.

process combined with OD resulted in the shortest drying time, that is, 238 min (Figure 4.19b), due to lower initial moisture content of approximately 7.38 kg/kg$_{d.b.}$, caused by preliminary OD and periodic cooling and heating. However, no significant difference was found in the duration of convective-intermittent drying between the samples with and without OD.

4.4.3.2 Intermittent Drying of Green Pepper

The effectiveness of convective-intermittent drying of green pepper was evaluated based on the comparison of drying kinetics obtained at a constant air temperature of 70°C and a variable air temperature of 58°C–72°C.

As shown in the graph in Figure 4.20a, convective drying of green pepper proceeded mainly in the CDRP (i.e., between 10 and 60 min of drying). It follows from the drying curve (50% moisture loss), as well as the sample temperature, which

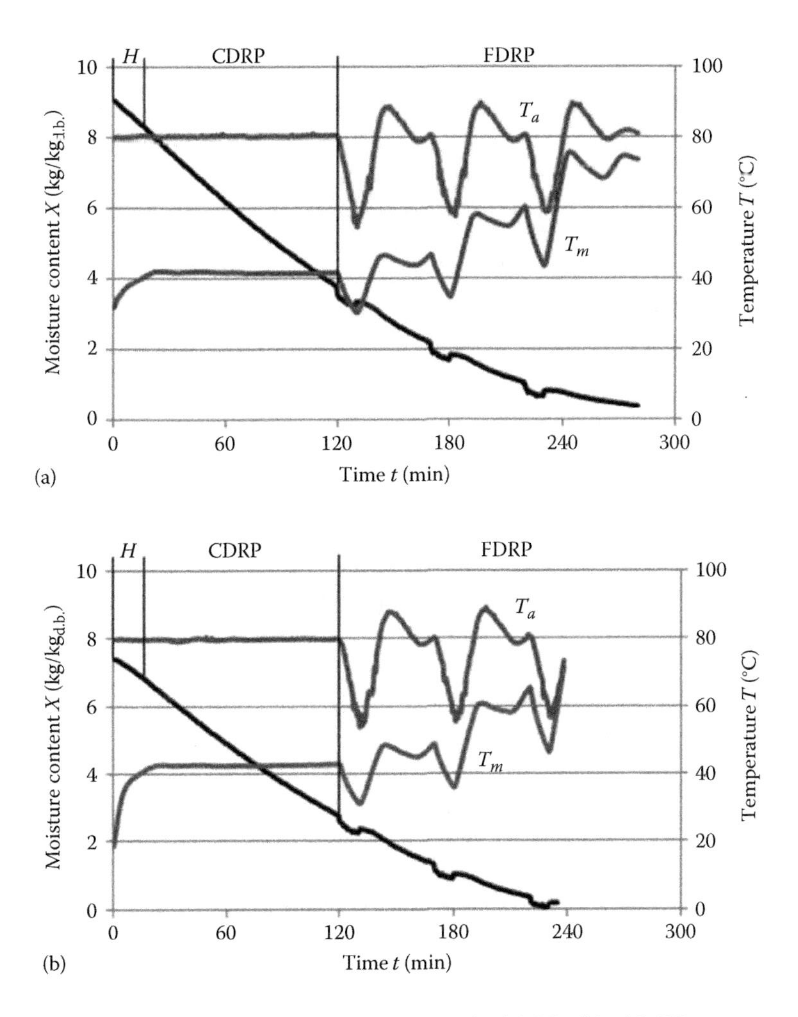

FIGURE 4.19 Intermittent drying of red beetroot: (a) with BL, (b) with OD.

remained at 43.5°C for a long time. Due to the relatively high drying rate, the dehydration of green pepper samples occurred fairly rapidly. The total drying time for constant conditions amounted to 188 min, on average. Due to a very high initial moisture content in green pepper of approximately 12.65 kg/kg$_{d.b.}$, short-lived cooling cycles in the intermittent mode were applied (Figure 4.20b). The cooling-heating periods were introduced after 60 min of drying (i.e., after CDRP), when the drying rate was slowed down. During intermittent drying, the air temperature was approximately in the range of 58°C–72°C, corresponding to decrease in material temperature to about 32°C (in the first cooling) and to 40°C (in the second cooling). When exposure to periodic cooling and heating was used, the overall drying time of green pepper was shortened to about 134 min, thus improving the economy of the process. Finally, this kind of intermittency in drying conditions contributed to a reduction in drying time of 29%, as compared to continuous convective drying.

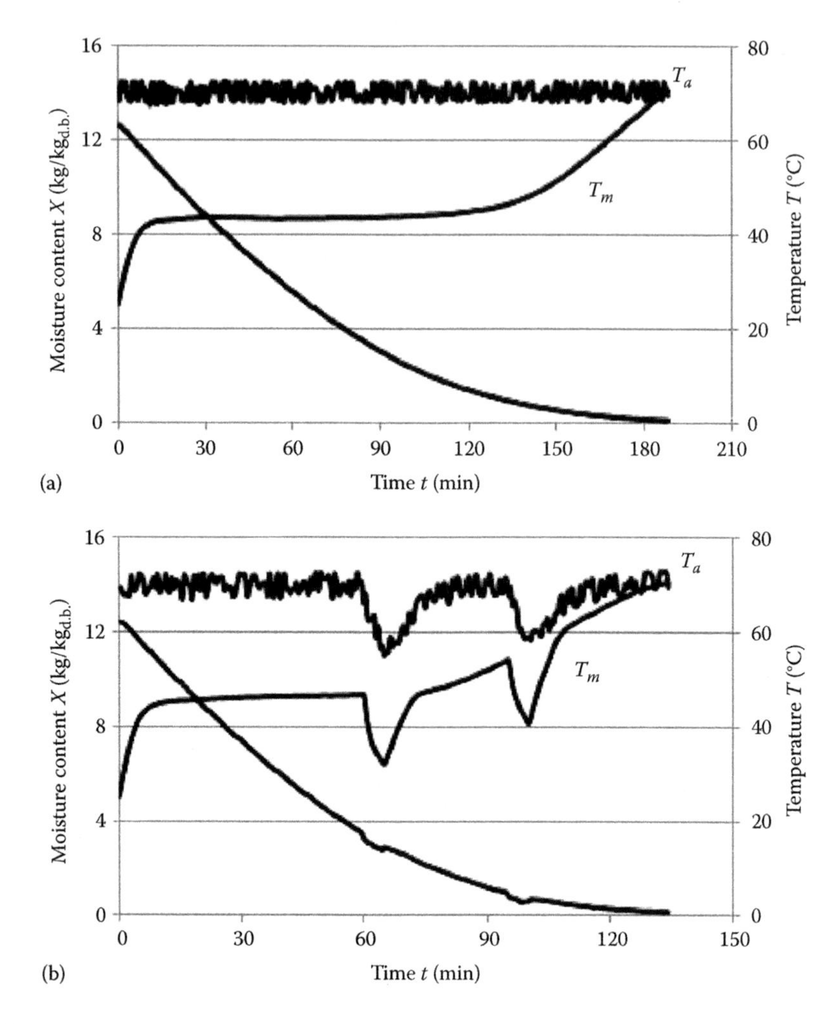

FIGURE 4.20 Drying of green pepper: (a) constant conditions, (b) intermittent conditions.

4.4.3.3 Summary

The main purpose of this research was to modify a commonly used drying method, that is, convective drying, also known as hot air drying, in order to eliminate a number of its drawbacks, that is, long drying time, material overheating, low efficiency of the process, and high operating costs. As a remedy for such disadvantages, intermittent drying conditions were proposed, resulting in higher drying efficiency of biological materials. Based on the obtained results, it was proved that application of changeable air temperature has a favorable impact on drying kinetics, as the intermittency of drying conditions reduces the temperature and moisture gradients and causes the distribution of these variables to become more uniform. In this way, the total drying time of high-moisture materials can be reduced by about 30%, in comparison to convective drying carried out in constant conditions. The application of

OD before intermittent drying is profitable due to the reduction in drying time as well as a better-quality product.

4.4.4 Quality and Energy Aspect in Intermittent Drying of Biological Materials

In view of the rising consumer demand, the final quality of dried products has become an extremely important factor of drying effectiveness. For biological materials such as vegetables and fruits, which are characterized by perishability and a very delicate internal structure sensitive to high temperature, quality is a priority. High quality of processed food products is associated with naturalness and freshness, which reflects retained natural values, sensory properties, and high nutritional content. Drying of biological materials rich in water is aimed at reduction in moisture content and mainly water activity, that is, shelf-life preservation for further consumption and storage. Evaporation of water from fruits and vegetables is a highly complex process, and therefore it is associated with many negative consequences, such as changes in structure, deterioration of sensory properties (color, flavor, aroma), changes in chemical composition, and degradation of bioactive components. Therefore, the primary challenge for dried foods is to reduce the water content (water activity) while preserving the quality (Woodroof and Luh 1986). The quality is defined as a set of important characteristics that determine the usefulness and acceptance of the dried product by the consumer. It includes mainly sensory parameters (texture, color, aroma, flavor), but primarily nutritional/biological value (retention of biologically active ingredients, e.g., vitamins and natural dyes) and microbiological stability, that is, water activity (Mitek and Słowiński 2006; Kumirska et al. 2010).

The amount of energy used for drying is one of the elements of its effectiveness. Past experience has demonstrated that common convective drying is highly energy consuming. One of the recently developed methods for the reduction of energy consumption is the application of intermittent drying. Due to tempering/relaxation periods consisting of, for example, periodic cooling or a complete single stop or multiple operation stops of the dryer, intermittent drying is one of the most promising solutions from the energy improvement point of view, often without increasing the substantial capital cost of the dryer (Gunasekaran 1999; Chin and Law 2010).

The following section describes the selected results of quality assessment of biological materials dried convectively in constant and intermittent conditions realized with periodic air temperature changes. The analysis of total color change, water activity, retention of natural dyes and vitamin C, and EC in drying of biological materials is presented below.

4.4.4.1 Total Color Change

Color is one of the most important quality indicators of both fresh and processed food products. Color depends mainly on the composition and content of natural dyes. The visual attributes of a food product such as attractive appearance and color mainly affect the human senses. The first quality factor analyzed after drying was total color change ΔE. The measurements were indicated by CIELab color space, where L^* is lightness, a^* is the color parameter from red to green, and b^* denotes the color

TABLE 4.2
Total Color Change

Material ΔE	Red Beetroot	Green Pepper
Convective drying	16.51 ± 0.36	8.42 ± 1.16
Convective-intermittent drying	9.44 ± 0.72 9.42 ± 0.10 with OD 5.15 ± 0.09 with BL	4.36 ± 0.10

parameter from yellow to blue. The value of ΔE was calculated using the following formula:

$$\Delta E = \sqrt{\left(\Delta L^*\right)^2 + \left(\Delta a^*\right)^2 + \left(\Delta b^*\right)^2} \tag{4.1}$$

Total color difference between fresh and dried red beetroot and green pepper is shown in Table 4.2.

The results of ΔE showed that the highest color change for the tested biological materials was observed after convective drying in constant conditions. The values of ΔE for the red beetroot samples dried using the convective-intermittent method with and without pretreatment (OD, BL) were significantly below the values obtained after the convective method. The main pigment in red beets (75%–95%) is reddish-violet betanin; however, it degrades when subjected to light, heat, and oxygen. Therefore, the osmosis and BL significantly improved the color and visual appearance of dried beetroots. Photos of dried red beetroots can be found in Kowalski and Szadzińska (2014). In the case of green pepper, the value of total color change for the samples dried in nonstationary conditions was almost twice as low as the samples dried in constant conditions.

4.4.4.2 Water Activity

Water activity is one of the most important quality parameters in the food industry as it determines microbial growth, and thus microbial stability. Most microorganisms do not develop at water activity below 0.6. Generally, microbial growth decreases in the following order: bacteria > yeast > molds. Since food stability usually decreases with an increase in water activity, a number of different pretreatments, for example, OD and addition of salt, have long been used to preserve foods. However, the value of water activity after osmosis does not ensure total microbiological stability. Therefore, in order to prolong the shelf-life of biological materials, they should be subjected to further drying to make them fully safe (Labuza et al. 1970; Sikorski 2007; Bonazzi and Dumoulin 2011; Moazzam 2012).

Water activity a_w in red beetroot and green pepper was measured before and after drying. Randomly selected fresh/dried samples were placed in a chamber of the water activity meter and kept there until equilibrium was reached. Since the moisture profile in the sample had to be aligned after drying, the measurements of a_w started after 3 h from the end of each drying process. The results of water activity are presented in Figure 4.21.

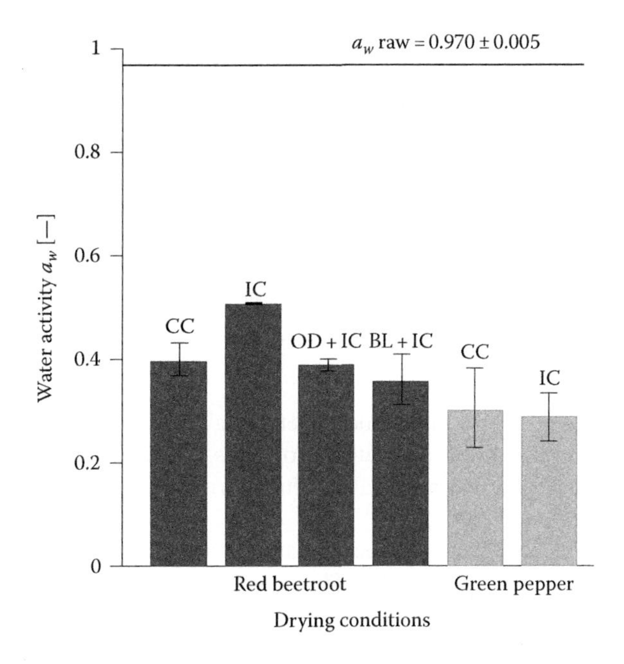

FIGURE 4.21 Water activity of fresh and dried red beetroot and green pepper. *Abbreviations:* CC, constant conditions; IC, intermittent conditions; OD, osmotic dehydration; BL, blanching.

As follows from the above graph, water activity in all raw biological materials amounted to 0.970 ± 0.005 (black line). For the dried samples, the value of a_w was less than 0.6, and thus bacteria, yeast, and mold growth was inhibited. In the case of green pepper samples dried in constant conditions, the water activity was comparable to those dried in intermittent conditions. In turn, the red beetroot samples dried with OD and BL followed by convective-intermittent drying were characterized by a lower water activity, as compared to constant conditions.

4.4.4.3 Retention of Natural Dyes and Vitamin C

Retention is understood as the retained number of bioactive compounds after processing. Fruits and vegetables are rich in vitamins, minerals, trace elements and terpene compounds, flavonoids, tannins, quinones, and phytoncides. The most frequent reason for the loss of many nutrients is thermal processing, including drying. Quality evaluation of dried products is usually carried out on the basis of thermolabile components content such as vitamins, for example, vitamin C. Vitamin C participates in many biochemical reactions, removes free radicals, affects immunity, and is one of the most important antioxidants. In addition, vitamin C possesses bacteriostatic and bactericidal properties and lowers the risk of cancer development. Natural dyes are unstable and easily degraded (oxidation) during processing (including drying) and storage due to the influence of oxygen, light, temperature, and pH, resulting in changes in color and smell. The color of fresh and processed fruits and vegetables is mainly due to carotenoids, chlorophylls, anthocyanins, and betalain pigments. Betalain pigments include reddish-violet betacyanins and yellow betaxanthins.

TABLE 4.3
Retention of Betanin and Vitamin C

Material	Red Beetroot	Green Pepper
Compound	Betanin (%)	Vitamin C (%)
Convective drying	31.40 ± 5.00	60.00 ± 4.00
Convective-intermittent drying	37.70 ± 5.00	75.00 ± 6.00
	77.40 ± 5.00 with OD	
	56.50 ± 5.00 with BL	

The source of these dyes is red beetroots. Betanin gives a specific and attractive red color and moreover has health-promoting properties, is a strong antioxidant, and has anticancer, antibacterial, and antiviral effects (Harmer 1980; Dreosti 1993; Asard et al. 2004; Prakash et al. 2004).

The retention of betanin in red beetroot and vitamin C in green pepper samples after drying is presented in Table 4.3. The content of betanin in red beetroots was determined using the Nilson method (Nilson 1970). Retention of vitamin C in green pepper was examined by the Tillmans method (Tillmans et al. 1932).

As shown in Table 4.3, convective drying causes a significant loss of betanin in red beetroot and vitamin C in green pepper. However, modification of drying conditions (intermittency), as well as application of pretreatment methods such as OD and BL, can preserve the content of these bioactive compounds up to 77%. The retention of another natural dye in food products, that is, β-carotene in carrots dried in intermittent conditions, can be found in Kowalski et al. (2013b).

4.4.4.4 Energy Aspect

Both drying time and energy consumption are of key importance in the cost structure of dried foods production. The problem of energy consumption in drying processes is extremely important, and the aim is to minimize it. In the studies on drying of green pepper and red beetroot, an attempt was made to characterize the energy aspect of convective-intermittent drying. The EC in drying tests of biological materials is given in Table 4.4.

TABLE 4.4
Total Energy Consumption

Material EC (kWh)	Green Pepper	Red Beetroot
Convective drying	0.40 ± 0.01	0.80 ± 0.01
Convective-intermittent drying	0.40 ± 0.01	1.00 ± 0.01
		0.90 ± 0.01 with OD
		1.00 ± 0.01 with BL

In the case of green pepper, the same value of EC was obtained for the processes carried out in constant as well as intermittent conditions. In drying of red beetroot, the lowest EC was observed for continuous convective drying. However, the EC in intermittent drying processes with and without pretreatment amounted to about 1 kWh. The increase in EC for the analyzed nonstationary drying here is closely related to work of the cooling system, that is, alternate switching on and off of the cooler.

4.4.4.5 Summary

The experiments on intermittent drying of biological materials described above proved that quality improvement is the result of periodic changes in material temperature and a shorter exposure to hot drying medium. All of the analyzed quality indicators of dried biological materials indicate much better quality of the products obtained after convective-intermittent drying, as compared to continuous convective drying. Color difference between the samples dried in constant and intermittent conditions follows from the reduction of the heating period, strictly due to periodic cooling and heating of drying medium. As a result, the dried biological material temperature decreases and increases, which affects the color preservation. It is also possible to maintain natural color due to pretreatments applied before drying, for example, BL and OD. Other quality parameters such as retention of natural dyes, that is, betanin in red beetroot, as well as retention of vitamin C in green pepper, also confirm that intermittent drying positively affects the content of nutrients. Furthermore, on the basis of EC, it was found that by carefully programming changes of drying conditions and at a comparable energy expenditure, one can get a better-quality product in a shorter drying time.

4.4.5 Intermittent Drying Controlled by Acoustic Emission Method

As described previously in this chapter, the nondestructive AE method was utilized in drying experiments. As the metal AE sensor needs to be placed in direct or indirect contact with the considered material in the drier, the use of this method in some hybrid drying processes becomes impossible, for example, when using microwaves. Nevertheless this control technique can be successfully used in a number of different drying processes. Figure 4.22 presents the example of AE descriptors collected during convective kaolin clay drying at 100°C.

The small increase of the value of descriptors at the beginning of the process is caused by thermal tensions and shrinkage of dry material. Nevertheless, they do not show the tendency to suddenly increase, which proves that drying is carried out in conditions that are safe for material. Such behavior in the CDRP can be confirmed by the sample observation presented in Figure 4.23, which visualizes the periods of material fracture. When the sample reaches critical moisture content and the process goes into the FDRP, an instant rise of AE descriptors appears, which explains the material fracture in this part of the process. The big advantage of AE in comparison with visual control is the possibility to collect the signals from the whole material and not only analysis of its surface appearance. This is why this tool can be used

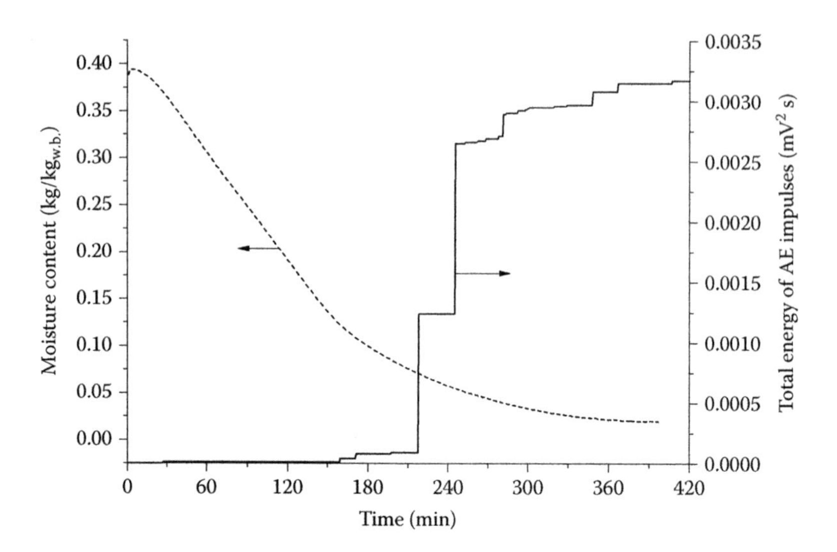

FIGURE 4.22 Drying curve and AE descriptor (total energy of impulses) for drying of kaolin clay at 100°C.

to design intermittent drying processes as well as online control of materials that undergo drying.

Quite low total energy of hits was achieved for stationary processes, and expected lower values for intermittent drying forced the improvement of the sensitivity of the AE device in nonstationary drying processes. Figure 4.23 shows the evolution of AE signals during convective drying with periodic temperature changes. Due to

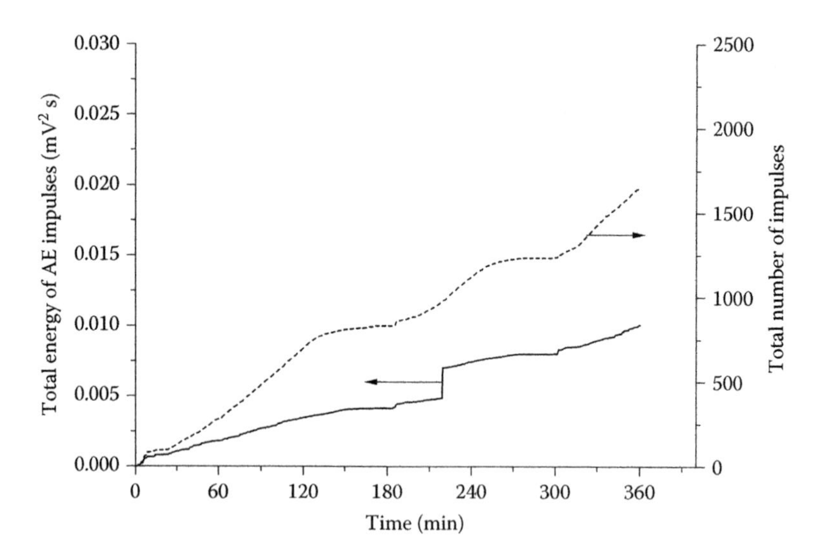

FIGURE 4.23 AE descriptors; total energy and total number of impulses for drying of kaolin clay with periodic air temperature changes.

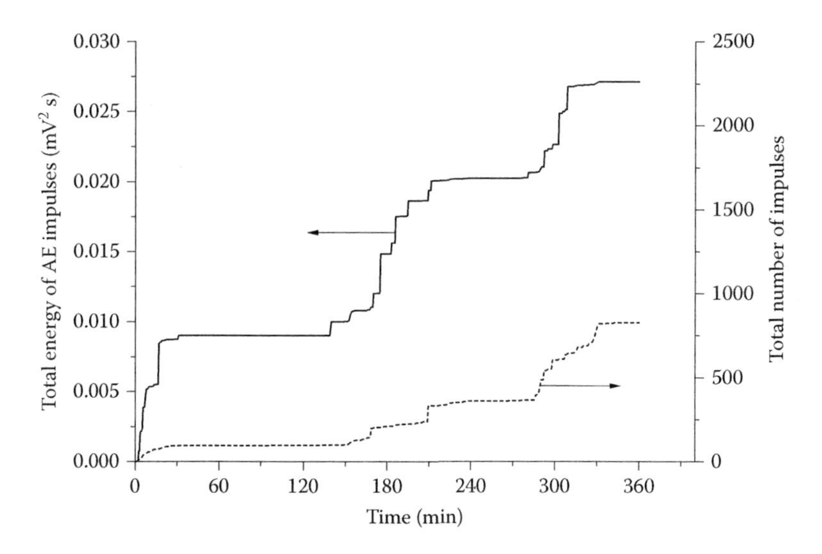

FIGURE 4.24 AE descriptors; total energy and total number of impulses for drying of kaolin clay with periodic air humidity changes.

higher system sensitivity, constant increase in AE summary hits and their energy can be observed; however, they do not show instant and stepwise changes except one. Increase in AE descriptors is stopped after application of relaxation periods, which reflects safe drying conditions and allows the reduction of drying stresses. Small stepwise increase of hits total energy in 180 and 300 min results from noises generated in the moments of drying parameter changes. This effect does not in any way influence the process control for such conditions, but it has to be taken into account during AE description.

Accurate investigation of the AE method in different intermittent drying processes unfortunately shows that this control method cannot be used successfully in all cases. The results of nonstationary drying with periodic humidity changes (Figure 4.24) are a good example of this. As the signals during the first part of the process and intensive drying periods come strictly from material, the other signals generated only in tempering periods consist of many other additional signals/noises. These signals appear only in relaxation periods and are not propagated in other ones, which leads to the conclusion that they depend on process parameter changes. Because the energy and number of hits rise significantly in humidifying periods, the signals can interfere with the proper signals and mask their real character. This is why in such periods this method of process control is not recommended.

4.5 MODELING OF INTERMITTENT DRYING

4.5.1 Equations of Drying Kinetics

Drying kinetics exposes the moisture content and the temperature of the drying body as a function of time. The mass and heat balances constitute the basis for construction of the respective equations for determination of the drying kinetics. The balance equations are of the form

$$-m_s \frac{dX}{dt} = A_m J_m + \Delta m \tag{4.2}$$

$$m_s \frac{d}{dt}\left[\left(c_s + c_l X\right)T\right] = A_T J_T - A_m l J_m + \Delta Q \tag{4.3}$$

where

- m_s is the mass of dry body
- X is the moisture content (mass of moisture per mass of dry body)
- c_s and c_l denote the specific heat for solid and liquid
- J_m and J_T denote the moisture and heat fluxes
- l is the latent heat of evaporation
- A_m and A_T are the surfaces of mass and heat exchange
- Δm denotes an increase of moisture mass due to the vapor condensation by air humidification
- ΔQ is the heat source specified for different materials in different drying methods, and also due to recovery of the latent heat by vapor condensation

In order to define the heat and mass fluxes, let us consider the contact of two layers of air having different vapor content x (molar ratio of vapor in air) and temperatures T. One layer is close to the dried body and is characterized by the temperature T (equal to that of dried body), the vapor concentration x, and the vapor chemical potential $\mu(T,x)$. The other layer constitutes the ambient air characterized by the temperature T_a, the vapor concentration x_a, and the vapor chemical potential $\mu_a(T_a,x_a)$.

It is assumed that $T_a > T$ and $\mu(T,x) > \mu_a(T_a,x_a)$. Therefore, the heat flux J_T is directed from the ambient air to the dried body and the mass flux J_m in the opposite direction. The process of heat and mass transfer is irreversible so entropy is produced during this process. The rate of entropy production per unit surface reads (Szarawara 1985; Berry et al. 2000)

$$\sigma = \left(\frac{1}{T} - \frac{1}{T_a}\right) J_T + \left(\frac{\mu}{T} - \frac{\mu_a}{T_a}\right) J_m \geq 0 \tag{4.4}$$

Assuming that the expression (4.4) is a positive defined quadratic form, the sufficient conditions to satisfy this expression require the heat and mass fluxes to be of the form

$$J_T = L_{11}\left(\frac{1}{T} - \frac{1}{T_a}\right) + L_{12}\left(\frac{\mu}{T} - \frac{\mu_a}{T_a}\right) \tag{4.5}$$

$$J_m = L_{21}\left(\frac{1}{T} - \frac{1}{T_a}\right) + L_{22}\left(\frac{\mu}{T} - \frac{\mu_a}{T_a}\right) \tag{4.6}$$

Equations 4.5 and 4.6 express coupled heat and mass transfer. In order to accomplish all possible coupled heat and mass transfer processes the coefficients with mixed indices ought to be equal to each other, that is, $L_{12} = L_{21}$, and the other should be $L_{11} > 0$ and $L_{11} L_{22} > (L_{12})^2$.

It is assumed that the coupling effects are very small and are therefore neglected in further considerations. Thus, the rate equations for heat and mass fluxes are reduced to the form

$$J_T = h_T\left(T_a - T\right) \tag{4.7}$$

$$J_m = L_{22}\left(\frac{\mu}{T} - \frac{\mu_a}{T_a}\right) \tag{4.8}$$

where h_T is termed the coefficient of convective heat transfer.

The chemical potential of vapor in air has the form (Szarawara 1985)

$$\mu\left(p, T, x\right) = \mu_0\left(p, T\right) + RT\ln x \tag{4.9}$$

If the body and air temperatures do not differ significantly, one can write (Berry et al. 2000)

$$J_m = L_{22}\left(\frac{\mu}{T} - \frac{\mu_a}{T_a}\right) = L_{22}\left(\frac{\mu_0\left(p, T\right)}{T} - \frac{\mu_0\left(p, T_a\right)}{T_a} + R\ln\frac{x}{x_a}\right) \approx h_m \ln\frac{x}{x_a} \tag{4.10}$$

where h_m is termed the coefficient of convective vapor transfer.

It is convenient to write the driving force of convective vapor transfer as

$$\ln\frac{x}{x_a} = \ln\frac{\varphi|_{\partial B}\, p_{wn}\left(T\right)}{\varphi_a p_{wn}\left(T_a\right)} \tag{4.11}$$

where

φ is the relative air humidity

p_{wn} is vapor partial pressure in the saturated state

The air relative humidity (RH) close to the dried sample surface depends on the drying body moisture content, so it reads

$$\varphi|_{\partial B} = \begin{cases} 1 & \text{for } X \geq X_{cr} \\ 1 - \left(1 - \varphi_a\right)\dfrac{X_{cr} - X}{X_{cr} - X_{eq}} & \text{for } X_{cr} \geq X \geq X_{eq} \end{cases} \tag{4.12}$$

In the above equations, φ_a and T_a denote the air relative humidity and temperature, that is, the assumed parameters of drying, and X_{cr} and X_{eq} are the critical and the equilibrium moisture content in the dried sample (parameters determined experimentally). Function $p_{wn}(T)$ is given in the literature (Strumiłło 1983) in the form of a table.

The governing equations in the final form are

$$-m_s \frac{dX}{dt} = A_m h_m \ln \frac{\varphi|_{\partial B}\, p_{wn}(T)}{\varphi_a p_{wn}(T_a)} + \Delta m \tag{4.13}$$

$$m_s \frac{d}{dt}\left[(c_s + c_l X)T\right] = A_T h_T (T_a - T) - A_m l h_m \ln \frac{\varphi|_{\partial B}\, p_{wn}(T)}{\varphi_a p_{wn}(T_a)} + \Delta Q \tag{4.14}$$

The initial conditions for the moisture content and temperature are assumed to be $X(t = 0) = X_0$ – the initial moisture content, and $T(t = 0) = T_0$ – the initial temperature.

The heat used for drying $\Delta Q = \{\Delta Q_{US}; \Delta Q_{MV}; \Delta Q_{IR}\}$ depends on the energy source and represents the additional amount of energy (e.g., ultrasound [*US*], microwave [*MV*]), and/or infrared [*IR*]). This is an auxiliary relation that has to be determined independently.

4.5.1.1 Microwave Heat Source

It is possible to determine the heat source ΔQ_{MV} in the case of microwave heating on the basis of Maxwell electromagnetic equations (Perré and Turner 1996; Suwannapum et al. 2010). In practical applications, the heat source in microwave drying can be written as (Kowalski et al. 2010)

$$\Delta Q_{MV} = (A + BX)\exp\left[-2\delta_{MV}(\boldsymbol{x} \cdot \boldsymbol{n})\right] \tag{4.15}$$

where

- A and B express the amount of microwave energy absorbed by the skeleton and the moisture
- $(\boldsymbol{x} \cdot \boldsymbol{n})$ denotes the distance of microwave propagation in n-direction
- δ_{MV} expresses the decay of microwave energy due to its absorption with distance, respectively

When the dried samples are small, the decay of microwave energy due to absorption can also be considered small, and therefore $\delta_{MV}(\boldsymbol{x} \cdot \boldsymbol{n}) \approx 0$.

4.5.1.2 Infrared Heat Source

The amount of infrared heat can be calculated using the following correlation (Islam et al. 2010; Kowalski et al. 2010):

$$\Delta Q_{IR} = \chi\varepsilon\sigma\left(T_R^4 - T^4\right)\exp\left[-2\delta_{IR}(\boldsymbol{x} \cdot \boldsymbol{n})\right] \tag{4.16}$$

where

- T_R is the temperature of the emitter
- χ is the shape factor related with geometry and surface conditions of the radiating and absorbing surface
- ε is the emissivity of the emitter system
- σ is the Stefan-Boltzmann constant
- δ_{IR} expresses the decay of infrared energy due to its absorption with distance, respectively

Infrared radiation mainly involves a surface effect, and the term describing the decay of infrared energy due to absorption can be considered a very small one, that is, $\delta_{IR}(\boldsymbol{x} \cdot \boldsymbol{n}) \approx 0$.

While the subject of radiation does not have any analogy in momentum or mass transfer, this mode of heat transfer is extremely important in drying practice. The role of geometry in radiation is a dominant one, and considerable effort and care must be exercised in determining the shape factor χ (Glouannec et al. 2002; Baehr and Stephan 2006). The use of the gray-body idealization results in a simplified approach to radiant energy exchange of great utility.

4.5.1.3 Ultrasound Heat Source

The ultrasound heat source is expressed as $\Delta Q_{US} = a_U \chi_U P_U$, where a_U [–] denotes the dimensionless absorption coefficient of the ultrasonic wave, χ_U [–] is the dimensionless working efficiency of the ultrasonic transducer, and P_U [W] is the power of the ultrasonic generator.

4.5.2 Intermittent Drying of Ceramics: An Example

Numerical calculations based on the model described earlier allow the reflection of the process kinetics for a number of different materials and process conditions. Their utilization is possible whenever the chemical and physical parameters of considered material as well as characteristic drying parameters are known for a given process. Appropriate description of drying conditions enables simulation not only of constant but also intermittent drying. Nevertheless, to model intermittent drying processes it is important to apply the model to constant drying conditions, to check the correctness of the chosen process parameters.

The intermittent drying conditions can be realized by changes of different parameters dependent on application of an additional energy source. In this section, consideration of an exemplary intermittent drying process based on numerical simulations will be presented. The presented convective nonstationary drying processes are realized by the periodic changes of the air temperature and/or humidity.

4.5.2.1 Drying in Variable Air Temperature

The variable air temperature in numerical calculations was accomplished by the following equation:

$$T_a(t) = \frac{T_{\max} + T_{\min}}{2} + \frac{T_{\max} - T_{\min}}{2} \cos\left[\frac{2\pi}{\lambda}(t - t_{ch})\right] \quad (4.17)$$

where

$T_{\min}$ and $T_{\max}$ denote minimum and maximum of the variable air temperatures
λ is the period of temperature variation
t_{ch} denotes the time at which the change of drying parameters begins

In numerical calculation t_{ch} equals exactly the same time as in experimental test, where the the first changes were applied before the critical time t_{cr} (for X_{cr}).

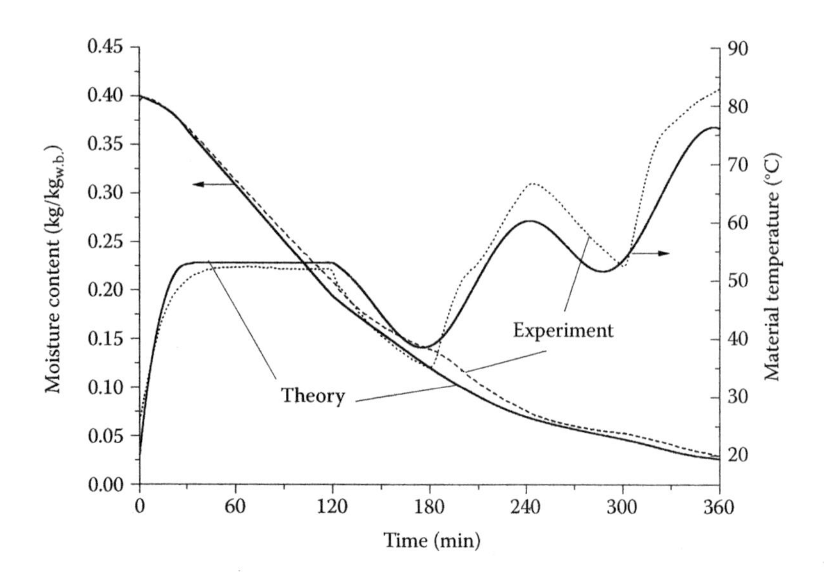

FIGURE 4.25 Comparison of the kinetic curves determined experimentally and numerically for kaolin clay dried with periodic air temperature changes.

The change of air temperature is automatically followed by the variation in the air humidity φ_a. The experimental measurement of these changes leads to the conclusion that the air relative humidity was changed exactly inversely to temperature changes while maintaining the same periodicity. In order to take this fact into account the air relative humidity can be written as follows:

$$\varphi_a(t) = \frac{\varphi_{max} + \varphi_{min}}{2} + \frac{\varphi_{max} - \varphi_{min}}{2} \cos\left[\frac{2\pi}{\lambda}(t - t_{ch}) + \pi\right] \tag{4.18}$$

Applying these periodically changing drying parameters in Equations 4.17 and 4.18 one can determine the variation of temperature and moisture content in the dried body. The mass and heat sources due to vapor condensation can be omitted by intermittent drying realized through the change of air temperature.

Figure 4.25 illustrates the results of numerical calculations of the drying curve and dried body temperature by periodically changing air temperature. The experimental curves are compared with the theoretical ones.

It can be seen that the numerical results obtained on the basis of the model presented above reveal very good adherence with experimental data. A small discrepancy between the theoretical and experimental results for intermittent drying follows from the slightly different character of periodic air temperature changes in the experimental tests and numerical assumption.

4.5.2.2 Drying in Variable Air Humidity

Periodic humidity changes are conducted by a humidifier, which delivers vapor into a drier. As dried material during the process is usually colder than 100°C, the vapor application can lead to reduction of drying rate not only by changes in driving force but also by condensation of vapor on colder material. When applying harmonic function of the form given by Equation 4.18 with values $\varphi_{min} = 4\%$ and $\varphi_{max} = 80\%$ it is possible to calculate drying kinetics for this process. Nevertheless, in the equations expressing the temperature and moisture changes it is also necessary to consider vapor condensation and sample temperature change due to recovery of the latent heat of condensation.

The increment of mass Δm^* condensed on the sample can be evaluated from the experiments, where it diminishes with the increase of sample temperature. In order to take this into account, exponential decrease of the condensed mass over time can be expressed in the form

$$\Delta m^* = \Delta m_0 \exp\left[-\delta \cdot T\right] \cos\left[\frac{2\pi}{\lambda}\left(t - t_{ch}\right)\right] \tag{4.19}$$

where

Δm_0 is the initial increment of condensed vapor

δ is the coefficient of reduction of condensing mass with increase of the body temperature T

Similarly, the increment of condensed heat can be determined as

$$\Delta Q^* = l \Delta m_0 \exp\left[-\delta \cdot T\right] \cos\left[\frac{2\pi}{\lambda}\left(t - t_{ch}\right)\right] \tag{4.20}$$

where l is the latent heat of evaporation.

Figure 4.26 presents results for drying at a temperature of 100°C and variable air humidity changing between 4% and 80%.

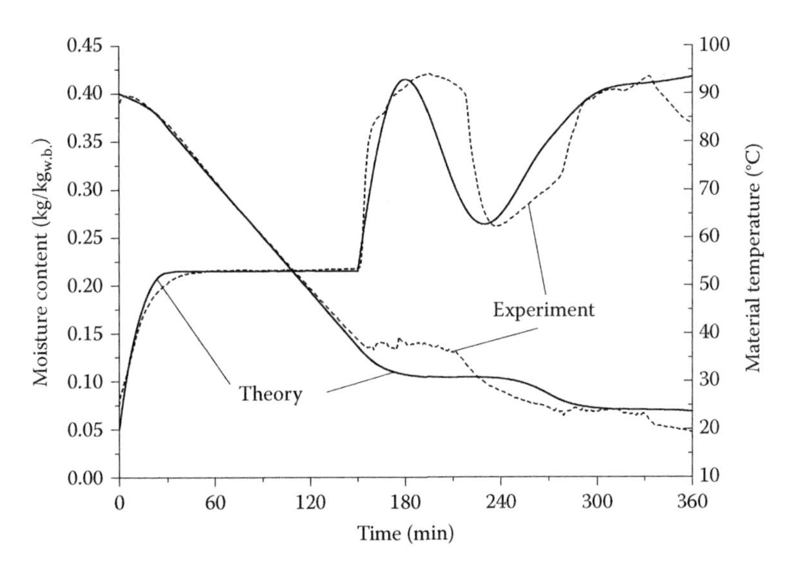

FIGURE 4.26 Comparison of the kinetic curves determined experimentally and numerically for kaolin clay dried with periodic air humidity changes.

It can be seen that the drying curves predicted numerically and those determined experimentally for the changing air humidity coincide with each other. The theoretical model quantitatively reflects the experimental curves very satisfactorily; however, a slight variation between the data follows from sinusoidal periodic changes taken for a model, which is not reflected ideally in the experiments.

The drying model describing the kinetics of intermittent drying may be helpful in the construction of nonstationary processes optimized with respect to drying time and energy consumption. The modeling of intermittent drying accomplished through periodically changeable temperature is easier than that with periodically changeable air humidity, because in the first case there is no effect of vapor condensation, which occurs during humidification of the air.

4.5.3 Intermittent Drying of Carrot: An Example

The model based on physical and chemical properties makes it possible to describe processes not only for engineering materials but also for biological products like fruits or vegetables. The variability of structure in these materials has a significant influence on drying kinetics where pretreatment, for example, OD, makes the calculations even more difficult. Nevertheless, appropriate determination of process and material parameters enables quite accurate numerical determination of process kinetics, which is presented later in this section. The developed theoretical drying kinetics is validated using the experimental data. Good adherence between results confirms the usefulness of the presented model and its possible application to the construction of controlled and optimized drying processes.

By intermittent drying, the influence of temperature changes on air humidity was taken into account by calculations of the drying kinetics curves. As the periodic changes of the drying air temperatures in experimental tests did not have a strictly sinusoidal character, their shape by mathematical modeling was described as being as close as possible to the experimental one. Their shape determined through mathematical modeling matched the shape of experimental curves. However, accuracy in description of all the parameters in intermittent drying was not completely satisfactory. Figure 4.27 shows the drying kinetics curves for drying carried out in intermittent conditions obtained on the basis of the mathematical model and experiment.

As shown in the above figures the drying curves show quite a good adherence, but the temperature curves are only qualitatively similar and shifted slightly over time. Such behavior of temperature curves shows that the model response is not fast enough to reveal the real temperature changes in dried material. The modeled air temperature during periodic changes tends to reach equilibrium value faster than the experimental one by the given drying conditions, and reacts with a retardation on the temperature changes. Another explanation for the discrepancy between the temperature curves could be the fact that the carrot samples were very thin and during cooling periods the cold air flowing near the sample could intensify the heat exchange and affect the temperature of the samples, which was not taken into account in the modeled temperature. The mathematical description considers only temperature changes without any additional effects that could occur from some external sources and influence the heat exchange and its intensification over time. However, it is worth

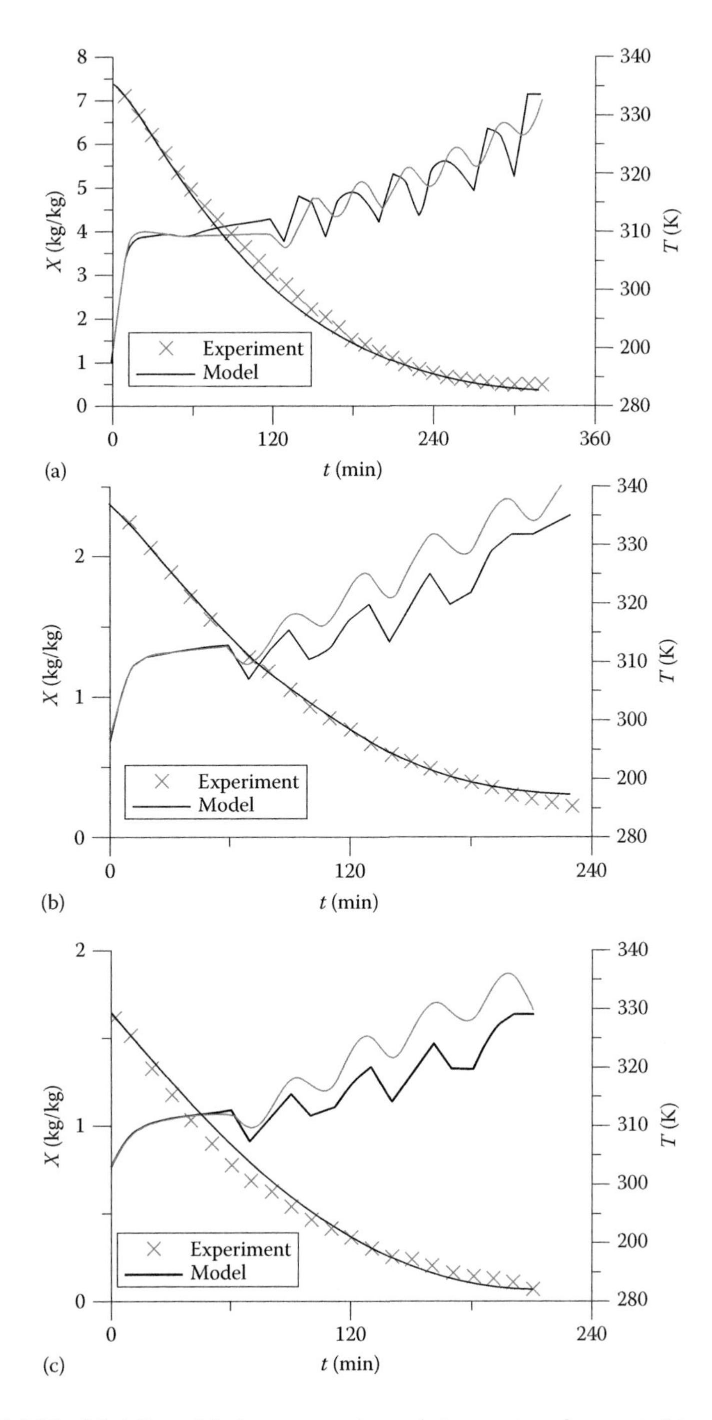

FIGURE 4.27 Modeling of drying curve and sample temperature for carrot: (a) convective-intermittent drying, (b) osmosis with convective-intermittent drying, (c) ultrasound-assisted osmosis with convective-intermittent drying.

noting that the differences in the experimental and model results do not significantly affect the drying curves, as the changes in temperature only shift over time, but still describe the changes on the same level.

4.5.4 Estimation of Drying Effectiveness

In the dryer being modified for the purpose of nonstationary (intermittent) drying, some additional energy sources like microwaves or ultrasound may be installed to enhance the heat and mass transfer by intermittent drying. One such source could be ultrasound waves, which, for example, involve air velocity oscillations which can increase the drying rate. In addition, a high-intensity airborne ultrasound causes microstreaming at the interfaces. Such turbulence reduces the dimension of boundary layer and accelerates moisture diffusion, which increases mass transfer and contributes positively to the drying process. Besides, the high-intensity ultrasound may induce "sponge effect," which involves alternative contractions and expansions of the material's skeleton, and thus due to alternating stress creates microscopic channels enabling ease of water movement.

As an example, the effect of ultrasound-assisted convective (CV-US) drying will be shown, where a similar effect can be achieved by only intermittent ultrasound assistance. The global model of drying kinetics is used to compare the theory with the experiments. In Figure 4.28, experimental (Exp) and numerically calculated (Model) moisture ratio and temperature curves, for the fastest drying schedules, are presented (Welty et al. 1976). It can be easily noted that both moisture reduction (*MR*) and temperature curves (*T*), simulated numerically, are coherent with the experimental data.

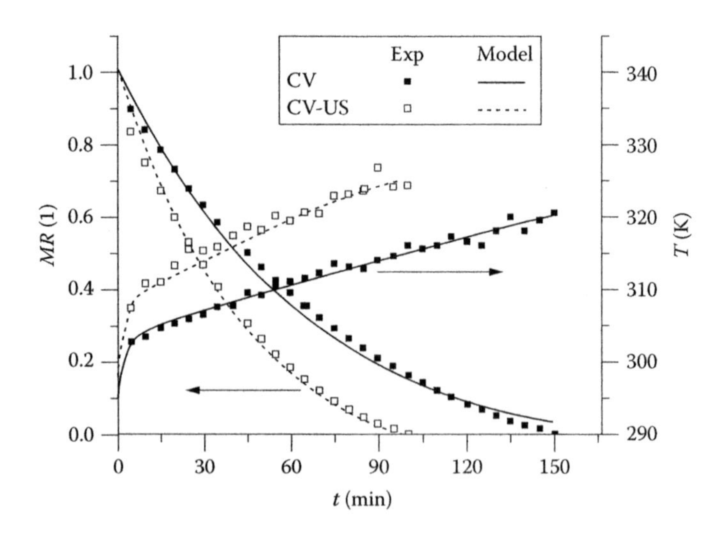

FIGURE 4.28 Comparison of the experimental (Exp) and numerical (Model) curves of moisture ratio and temperature, obtained during drying at 323 K with 2 m/s.

TABLE 4.5
Kinetic Parameters Acquired from Numerical Calculations

Parameters	CV	CV-US
T_a (K)	323	323
v_a (m/s)	2	2
h_m (kg/m$^2 \cdot$ h)	0.64	1.26
h_T (W/m $\cdot$ K)	16.50	26.04

In order to make use of this model for complex drying of biological materials, the physical heat (h_T) and mass (h_m) transfer coefficients have to be determined first. For this purpose, a new optimization methodology of assessment of the model heat and mass transfer coefficients was used, based on adjusting the drying kinetics plots determined numerically to the experimentally established data. The parameters of drying kinetics presented in Table 4.5 allowed the assessment of the influence of US enhancement on mass and heat transfer processes. One can see that both the heat h_T and the mass transfer coefficient h_m increased due to US enhancement of convective drying. This proves that application of US has a beneficial effect on heat and mass transfer during drying. The most probable explanation for the increase of h_m and h_T coefficients is the turbulence effect at boundary layer and US action on material structure (increase of porosity, micro channels, etc.). An intensive agitation of air just above the material's surface, due to changeable pressure (compression and depression) induced by US, leads to strengthening of heat and mass transfer processes.

Drying rate D_r (g/h) expresses the rate of moisture decrease in the drying material as a function of time, that is

$$D_r(t) = m_s \frac{dX}{dt} = -A_m h_m \ln \frac{\varphi|_{\partial B}\, p_{vs}(T_m)}{\varphi_a p_{vs}(T_a)} \tag{4.21}$$

and the average drying rate $D_{r,ave}$ is determined as

$$D_{r,ave} = \frac{1}{t_e} \int_0^{t_e} D_r(t)\, dt \tag{4.22}$$

where t_e is the drying time at which the moisture content reaches equilibrium.

Based on Figure 4.28, one can state that for drying at temperature 323 K and air velocity v = 2 m/s the average drying time for convective drying equals $t_e = t_e^{NU} \approx 150\,\text{min}$, and for CV-US drying is $t_e = t_e^{U} \approx 98\,\text{min}$.

The calculation of moisture mass reduction is attributed to 26.648 g moisture removed from material during the process. Dividing this total mass by the drying time, one can estimate the average drying rate in a given drying process. The average drying rate by drying at temperature 323 K and air velocity v = 2 m/s amounted to $D_{r,ave}^{NU} = 10.66\,\text{g/h}$ for pure convective drying, and $D_{r,ave}^{U} = 16.35\,\text{g/h}$ for ultrasonic-assisted convective drying for given material.

Drying rate enhancement D_rE and the ratio of drying rate enhancement AD_rE are used to evaluate the effectiveness of CV-US drying. In the case of drying at temperature 323 K and air velocity $v = 2$ m/s, one obtains

$$D_rE = D^{U}_{r,ave} - D^{NU}_{r,ave} = 16.35 - 10.66 = 5.69\,\mathrm{g/h} \tag{4.23}$$

$$AD_rE = \frac{D^{U}_{r,ave} - D^{NU}_{r,ave}}{D^{NU}_{r,ave}} \cdot 100\% = \frac{5.69}{10.66} \cdot 100\% = 53.38\% \tag{4.24}$$

Taking into account the above results, one can state that ultrasound enhancement of carrot drying accelerated the drying process in a similar way in both cases, as the ratio of the drying rates amounted to about 53.38% and 48%, respectively.

To describe quantitatively the components of ultrasound action (C) such as "heating effect" (T), "vibration effect" (v), and "synergistic effect" (s), the difference in mass balance equations for CV-US and CV processes is used. As an example, the case of drying at temperature 323 K and air velocity $v = 2$ m/s is considered. Thus, the components of C such as T, v, and s read

$$CD_rE_{T,eff} = \frac{A_m h_m}{D_rE} \ln \frac{p_{vs}(T_m + \Delta T_m)}{p_{vs}(T_m)} \cdot 100\% = \frac{19.8 \cdot 0.64}{5.69} \ln \frac{1296.0}{1173.6} \cdot 100\% \approx 22\% \tag{4.25}$$

$$CD_rE_{v,eff} = \frac{A_m \Delta h_m}{D_rE} \ln \frac{\varphi|_{\partial B}}{\varphi_a} \cdot 100\% = \frac{19.8 \cdot 0.62}{5.69} \ln \frac{0.75}{0.55} \cdot 100\% \approx 67\% \tag{4.26}$$

$$CD_rE_{s,eff} = \frac{A_m \Delta h_m}{D_rE} \ln \frac{p_{vs}(T_m + \Delta T_m)}{T_a} \cdot 100\% = \frac{19.8 \cdot 0.62}{5.69} \ln \frac{1296.0}{1233.5} \cdot 100\% \approx 11\% \tag{4.27}$$

On the basis of the above calculations, it can be concluded that the "vibration effect" generated by ultrasound contributed the most to the increase in drying efficiency in the case of drying at temperature 323 K and air velocity $v = 2$ m/s. Thus, it was found that ultrasound has a positive influence on the acceleration of the convective drying process. Application of the US caused intensification of the heat and mass transport and contributed to shortening of drying time.

In a similar way, the effectiveness of different drying processes carried out in intermittent conditions, as well as hybrid methods, can be calculated.

4.6 CONCLUDING REMARKS

The objective of this chapter was to provide a concise introduction to selected new advanced drying technologies like intermittent drying that are briefly presented elsewhere in the literature. The interested reader can find more information here as well as descriptions of the innovative designs and operational modifications in the advanced drying technologies. This chapter presents applications of intermittent drying including hybrid methods for engineering materials and for food processing technology.

Intermittent drying is an alternative technology to continuous heat-supply drying as it could improve the product quality and the energy efficiency of drying processes due to reduction of both the energy consumption and drying time. In this study the application of intermittent drying technology to drying of engineering and biomaterials was considered for different drying air temperatures and humidity, and especially for proper arrangement of drying cycles.

The industrial interest in intermittent and hybrid drying enhanced with ultrasound and microwaves has revived during recent years because such a technology may represent a flexible alternative for more energy-efficient drying. A challenge is the application of high-intensity ultrasound to industrial processing and the design and development of specific power ultrasonic systems for large-scale operation. In the area of ultrasonic processing of multiphase media the development of a new family of power generators with extensive radiating surfaces has contributed significantly to the implementation at industrial scale of several applications in sectors such as engineering, food industry, environment, and manufacturing.

This chapter discusses the variety of drying materials and drying equipment with additional devices. To quantify the objective functions like "heating effect," "vibration effect," and "synergistic effect," a drying model was developed for considering mass and heat transfer and showing non-negligible external resistances, and transport properties that are moisture and temperature dependent. The optimization results show that there is an advantage of intermittent over continuous drying as regards material properties and drying time under the operating conditions and constraints that were considered. The analysis also indicates a decrease in the energy consumption and improvement of the quality of dried products.

Drying of heat-sensitive biological products is a complex process, which can alter a variety of mechanical, chemical, and biochemical characteristics of these products after drying. For example, thermal dehydration of biomaterials can affect one or more of their biochemical (viability, atrophy of cells), enzymatic (stability, catalytic ability), chemical (decrease of nutritive value, degradation of vitamins), physical (solubility, rehydration, shrinking, loss of aroma), and mechanical properties (elasticity, hardness, grindability, fragility).

Proper selection of drying technology and operating parameters is therefore important to obtain a good-quality product. One of the recommended techniques for drying foods and other biomaterials with innovative drying is intermittent drying. Here the conventional hot air drying down to the critical moisture content is followed by a tempering period in ambient air for an extended period (several hours or days). During tempering, some evaporation of moisture occurs utilizing the sensible heat of the product accumulated during the period of convective drying. At the same time, moisture from the wet core diffuses to the material surface so that after tempering the moisture distribution within the drying material becomes more uniform. This brings drying conditions to the original ones, although at much lower initial moisture content. Then, finishing drying to the final moisture can be carried out with reduced risk of thermal damage of the material because evaporation withdraws from the material surface, and the temperature of drying air can be reduced as well. This method of drying is particularly suitable for biomaterials because moisture from such materials is difficult to remove by conventional drying methods.

Independently of the conventional or intermittent drying there is always a risk of some quality loss. From the consumer point of view, the nutritional value of dried products and rehydrated foods is the main concern, followed by sensory properties such as color, texture, and aroma. In concluding the present consideration one can state that

- Tempering-intermittent drying can reduce degradation of biocomponents and shorten the drying time.
- The mathematical model of drying kinetics fits the experimental data and enables the assessment of the effectiveness of biological materials dried in intermittent conditions.

REFERENCES

Asard, H., May, J. M., and Smirnoff, N. 2004. *Vitamin C Functions and Biochemistry in Animals and Plants*. New York: BIOS Scientific Publishers.

Baehr, H. D. and Stephan, K. 2006. *Heat and Mass Transfer*. Berlin, Germany: Springer–Verlag.

Berry, R. S., Kazakov, V. A., Sieniutycz, S., Szwast, Z., and Tsirlin, A. M. 2000. *Thermodynamic Optimization of Finite–Time Processes*. New York: John Wiley & Sons.

Bon, J. and Kudra, T. 2007. Enthalpy-driven optimization of intermittent drying. *Drying Technology* 25:523–532.

Bonazzi, C. and Dumoulin, E. 2011. Quality changes in food materials as influenced by drying processes. In *Modern Drying Technology*, Vol. 3: Product Quality and Formulation, eds. E. Tsotsas and A. S. Mujumdar, pp. 10–16. Weinheim, Germany: Wiley-VCH Verlag GmbH & Co. KGaA.

Chin, S. K. and Law, C. L. 2010. Product quality and drying characteristics of intermittent heat pump drying of *Ganoderma tsugae* Murrill. *Drying Technology* 28(12):1457–1465.

Chua, K. J., Mujumdar, A. S., and Chou, S. K. 2003. Intermittent drying of bioproducts—An overview. *Bioresource Technology* 90(3):285–295.

Chua, K. J., Mujumdar, A. S., Hawlader, M. N. A., Chou, S. K., and Ho, J. C. 2001. Convective drying of agricultural products. Effect of continuous and stepwise change in drying air temperature. *Drying Technology* 19(8):1949–1960.

Dreosti, I. E. 1993. Vitamins A, C, E and β-carotene as protective factors for some cancers. *Asia Pacific Journal of Clinical Nutrition* 2(1):21–25.

Glouannec, P., Lecharpentier, D., and Noel, H. 2002. Experimental survey on the combination of radiating infrared and microwave sources for the drying of porous material. *Applied Thermal Engineering* 22:1689–1703.

Gunasekaran, S. 1999. Pulsed microwave-vacuum drying of food. *Drying Technology* 17:395–412.

Harmer, R. A. 1980. Occurrence, chemistry and application of betanin. *Food Chemistry* 5(1):81–90.

Herrithsch, A., Dronfield, J., and Nijdam, J. 2008. Intermittent and continuous drying of red-beech timber from the green conditions. Paper presented at the *16th International Drying Symposium*, Hyderabad, India, November 9–12, 2008.

Ho, J. C., Chou, S. K., Chua, K. J., Mujumdar, A. S., and Hawlader, M. N. A. 2002. Analytical of cyclic temperature drying: Effect on drying kinetics and product quality. *Journal of Food Engineering* 51:65–75.

Islam, Z. U., Dhib, R., and Dahman, Y. 2010. Modelling of infrared drying of polymer solutions. *Chemical Product and Process Modelling* 5(1):1–19.

Jumah, R., Al-Kteimat, E., Al-Hamad, A., and Telfah, E. 2007. Constant and intermittent drying characteristics of olive cake. *Drying Technology* 25(9):1421–1426.

Katekawa, M. E. and Silva, M. A. 2006. A review of dying models including shrinkage effects. *Drying Technology* 24(1):5–20.

Knorr, K. D., Zenker, M., Heinz, V., and Lee, D.-U. 2004. Applications and potential of ultrasonics in food processing. *Trends in Food Science & Technology* 15(5):261–266.

Konopacka, D., Jesionkowska, K., Klewicki, R., and Bonazzi, C. 2009. The effect of different osmotic agent on sensory perception of osmo-treated dried fruit. *Journal of Horticultural Science and Biotechnology*, ISAFRUIT Special Issue:80–84.

Kowalski, S. J. 2003. *Thermomechanics of Drying Processes*. New York: Springer-Verlag.

Kowalski, S. J. 2010. Control of mechanical processes in drying. Theory and experiment. *Chemical Engineering Science* 65(2):890–899.

Kowalski, S. J., Banaszak, J., and Rajewska, K. 2013a. Acoustic emission in drying materials. In *Acoustic Emission—Research and Application*, ed. W. Sikorski. InTech, Tijeka, Croatia. http://www.intechopen.com/books/acoustic-emission-research-and-applications/acoustic-emission-in-drying-materials.

Kowalski, S. J. and Mielniczuk, B. 2007. Analysis of effectiveness and stress development during convective and microwave drying. *Drying Technology* 26(1):64–77.

Kowalski, S. J. and Mierzwa, D. 2013. Influence of osmotic pretreatment on kinetics of convective drying and quality of apples. *Drying Technology* 31:1849–1855.

Kowalski, S. J., Musielak, G., and Banaszak, J. 2010. Heat and mass transfer during microwave–convective drying. *American Institute of Chemical Engineers Journal* 56(1):24–35.

Kowalski, S. J., Musielak, G., and Rybicki, A. 1997. The response of dried materials to drying conditions. *International Journal of Heat and Mass Transfer* 40(5):1217–1226.

Kowalski, S. J. and Pawłowski, A. 2011a. Intermittent drying of initially saturated porous materials. *Chemical Engineering Science* 66(9):1893–1905.

Kowalski, S. J. and Pawłowski, A. 2011b. Energy consumption and quality aspect by intermittent drying. *Chemical Engineering and Processing* 50:384–390.

Kowalski, S. J. and Pawłowski, A. 2015. Intensification of apple drying due to ultrasound enhancement. *Journal of Food Engineering* 156:1–9.

Kowalski, S. J., Pawłowski, A., Szadzińska, J., Łechtańska, J., and Stasiak, M. 2016. High power airborne ultrasound assist in combined drying of raspberries. *Innovative Food Science and Emerging Technologies* 34:225–233.

Kowalski, S. J., Rajewska, K., and Rybicki, A. 2000. Destruction of wet materials by drying. *Chemical Engineering Science* 55(23):5755–5762.

Kowalski, S. J. and Rybicki, A. 1994. Interaction of thermal and moisture stresses in materials dried convectively. *Archives of Mechanics* 46(3):251–265.

Kowalski, S. J. and Rybicki, A. 2007. Residual stresses in dried bodies. *Drying Technology* 25(4):629–637.

Kowalski, S. J. and Szadzińska, J. 2012. Non-stationary drying of ceramic-like materials controlled through acoustic emission method. *Heat and Mass Transfer* 48(12):2023–2032.

Kowalski, S. J. and Szadzińska, J. 2014. Kinetics and quality aspects of beetroots dried in non-stationary conditions. *Drying Technology* 32:1310–1318.

Kowalski, S. J., Szadzińska, J., and Łechtańska, J. 2013b. Non-stationary drying of carrot: Effect on product quality. *Journal of Food Engineering* 118(4):393–399.

Kumar, C., Karim, M. A., and Joardder, M. U. H. 2014. Intermittent drying of food products: A critical review. *Journal of Food Engineering* 121:48–57.

Kumirska, J., Gołębiowski, M., Paszkiewicz, M., and Bychowska, A. 2010. *Analiza żywności*. Gdańsk, Poland: Wydawnictwo Uniwersytetu Gdańskiego (in Polish).

Labuza, T., Tannenbaum, S., and Karel, M. 1970. Water content and stability of low moisture and intermediate-moisture foods. *Food Technology* 24(5): 543–550.

Lewicki, P. P. and Lenart, A. 2006. Osmotic dehydration of fruits and vegetables. In *Handbook of Industrial Drying*, ed. A. S. Mujumdar, pp. 665–681. Boca Raton, FL: CRC Press.

Łykow, A. V. 1968. *Tieoria suszki*. Moscow, Russia: Energia.

Mitek, M. I. and Słowiński, M. 2006. *Wybrane zagadnienia z technologii żywności*. Warsaw, Poland: SGGW (in Polish).

Moazzam, R. K. 2012. Osmotic dehydration technique for fruits preservation—A review. *Pakistan Journal of Food Science* 22(2):71–85.

Mujumdar, A. S. 2007. *Handbook of Industrial Drying*. Boca Raton, FL: CRC Press.

Mujumdar, A. S. and Law, C. L. 2010. Drying technology: Trends and applications in postharvest processing. *Food and Bioprocess Technology* 3(6):843–852.

Nilson, N. 1970. Studies into pigments in beetroot. *Lantbrukshogskolans Annaler* 36:179–219.

Nowacka, M., Wiktor, A., Śledź, M., Jurek, N., and Witrowa-Rajchert, D. 2012. Drying of ultrasound pretreated apple and its selected physical properties. *Journal of Food Engineering* 113:427–433.

Ong, S. P., Law, C. L., and Hii, C. L. 2012. Optimization of heat pump–assisted intermittent drying. *Drying Technology* 30(15):1676–1687.

Pan, Y. K., Zhao, L. J., Dong, Z. X., Mujumdar, A. S., and Kudra, T. 1999. Intermittent drying of carrot in a vibrated fluid bed: Effect on product quality. *Drying Technology* 17:2323–2340.

Perré, P. and Turner, I. W. 1996. A complete coupled model of the combined microwave and convective drying of softwood in an oversized waveguide. Paper presented at the *10th International Drying Symposium*, Krakow, Poland, July 30–August 2, 1996.

Prakash, S., Jha, S. K., and Datta, N. 2004. Performance evaluation of blanched carrots dried by three different driers. *Journal of Food Engineering* 62:305–313.

Rawson, A., Tiwari, B. K., Tuohy, M. G., O'Donnell, C. P., and Brunton, N. 2011. Effect of ultrasound and blanching pretreatments on polyacetylene and carotenoid content of hot air and freeze dried carrot discs. *Ultrasonics Sonochemistry* 18:1172–1179.

Sherer, G. 1990. Drying theory. *Journal of American Ceramic Society* 73(1):3–14.

Sikorski, Z. E. 2007. *Food Chemistry. Food Ingredients*. Warsaw, Poland: WNT (in Polish).

Strumiłło, C. Z. 1983. *Fundamentals of the Theory and Technology of Drying*. Warsaw, Poland: WNT (in Polish).

Suwannapum, N., Rattanadecho, P., and Vongpradubchai, S. 2010. Analysis of microwave heat-mass transport and pressure built up induced inside unsaturated porous media subjected to microwave energy using a single (TE10) mode cavity. *Drying Technology* 29(9):1010–1024.

Szarawara, J. 1985. *Chemical Thermodynamics*. Warsaw, Poland: WNT (in Polish).

Thomkapanich, O., Suvarnakuta, P., and Devahastin, S. 2007. Study of intermittent low-pressure superheated steam and vacuum drying of a heat-sensitive material. *Drying Technology* 25(1):205–223.

Tillmans, J., Hirsch, P., and Jackisch, J. Z. 1932. Untersuch. *Lebensmittel* 63:241–276.

Welty, J. R., Wicks, C. E., and Wilson, R. E. 1976. *Fundamentals of Momentum, Heat and Mass Transfer*. New York: John Wiley & Sons.

Woodroof, J. G. and Luh, B. S. 1986. *Commercial Fruit Processing*. Westport, CT: AVI Publishing Company.

Zhang, D. and Mujumdar, A. S. 1992. Deformation and stress analysis of porous capillary bodies during intermittent volumetric thermal drying. *Drying Technology* 10(2):421–443.

5 Effects of Process Conditions of Intermittent Drying on Quality of Food Materials

Nghia Duc Pham, Sami Ghnimi, A.M. Nishani Lakmali Abesinghe, Mohammad U.H. Joardder, Tony Petley, Scott Muller, and M. Azharul Karim

CONTENTS

5.1 INTRODUCTION

Nowadays, there has been significant interest in producing appealing and healthy processed food. Several studies emphasize on the benefits of agricultural products in preventing and controlling different diseases such as cataract, cancer, scurvy, and heart disease (Alasalvar and Shahidi 2012, Grosch and Schieberle 2009, Hui and Barta 2006, Joshi et al. 2011, Kumar et al. 2016b). As living standards keep improving, people are becoming increasingly concerned about their well-being and thus favor foods with potential health benefits (Röhr et al. 2005). Food is essential for human sustainability and development, by its contribution to macronutrients (such as fiber, polysaccharide, carbohydrate, protein, lipid) and micronutrients (namely, vitamins, minerals, and antioxidants such as vitamin C, β-carotene, lycopene, lutein, flavonoids, and anthocyanins) (Alasalvar and Shahidi 2012, Hui and Barta 2006). The great diversity of food materials leads to the wide variety of sensory characteristics in term of taste, aroma, color, and texture and different ways of preparation make them widely accepted by consumers.

In most drying processes, several physical, chemical, nutritional, and microbial modifications take place in plant-based food that might result in changing the nutritive and organoleptic quality attributes of the product (Kumar et al. 2012a, Vega-GÁLvez et al. 2010, Mujumdar and Jangam 2011). Hence, greater attention is necessary to attain information regarding how the quality of food changes during drying in order to make process selection, estimation, characterizing, and improving the quality of a dried product.

Many drying methods are investigated in the literature, which reduce heating time and energy usage while preserving quality. Different dehydration techniques have been studied for fruits and vegetables, and each approach has its own advantages and disadvantages (Abbasi Souraki and Mowla 2008, Gao et al. 2012, Kumar et al. 2012b). Intermittent drying is a relatively new and innovative technique in which the drying conditions and duration of supplying heat input, airflow, humidity, and pressure are wisely adjusted to minimize quality degradation while enhancing drying effectiveness.

This chapter begins by providing an overview of the effects of variably controlled conditions of different intermittent drying methods on dried food quality in terms of nutritional, chemical, and microbiological properties. Then, the latter part of the chapter evaluates current food quality modeling that can be applied to intermittent drying to improve drying performance and reducing unnecessary amount of experimental works.

5.2 DIFFERENT PRACTICES OF INTERMITTENT DRYING AND THEIR EFFECTS ON PRODUCT QUALITY

Several research works have been undertaken to enhance dehydrated food quality. One of them is intermittent drying, which can be applied by controlling heat source, airflow velocity, temperature, relative humidity, or drying pressure individually or simultaneously. The ultimate target is to fine-tune the drying conditions to achieve high drying efficiency without compromising product quality. It can also be done by integrating different heat sources such as convective drying, infrared drying, or dielectric heating (Fernandes et al. 2011). The application of intermittent drying in fruits and vegetable has proved to be able to reduce enzymatic and nonenzymatic browning, inhibit vitamin E and C loss, and retain better carotenoid content (Chou et al. 2000, Chua et al. 2000b, Jumah et al. 2007, Thomkapanich et al. 2007). Table 5.1 provides an overview of several intermittent drying practices in research to improve product quality.

TABLE 5.1
Summary of Some Intermittent Drying Works in Literature

Study	Material	Drying Method	Operating Conditions	Quality Attributes
Mishkin et al. (1984b)	Potatoes	Air drying	Air temperature 40°C–80°C, final moisture 0.05 g/g solids	Vitamin C content
Pan et al. (1999)	Carrots	Intermittent vibrated fluid bed	Vibration frequency from 0 to 20 Hz, air temperature 100°C–130°C	Rehydration ratio, β-carotene
Kaminski and Tomczak (2000)	Silica gel, green peas, potatoes, cabbage	Vibrofluidized drying	Temperature 60°C–120°C, velocity 30–50 m/s	Vitamin C
Ho et al. (2001)	Potatoes	Intermittent two-stage heat pump dryer	Various conditions	Nonenzymatic browning, vitamin C content
Khraisheh et al. (2004)	Potatoes	Microwave convective drying	Air temperature 30°C, 40°C and 60°C, velocity of 1.5 m/s, and microwave power levels between 90 and 650 W	Shrinkage, rehydration, vitamin C
Da Silva et al. (2005)	Camu-Camu	Hot air drying	Air temperature 50°C, 60°C, 70°C at velocity 1.5 m/s	Vitamin C
Nicoleti et al. (2007)	Persimmon	Hot air drying	Air temperature 40°C–70°C, at air velocity 0.8–0.2 m/s	Vitamin C
Di Scala and Crapiste (2008)	Red pepper	Hot air dryer	Hot air temperature 50°C, 60°C and 70°C, at air velocity 0.2 and 1.2 m/s	Vitamin C and carotenoids content
Saxena et al. (2010)	Jackfruit bulb slices	Hot air drying	Air temperature 50°C, 60°C, and 70°C	Color and β-carotene
Ong et al. (2012)	Salak fruit	Heat pump	*Initial stage*: heat pump at 26°C, 37°C; hot air at 40°C, 50°C *Later stage*: hot air 50°C, 70°C, 90°C	Vitamin C and total phenol content

(*Continued*)

TABLE 5.1 (*Continued*)
Summary of Some Intermittent Drying Works in Literature

Study	Material	Drying Method	Operating Conditions	Quality Attributes
Botha et al. (2011)	Pineapple	Osmotically microwave-assisted air drying	Osmotic dehydration at 40°C in 55Brix sucrose for different times Various power density and time.	Appearance, moisture content, soluble solids content, water activity, firmness, color, and volume
Demiray et al. (2013)	Tomatoes	Hot air drying	Air temperature 60°C–80°C, at air velocity 0.2 m/s and 20% relative humidity	Lycopene, β-carotene, ascorbic acid
Karaaslan et al. (2014)	Pomegranate arils	Blanching Vacuum drying	Blanching at 80°C for 2 min Drying at 55°C, 65°C, 75°C, 85 kPa vacuum	Phenolic compounds and anthocyanin
Aghilinategh et al. (2015)	Apple	Intermittent microwave convective	Hot air temperature (40°C–80°C), velocity (1–2 m/s), pulse ratio PR (2–6), and microwave power (200–600 W)	Rehydration capacity, bulk density, phenolic content

The following sections will present the effect of different intermittent drying applications on processed food quality when the heating source, airflow, drying pressure, and relative humidity of drying medium were applied variably and periodically.

5.2.1 Heating Source Supplement

Continuous supply of heat during drying is the traditional dehydration technique that has proved to be sufficiently productive, versatile, and easy to use. However, this drying method often has several disadvantages, such as high drying temperature and lengthy processing time, which result in products with low essential important nutrients, tougher texture, and hard product surface due to case-hardening. During continuous hot air drying process, the loss of functionality in cell membranes causes significant changes in sensory and nutritional quality (Torreggiani 1993). Several studies have reported modified drying regimes to overcome the difficulties in convective dryers and to promote the organoleptic and nutrient properties of the dried products (Jangam 2011). The application of periodically supplying heat during drying is the most dominant method in the intermittent drying strategy.

A substantial amount of research has been done in exploring the benefit of the intermittent technique in convective drying for enhancing dried food quality and energy efficiency. In their early work, Chua et al. (2001) employed different patterns for periodic temperature changes to dry guava samples to yield higher ascorbic acid retention compared to conventional hot air drying method. There was a 20% higher retention of ascorbic acid achieved without significant increase in drying time. Continuing previous work with the two-stage heat pump dehydration method, Chua et al. (2001) tested the change in drying kinetic and color of banana, employing stepwise control of drying temperature. They reported that with proper initial temperature and cyclic pattern, the drying time could be reduced while also the color could be improved. The moisture layer formed during relaxation time minimizes color and nutrient degradation. Chong et al. (2014) compared variable temperature heat pump drying with microwave vacuum heat pump drying, constant convective heat pump drying, and microwave vacuum convective drying. The best result was achieved by microwave vacuum heat pump drying and the step-up temperature heat pump drying retained the second highest amount of total phenolic content. The cyclic temperature heat pump drying can retain antioxidant levels better than the convective microwave drying method.

Intermittent energy supply is also used for some drying methods which are commonly believed as unfavorable treatments to food quality, such as radio frequency, microwave drying, infrared drying, or superheated steam drying. The main drawbacks of supplying radio frequency, microwave, and infrared continuously during the whole drying period are uneven temperature and moisture distribution inside the food material, which results in food quality deterioration (Kumar et al. 2016a, Soysal et al. 2009a, Vadivambal and Jayas 2010). Researchers investigated several combinations of drying methods, for example, microwave-assisted convective drying and freeze-drying to name a few. Among various techniques used for improving the performance of thermal dehydration, intermittent microwave convection (IMWC) drying has drawn special attention due to its reduced negative impact on the quality of dried products (Joardder et al. 2015b, Kowalski and Pawłowski 2011, Kumar et al. 2014a). IMWC is advantageous in terms of quality of dried food, which has been reported in several studies. For instance, convective-assisted microwave drying with variable power levels of osmotically treated pineapple was studied by Botha et al. (2012). They concluded that a high drying rate can be achieved without burning the pineapple samples when they applied variable microwave level with low temperature of drying medium.

Soysal et al. (2009b) noted that intermittent convective-assisted microwave drying of red pepper produced better sensory quality, color, texture, and general quality, than constant convective-assisted microwave drying and conventional drying. Besides attractive product appearance, red peppers are also known for having a high amount of important nutrient components, such as vitamin C, β-carotene, vitamin K, E, and folate, but they were not investigated in the study.

A complex combination and comparison in terms of physical appearance, antioxidant activity, and phenolic content of apple under different drying methods were undertaken by Chong et al. (2014). It was pointed out that heat pump and vacuum microwave were the most efficient drying methods in retaining phenolic content, antioxidant activity, and the best sensorial quality score.

Aghilinategh et al. (2015) analyzed the effect of convective-continuous microwave with intermittent microwave–convective drying on physical properties including color change, rehydration capacity, and bulk density and total polyphenol. Their research pointed out that higher microwave level and air velocity can provide high phenolic retention, but the increase in hot air temperature and power ratio has a negative impact on phytochemical properties. They explained that intensive heating due to high temperature and lengthy microwave radiation would severely break the cell compartment, which releases the degradative enzymes and accelerates the oxidation reaction.

5.2.2 Airflow

In many intermittent drying experiments, the velocity of drying medium was controlled to match with drying conditions. This strategy was recorded as a useful technique during the initial warming-up stage and constant drying period to remove free moisture accumulated near the surface of samples. Reyes et al. (2012) studied the effect of pulsed fluidized bed drying kinetics, moisture diffusivity, antioxidant capacity, and the retention of polyphenols in ordinary and selenium-treated broccoli florets. The pulsed gas supply during the course of the drying process helped to improve uniformity and reduce pressure drop. However, its effect on nutrient quality properties was not mentioned.

Effects of pulsed gas in fluidized bed starch particles were evaluated in Dacanal et al.'s (2016) research. The use of pulsed air resulted in a better dryness uniformity, preventing powder over-wetting, higher porosity, easy to flow, less cohesiveness (Joardder et al. 2015a). The retention of β-carotene was analyzed in intermittent vibrated fluidized bed drying of carrot by Pan et al. (1999). They indicated that the degradation of β-carotene during intermittently vibrated fluidized bed drying was reduced in the tempering period. However, the efficiency of this method was not evaluated, and it can be clearly seen that the drying time was substantially increased.

Pan et al. (1998) also investigated the drying effect on some quality parameters of squash slices using vibrofluidized bed drying. By applying the intermittent technique, the degradation of β-carotene, drying time and energy usage were reduced. A comparison was made between conventional and intermittent convective dried squash slice that highlighted the benefit of the intermittent technique.

The benefit of intermittent spouting in two-dimensional spouted-bed cardamom drying was investigated by Balakrishnan et al. (2011). While there was an increase in total drying time, a reduction of 25% of energy consumption and an increase of 10%–15% of oleoresin content was reported compared to continuous drying.

A newly designed fluidized bed dryer by Vega-Valencia et al. (2014) was investigated to test the energy efficiency, the reservation of invertase and the rehydration characteristic of enriched broccoli. Drying time was significantly reduced, while the moisture diffusivity increased. The product had high retention of enzymatic activity and a high degree of rehydration ability.

5.2.3 Pressure of Drying Environment

The pressure in the drying chamber plays an essential role in the early steps of dehydration, from cost estimation, the designing step to choosing drying conditions.

There are mainly three categories, so far as pressure is concerned, in which drying processes take place: vacuum pressure, atmospheric pressure, and high-pressure drying. During the drying process, the sample is heated by contact with drying medium from heat source or by radiation. Due to the vapour pressure difference between the outer environment and inside the food, the moisture will move from inside to outside, and it will be carried away by drying medium.

Drying under vacuum conditions provides a mild heating condition as the moisture evaporation reduces the temperature. Moreover, the oxidation of the sample decreases since less air is available inside vacuum drying chamber. Another advantage is that the vacuum condition increases the pressure gradient between drying medium and product and thus the external mass transfer is enhanced.

Superheated steam drying is a dehydration method in which all the drying medium is steam; hence, the harmful oxidation reaction is significantly reduced. However, the high temperature of the drying medium in this method is unfavorable for the retention of health-related compounds in most plant-based foods that are heat-sensitive. The application of pressure variance during drying was conducted in some research works. However, the outcome was not as expected. The nonstationary pressure during superheated steam drying in Thomkapanich et al.'s (2007) experiments showed that the degree of ascorbic acid retention was lower than convective drying. Therefore, its application was not deemed suitable to dehydrate the thermolabile and easily oxidized products (Karim et al. 2014).

5.2.4 Relative Humidity of Drying Medium

The relative humidity is the ratio between actual vapor pressure and saturated vapor pressure. The lower the relative humidity, the higher is the dehydration capability of the drying medium. When the drying medium is heated, it facilitates the drying rate due to low relative humidity and increase in product temperature. The relative moisture gradient between the food and the surrounding the drying medium enhances the drying process. There are some reports in the literature that the relative humidity is controlled periodically during drying of food. The varying humidity of the drying medium prevents high moisture gradient between the inner and outer food surface, which may result in the surface becoming overdried when the moisture migration by diffusion from sample surface to surrounding environment is substantially higher than by thermal diffusion from inside to outer sample surface. Plus, adjusting the humidity of the drying medium is often applied in combination with other processing parameters as this controlling technique is mostly valuable during the early drying stage when most of the moisture of food is in free and/or loosely bound water. As the drying progresses, the potential of this method is reduced. Some research works investigating the effect of moisture content of drying air on nutrient quality showed that drying under relative high moisture conditions could retain a higher amount of nutrient due to the reduction of oxygen in drying medium (Kaya et al. 2010, Qing-guo et al. 2006).

In the following section, the change of nutrient quality over the course of intermittent drying has been presented in order to illustrate how intermittency affects nutritive quality of dried food.

5.3 NUTRITIONAL QUALITY CHANGES DURING INTERMITTENT DRYING

Previous studies were mainly focused on lengthening the storage life and reducing energy consumption, while health-related quality was being compromised. Recently, several studies have been directed toward producing high-quality dehydrated foods by taking advantage of intermittent drying technologies, enhancing and optimizing existing drying methods (Sablani 2006).

Under the heat treatment during thermal drying, many physical and biochemical reactions can be induced, resulting in the disruption and breakage of plant cell wall and release of the compounds inside the cell. Once these health-promoting compounds are exposed to the drying environment, the nutrients, which are usually stable within the original fruit matrix, are highly vulnerable to deterioration (Faulks and Southon 2005, van Schie and Young 2000). Because of their thermolabile characteristic, most nutrients deteriorate markedly during the thermal dehydration process (Joardder et al. 2015b, Sablani 2006). In general, the use of continuous hot air supply in the conventional convective drying methods shows a negative effect on the retention of nutrients, as high temperature accelerates destructive reaction rate. Moreover, prolonged exposure in the drying chamber worsens the quality of dried food due to heat stress and other degradative reactions. Some reactions provide appealing color and flavor, while others can cause undesirable physical and nutritional characteristics (Joardder et al. 2015b). All the degradative reactions are related to the change of food structure, temperature, and moisture content (or molecular mobility, or water activity), which are variable in different positions and drying stages (Tsotsas and Mujumdar 2011). The literature points out that the application of the intermittent technique in drying provides higher retention of nutrient components in product compared to conventional drying method, as outlined in Table 5.2.

5.4 CHEMICAL QUALITY CHANGES DURING INTERMITTENT DRYING

Browning, lipid oxidation, antioxidant activity, bioactive retention, and modification of flavor in foods can occur during drying. Many investigators have employed intermittent heating schemes in their research, which automatically result in significant increase in the quality of food products, as shown in Table 5.3.

5.4.1 Lipid Oxidation and Fatty Acid Composition

Lipids in foods serve several important functions including aroma, flavor, color, and texture. The oxidation process of lipid generates products that can adversely affect these quality attributes. Rancidity and off-flavors often occur in fatty foods when they have low moisture content. Lipid oxidation is also responsible for the degradation of fat-soluble health-promoting compounds and pigments, especially in dehydrated foods. The type of fatty acid is a major factor that influences the lipid oxidation rate. Lipid oxidation is a significant problem with food that contains high amount of oil and unsaturated fatty acids. Fu et al. (2016) studied the effect of intermittent oven

TABLE 5.2
Summary of Nutrient Quality Changes Observed in Different Intermittent Drying Studies

Study	Materials	Drying Method	Conclusion on Product Quality
Chua et al. (2000a)	Guavas	Two-stage heat pump dryer	Higher retention of vitamin C compared to isothermal conditions.
Ramallo et al. (2010)	Yerba mate	Convective pilot drier	Tempering time influenced caffeine concentration but had low influence on the physical product quality.
Mishkin et al. (1984b)	Potatoes	Air drying	Optimization of vitamin C can be archived by selecting an appropriate drying time and temperature.
Zhao et al. (1999)	Carrot	Vibrofluidized bed dryer	Tempering-intermittent drying can reduce degradation of β-carotene and shorten the drying time, but does not affect rehydration capacity.
Chua and Chou (2005)	Carrot	Microwave compared with infrared drying	Intermittent MW drying can significantly reduce color change compared with convective or intermittent infrared drying.
Thomkapanich et al. (2007)	Banana	Low-pressure superheated steam and vacuum drying	Higher level of ascorbic acid retention, especially at longer tempering (off) periods. The product colors in intermittent pressure drying were worse than in continuous drying. Product texture, the shrinkage were not significantly different.
Soysal et al. (2009b)	Red pepper	Microwave convective	Better sensory quality, color, texture, and appealing appearance, than continuous microwave–convective drying and commercial drying.
Soysal et al. (2009a)	Oregano	Microwave convective	No significant change in essential oil content, but unacceptable color.
Zhu et al. (2010)	Apple	Infrared dry-blanching and intermittent heating	No severe surface discoloration, faster inactivation of enzymes, when prolonged heating with intermittent heating mode.
Arikan et al. (2011)	Carrot	Intermittent microwave convective	Higher β-carotene loss compared to convective drying product, especially with higher power level applied.
Balakrishnan et al. (2011)	Cardamom	Spouted bed	Increase oleoresin retention.
Ong et al. (2012)	Salak fruit	Heat pump	Shorter drying time and better product quality compared to hot air constant drying total color change, ascorbic acid content, and total phenolic content.

(Continued)

TABLE 5.2 (*Continued*)
Summary of Nutrient Quality Changes Observed in Different Intermittent Drying Studies

Study	Materials	Drying Method	Conclusion on Product Quality
Szadzińska (2014)	Green pepper	Intermittent convective drying	Higher retention of vitamin C.
Chong et al. (2014)	Apple	Combined heat pump, vacuum–microwave	Best quality in physical appearance, antioxidant activity, and phenolic content.
Aghilinategh et al. (2015)	Apple	Microwave–convective	Minimum physical characteristic, energy usage, and phenolic content can be obtained by increasing the microwave power and decreasing air temperature.

TABLE 5.3
Chemical Quality Attributes Investigated in Intermittent Drying

Chemical Attributes	References	Type of Intermittency	Food Material
Lipid oxidation Fatty acids	Fu et al. (2016)	Oven drying	Walnut
β-carotene retention	Kowalski et al. (2013)	Nonstationary air drying	Carrot
Betanin retention	Pan et al. (1999a,b)	Tempering-drying	Squash
	Kowalski and Szadzińska (2014b)	Air drying	Beetroots
Volatile flavor compounds	An et al. (2016)	Microwave–convective drying	Ginger
Essential oil yield	Soysal et al. (2009a)	Microwave–convective drying	Oregano
Enzymes inactivation	Zhang et al. (2014)	Immersion	Cucumber
• CAT,[a] POD[a]	Zhu et al. (2010)	IR heating	Apple
• PPO,[a] POD[a] • LOX[a]	Fu et al. (2016)	Oven drying	Walnut
Nonenzymatic browning	Zhu et al. (2010)	IR heating	Apple
	Ho et al. (2001)	Air drying	Potato
Sugar content, caffeine content	Ramallo et al. (2010)	Cross-flow air dryer	Yerba mate

[a] CAT, Chloramphenicol acetyltransferase; POD, peroxidase; PPO, polyphenol oxidase; LOX, lipoxygenase.

drying on lipid oxidation of walnuts. Authors demonstrated that intermittent oven drying provided walnut products with the lowest lipid oxidation values. Moreover, the amount of linoleic acid in walnut was better retained in intermittent oven drying and sun drying than those dried in the oven. Therefore, intermittent drying can be applied to improve the quality of dehydration of fatty foods.

5.4.2 Pigment Retention

There are many types of pigments such as β-carotene, the main pigment in carrots, which are responsible for the color of food material. These pigments have an important impact on the food quality and biological properties such as pro-vitamin A and antioxidant activity. Betanin is the main pigment in beetroots and has different bioactivities including antioxidant and antibacterial activity. Recently, Kowalski et al. (2013) have demonstrated that the periodical changes in the drying temperature retain much higher β-carotene, with approximately 73%–92% retention, than in the case of unstructured changes. Kowalski and Szadzińska (2014b) examined the effect of nonstationary drying parameters on the betanin retention in beetroots. The authors demonstrated the benefit of intermittent convective drying in retention and preservation of pro-health natural pigments such as betanin. The stationary convective drying causes a high loss of betanin with a maximum loss (68%) at the drying temperature of 80°C.

5.4.3 Volatile Flavor Compounds

Volatile flavor compounds are significantly affected by the process conditions over the course of drying. An et al. (2016) investigated the effect of microwave, convective, infrared, and intermittent microwave–convective and freeze drying on the major volatile compounds of ginger. The amount of sesquiterpenes compounds significantly increased, while monoterpenes decreased considerably. The authors concluded that the IMWC drying process retained lower volatile compounds compared to the other drying processed as the microwaves accelerate destruction of cell wall to the release of volatile compounds with the assistant of hot air to elevates the migration of those compounds to the sample surface, which increases the reduction of volatile compounds. IMWC is not recommended in ginger due to significant reductions in volatile content. More research has to be conducted to improve the effect of IMWC drying of aromatic plants (An et al. 2016).

In addition to these, food matrix significantly controls the release of flavors during consumption and at all steps of food processing (Tokuşoğlu and Swanson 2014, Lafarge et al. 2008, Druaux and Voilley 1997). Therefore, the final product with minimal structural damage would retain maximum sensory quality.

5.4.4 Enzyme Inactivation

The quality of fresh fruits and vegetables rapidly declines during storage, mainly caused by disease infection and severe water transpiration. Plant cells are equipped with antioxidant defense systems, which usually include some antioxidant enzymes such as chloramphenicol acetyltransferase (CAT) and peroxidase (POD). The enzymes help to eliminate oxidative damage in senescence and regeneration of ascorbate and glutathione metabolites. Few studies have demonstrated the effects of intermittent heat treatment on antioxidant enzymes during storage. Meanwhile, there have been few outcomes about heat treatment effect from the perspective of heat transfer characteristics during heat treatment (Zhang et al. 2014). Zhang et al. (2014) investigated the effect of intermittent heat treatment on antioxidant enzymes in cucumber, namely, CAT and POD activity. CAT and POD activities, for intermittent heat treatment, were significantly higher than continuous heat treatment, which enhanced their activity and reinforced the capacity of scavenging peroxide. The intermittent heat treatment improved the ability to protect cells from oxidative injury, providing better sensory quality for intermittent heat treatment of cucumbers. Zhu et al. (2010) demonstrated that simultaneous infrared dry-blanching with intermittent heating can be used to manufacture high-quality plant-based food, when they investigated the effect of intermittent heating on residual oxidase and peroxidase in apple slices.

5.4.5 Sugar and Caffeine Content

Sugar and caffeine content is of paramount importance for many products such as yerba mate and coffee. The benefit of tempering in the dehydration of Yerba mate, a very popular tea, was investigated by Holowaty et al. (2012). They found that loss of caffeine content was 10% in intermittent drying, whereas it was about 30% in continuous drying. Conversely, sugar content was not influenced by tempering. Perhaps this is because caffeine content is more sensitive than sugar content when exposed to higher temperatures or temperature needs to be elevated to a certain level to remove the sugar. Some continuous drying studies also reported no changes in sugar concentration during drying. In the case of sugar accumulation on the product surface of the sample, it is very important to understand whether water transportation takes place as vapor form or as liquid phase which could convey sugar from the inside to the outer surface of the sample. If water migrates in liquid phase from inside to surface, soluble sugar migrates to the surface and develops sugar crust. Therefore, further investigation is necessary to explore whether intermittent drying has any impact on the sugar content of food items (Kumar et al. 2014b).

5.5 INTERMITTENT DRYING AND MICROBIOLOGICAL SAFETY

Drying is a popular preservation method used to inhibit the development of microbial organisms and delay the onset of some deteriorative biochemical reactions in biomaterials such as food and agricultural products. There are two major effects of drying toward the preservation of biomaterials: The reduction of surviving microorganisms to a safe level, which inhibits deterioration and the reduction of water activity (a_w) that avoids microbial growth; The minimization of destructive degradative reactions causing deterioration. Water activity contributes to chemical and biochemical reactions and cellular and biological matrices in foods (Chirife and Fontana 2008). A high water activity facilitates the growth of microbiological organisms and biochemical reaction in food material, which can reduce the shelf-life time of products. Microbiologically safe water activity is shown in Table 5.4.

In particular, the most critical microorganisms in dried foods are molds, such as *Aspergillus* spp., which can start growing from a water activity of 0.70. Mold growth is

TABLE 5.4
Safe Water Activity Level of Some Microorganisms in Food

Microbial Organisms Type	Water Activity Threshold	References
Bacteria	0.85	Perera (2005)
Yeasts	0.7	
Molds	0.65	
Spore-forming organisms	0.93	Lopez-Velasco et al. (2011), Sikorski (2007), and Valdramidis et al. (2006)

significant in dried biomaterials because it can decrease product shelf life. Moreover, some mold species can produce mycotoxins and the toxic secondary metabolites can be produced by some molds (Jay 2012). Generally, when water activity is less than 0.6, the material is considered as microbiologically safe (Kowalski and Szadzińska 2014a).

However, food quality can deteriorate due to the loss of moisture, which results in significant changes in the physical properties of dried foods (Bernstein and Noreña 2014, Setiady et al. 2009, Telis et al. 2005). Intermittent drying is one of the potential dehydration techniques used to minimize the adverse effects of the continuous stationary drying process with the introduction of the resting time or relaxation periods. During the intermittent drying process, drying process parameters are controlled variably during the course of the drying time. In the tempering period, there is a homogenization process of moisture and temperature distribution (Carmo et al. 2012, Dong et al. 2009). Cihan et al. (2008) examined the effect of tempering on the drying kinetics of rough rice (*Oryza sativa*) and reported similar results that drying rates are proportionate to the length of tempering periods.

Drying methods and process conditions critically affect the heat resistance of microbial organisms and their activity. Apart from that, it is strictly dependent on the microbiological quality of the raw materials because contamination often occurs during the preliminary processing step on the field before the main production chain, and there may be a health risk because of the survival of pathogens at low water activity values (e.g., *Salmonella* spp. and enterohemorrhagic *Escherichia coli*) or presence of toxins (Pittia and Antonello 2016). In fact, this effect can become more severe if drying is not properly managed. Prolonged tempering time can accelerate the microbial activity in the biomaterial. Therefore, the knowledge of ideal relaxing duration for different biomaterials is essential in terms of the microbiological safety of intermittent drying.

Several studies have proved prospective benefits of intermittent heat treatment on product quality. Kowalski et al. (2013) and Szadzińska (2014) studied the effects of variable drying air temperature on the quality of dehydrated carrot (*Daucus carota* L.) and green pepper (*Capsicum* L.), respectively. Both experiments proved that convective–intermittent drying efficiently minimizes the adverse effects of convective drying as well as increases the shelf life. The lowest water activity values $a_w = 0.288 \pm 0.045$ for green pepper and $a_w = 0.49 \pm 0.06$ for a carrot were observed for intermittent conditions (70°C) with 5 min heating and 30 min cooling cycles. For most intermittent drying experiments, water activity was maintained under 0.6, which indicates the stability against microbiological deterioration upon storage.

Due to the volumetric heating characteristic, intermittent supply of microwave energy during drying may improve the microbiological safety of biomaterials (Gunasekaran 1999, Joardder et al. 2013, Kumar et al. 2015). Soysal et al. (2009b) and Beaudry et al. (2003) reported the effect of intermittent convective-assisted microwave drying on the quality parameters of red pepper (*Capsicum* L.) and partially dehydrated cranberries (*Vaccinium macrocarpon*), respectively.

Pretreatment techniques such as osmotic treatment, blanching, and ultrasonic treatments (Frias et al. 2010, Gornicki and Kaleta 2007) are useful, as well as emerging technologies that can be used to improve the stability of biomaterials by further removal of moisture and by making it unavailable for microorganisms. Kowalski and Szadzińska (2014b) investigated the influence of heating pattern of intermittent convective drying and the effect of pretreatments such as osmotic drying and blanching on the quality of beetroots (*Beta vulgaris* L.). According to their results, the lowest water activity values were achieved with stepwise changing temperature (10 min on and 40 min off), preceded by osmosis ($a_w = 0.39 \pm 0.01$) and blanching ($a_w = 0.36 \pm 0.06$).

Kowalski et al. (2015) obtained similar kind of results for their investigation on the combined effect of intermittent convective drying and ultrasonic pretreatment on the stability of carrot (*Daucus carota* L.). They revealed that ultrasonic-assisted osmotic dehydration under intermittent conditions could reduce the water activity up to 0.328 ± 0.02, which can inhibit microbial growth; hence, the storage life of the dried carrot was extended up to 32% compared to the fresh biomaterial (Kowalski et al. 2015).

Despite the substantial improvement in product quality, it was noted that there is limited research available on the effects of process parameters of intermittent drying as well as its combination with pretreatment on the destruction of pathogens and spoilage flora in biomaterials.

5.6 FOOD QUALITY PREDICTION IN INTERMITTENT DRYING

Foods are complex in nature due to its variable characteristics in structure, composition, which are also unique among different varieties, cultivating the field, even the difference occurs with fruit/vegetable in the same tree. The difference can also be found in different parts in the same body of food. Many attempts have been made by scientists and technologists to understand and predict the behaviors of foods during processing. An in-depth understanding of fundamental changes in structure and nutrient can assist in optimizing the performance of drying process while minimizing the severe effect of heat on important quality attributes of dried food.

Intermittent drying is a promising drying control method that has a potential to increase the energy efficiency while maintaining product quality. Many attempts have been also made in the literature to predict the intermittent drying performance.

Table 5.5 provides an overview of frequently employed predictive models used by researchers, along with experimental conditions in modeling as well as the different quality indices of various foods when drying. Empirical and experimental-based models are often applied to predict changes in color, vitamins, and other phytochemical components of dried food. These models demonstrate relationships between the amount of one or a few nutritional quality and processing conditions used in the drying process. Artificial Neural Network (ANN) and Response Surface Methodology

TABLE 5.5
List of Quality Estimation Models

Model	Description	References
Kinetic Arrhenius model	$\frac{dQ}{dt} = -k \cdot Q^n$ Q is the quality index of a color value, nutritional content at a certain time to the initial content. n is the order of the reaction. The reaction rate constant k is the function of a process condition. In literature, quality reduction rate usually follows zero-, first-, or second-order reaction. Arrhenius equation $k = k_o \cdot \exp\left(-\frac{\Delta E_a}{R \cdot T}\right)$ where R is the universal gas constant: 8.134 (J/kmol), k_o is preexponential factor, and ΔE_a is activation energy.	Arslan and Özcan (2011), Devahastin and Niamnuy (2010), and Di Scala and Crapiste (2008)
Weibull	The model is usually used to describe the collapse of a system subjected to stress conditions over time: $\frac{Q_t}{Q_o} = \exp\left[-\left(\frac{t}{\alpha}\right)^{\beta}\right]$ (5.14) Q_t and Q_o are quality value at a certain time and the initial time of the process, respectively. α is the kinetic reaction constant. β: shape factor, i.e., $\beta = 1$, then the model become first order kinetics; $\beta > 1$ reaction rate increases with time, the curve adopts a sigmoidal shape; if $\beta < 1$ the reaction rate reduces with time, and deleterious rate higher than the observed exponential at the initial time.	Jiang et al. (2014), Uribe et al. (2011), and Zheng and Lu (2011)
Williams Landel Ferry (WLF)	$\log\frac{Q_t}{Q_o} = -\int_o^t \frac{10^{\frac{C_1 C_2 (T-T_r)}{[C_2+(T_g-T_r)][C_2+(T-T_g)]}} dt}{D_r}$ (5.15) T_r is the reference temperature; T_g is the glass temperature; D_r is the thermal death time at T_r; C_1, C_2 are WLF constants; Q_t and Q_o are quality value at a certain time and the initial time of the process, respectively.	Frías and Oliveira (2001) and Nicoleti et al. (2007)
Eyring and Polanyi model	Reaction rate: $k = \frac{k_B}{h} T \times e^{-\frac{\Delta G^*}{RT}} = \frac{k_B}{h} T \times e^{-\frac{\Delta H^* - T\Delta S^*}{RT}}$ (5.16) where ΔH^* is the enthalpy of activation (kJ/mol), ΔS^* is the entropy of activation (J/mol K), k_B is the Boltzmann constant (1.381×10^{-23} J/K), h is the Planck constant (6.626×10^{-34} J s)	Barsa et al. (2012) and Karaaslan et al. (2014)

(Continued)

TABLE 5.5 (*Continued*)
List of Quality Estimation Models

Model	Description	References
Multilayer neural networks (ANN)	The multilayer network was modeled based on a multilayer feed-forward algorithm that maps sets of input data onto a set of appropriate outputs, utilizes a supervised training strategy called back propagation method.	Devahastin and Niamnuy (2010), Fathi et al. (2011), Kaminski and Tomczak (2000), and Razavi et al. (2007)

(RSM) are empirical models that are often applied for modeling and optimizing of food processing. RSM and ANN use the input variables, alone or in combination, and predict the response of the process. They allow food processing researchers to investigate the process of interest efficiently (Madamba 2002). However, as RSM and ANN are purely empirical models containing no chemical/physical mechanism, they are mostly applicable in the range of experimental conditions (Kumar et al. 2014b).

Kaminski and Tomczak (2000) examined the capability of first-order reaction model, multilayer perceptron (MLP) model, and combined model to predict changes in vitamin C level in some type of vegetables during vibrofluidized bed drying. The first-order kinetics models provided a better fit in testing for silica gel and potatoes. For cabbage, combined model performed better than the others, while MPL model was more suitable for potatoes. The goodness of the result depends on the food matrix. Razavi et al. (2007) employed ANN to predict the color change of combined osmotic treatment and hot air drying of pumpkins. The authors claimed that the model delivers better fit than empirical and semiempirical models to simulate quality changes of some types of fruits and vegetable during drying (Alli et al. 2001). However, the predictions of ANNs models are greatly dependent on the number of neurons, and ANNs are subject to overfitting or memorization instead of generalization.

Kinetic models, which are a more fundamental modeling approach obtained from energy and mass relation, are widely adopted to simulate the reduction of nutrient content and color of fruits and vegetables (Gornicki and Kaleta 2007) in thermal food processing. These kinetic models have been successfully applied in many drying pieces of research to predict the changes of the quality index with time. The reaction rate in the kinetic model is widely applied to the Arrhenius equation.

The change of quality index of dried products is often described as an nth-order reaction kinetic approach (Valdramidis et al. 2012):

$$\frac{dQ}{dt} = -k \cdot Q^n \tag{5.1}$$

where

Q is the quality index (represent the relative value of color value, nutritional content at a certain time to the initial content)

n is the order of the reaction

The reaction rate constant k is the function of dehydration condition, that is, moisture content X_m, and sample temperature T_m, which depends on the local position in the drying material (Kaminski and Tomczak 2000):

$$k = f\left(X_m\left(t\right), T_m\left(t\right)\right) \tag{5.2}$$

Equation 5.1 can be solved by integration with reference to Equation 5.2 if variables in Equation 5.2 are known.

By integration with time, the quality index values can be obtained as follows:

$$Q^{1-n} = Q_o^{1-n} + \left(n-1\right)k \cdot t \quad \text{for } n \neq 1 \tag{5.3}$$

where Q_o is the quality index at the initial condition.

In food quality prediction in research, quality change rate under heat treatment are usually recorded to follow a zero-order, first-order, or second-order reaction and the decomposition rates are given as in Equations 5.4 through 5.6 respectively:

$$\frac{dQ}{dt} = -k \tag{5.4}$$

$$\frac{dQ}{dt} = -k \cdot Q \tag{5.5}$$

$$\frac{dQ}{dt} = -k \cdot Q^2 \tag{5.6}$$

After integration, Equations 5.4 through 5.6 become:

$$Q = Q_o - k \cdot t \tag{5.7}$$

$$Q = Q_o \exp\left(-kt\right) \tag{5.8}$$

$$Q = \frac{Q_o}{1 + Q_o \cdot k \cdot t} \tag{5.9}$$

As food quality changes owing to the biochemical reaction and physical alternation strongly depend on food matrix characteristics, no generic modeling by theoretical approach can be applied. The relationship between the reaction rate constant k and process conditions is usually nonlinear. In the literature, the reaction rate constant, k, is often calculated from the Arrhenius equation, which depends on temperature, as the following equation (Goula and Adamopoulos 2006):

$$k = k_o \exp\left(-\frac{\Delta E_a}{RT}\right) \tag{5.10}$$

Equation 5.10 can also be rewritten to:

$$\ln k = \ln k_o - \frac{\Delta E_a}{RT} \tag{5.11}$$

where

R is the universal gas constant: 8.134 (J/kmol)

ΔE_a is the activation energy

k_o is the preexponential factor that can be considered as functions of moisture content, as in Mishkin et al. (1984a) report:

$$k_o = A_0 + A_1 \cdot M + A_2 M^2 \tag{5.12}$$

$$\Delta E_a = B_0 + B_1 \cdot M + B_2 M^2 \tag{5.13}$$

where

M is the moisture content in dry basis (g/g solid mass)

A_0, A_1, A_2, B_0, B_1, B_2 are constants estimated by suitable fitting method

A suitable reaction rate can be calculated by a two-step linear or one-step nonlinear regression analysis from experimental data to obtain quality index Q.

The degradation of nutrient and color changes predicted by kinetic models are mostly found to obey zero-, first-, and second-order reaction. Chua et al. (2000a) applied the kinetic model to predict the degradation of ascorbic acid of grapes pieces dried in nonstationary conditions by using the equivalent temperature concept. The Weibull model is an expanded version of Arrhenius models that were originally used to predict the collapse of the structure, and microbiological inactivation has been increasingly used by researchers to model color and chemical deterioration (Boekel 2008). This model was already successfully applied in several studies to describe the kinetics of nutrient, enzymatic, or microbial inactivation processes.

The prediction of nutrients and color change of dehydrating fruits and vegetables can also be implemented by applying different types of models. Williams–Landel–Ferry (WLF) equation is derived from polymer science and the newly imported Eyring–Polanyi model from physical chemistry (Barsa et al. 2012). The WLF model uses the glass transition temperature (T_g) and thermal death time (DT) to describe the kinetics of some nutrient deterioration of food under thermal treatments. As the Arrhenius equation is not suitable to describe the degradation of food in the rubbery condition, the kinetics of food modification may be affected by food molecular mobility, free volume, molecular relaxation time, and glass transition temperature, T_g.

The kinetic temperature dependence of physical and chemical properties in this state is mostly described by the WLF model. The correlation between the reaction rate k and the reciprocal of the drying temperature ($1/T$) in the Arrhenius equation has to be linear, which is an issue in some real cases. The Eyring–Polanyi model has been recently suggested to substitute the Arrhenius equation where the plot of ln [$k(T)$] versus $1/T$ is curvy (Başlar et al. 2014). The Eyring–Polanyi has been applied in modeling total phenol retention and total anthocyanins in dried pomegranate arils (Karaaslan et al. 2014); thermal decomposition process of anthocyanins in blood orange, blackberry, and roselle (Cisse et al. 2009), chlorophyll and vitamin C in nectar (Diop Ndiaye et al. 2011).

A combination of the empirical and kinetic model with the assistance of a mathematical model is believed to have the advantage that has the strengths of each approach in predicting the response quality indexes and suggesting an optimum solution for drying fruit (Table 5.5).

5.7 CONCLUSION

The application of intermittent food drying is gaining more interest because of its better drying performance, energy saving, and product quality improvement. The intermittent process can be made by varying drying conditions such as a heat source, airflow, pressure, humidity, and effective drying time depending on types of sample. Among many intermittent drying researches, most of the practice has been done in convective and microwave drying due to the fact that the conventional application of those drying methods often results in severe quality degradation and high power consumption. The application of the intermittent drying technique can be conducted on drying methods in tandem or simultaneously to utilize the advantages of each drying method. However, its high drying performance can only be achieved by acquiring an in-depth understanding of the effects of process parameters on the food quality attributes and the mechanism of quality degradation. A robust and accurate model that is flexible to be adapted for different products is necessary for better prediction of food quality.

REFERENCES

Abbasi Souraki, B. M. and Mowla, D. (2008). Experimental and theoretical investigation of drying behaviour of garlic in an inert medium fluidized bed assisted by microwave. *Journal of Food Engineering*, *88*(4), 438–449.

Aghilinategh, N., Rafiee, S., Hosseinpur, S., Omid, M., and Mohtasebi, S. S. (2015). Optimization of intermittent microwave–convective drying using response surface methodology. *Food Science & Nutrition*, n/a–n/a doi:10.1002/fsn3.224.

Alasalvar, C. and Shahidi, F. (2012). *Hui: Food Science and Technology: Dried Fruits: Phytochemicals and Health Effects*. Somerset, NJ: John Wiley & Sons.

Alli, I., Ramaswamy, H., and Chen, C. (2001). Prediction of quality changes during osmo-convective drying of blueberries using neural network models for process optimization. *Drying Technology*, *19*(3), 507–523.

An, K. J., Zhao, D. D., Wang, Z. F., Wu, J. J., Xu, Y. J., and Xiao, G. S. (2016). Comparison of different drying methods on Chinese ginger (*Zingiber officinale* Roscoe): Changes in volatiles, chemical profile, antioxidant properties, and microstructure. *Food Chemistry*, *197*, 1292–1300.

Arikan, M. F., Ayhan, Z., Soysal, Y., and Esturk, O. (2012). Drying characteristics and quality parameters of microwave-dried grated carrots, *Food and Bioprocess Technology*, 5(8), 3217–3229.

Arslan, D. and Özcan, M. M. (2011). Drying of tomato slices: Changes in drying kinetics, mineral contents, antioxidant activity and color parameters [Secado de rodajas de tomate: cambios en cinéticos del secado, contenido en minerales, actividad antioxidante y parámetros de color]. *CyTA—Journal of Food*, *9*(3), 229–236.

Balakrishnan, M., Raghavan, G. S. V., Sreenarayanan, V. V., and Viswanathan, R. (2011). Batch drying kinetics of cardamom in a two-dimensional spouted bed. *Drying Technology*, *29*(11), 1283–1290.

Barsa, C., Normand, M., and Peleg, M. (2012). On models of the temperature effect on the rate of chemical reactions and biological processes in foods. *Food Engineering Reviews*, *4*(4), 191–202.

Başlar, M., Karasu, S., Kiliçli, M., Us, A. A., and Sağdiç, O. (2014). Degradation kinetics of bioactive compounds and antioxidant activity of pomegranate arils during the drying process. *International Journal of Food Engineering*, *10*(4), 839–848.

Beaudry, C., Raghavan, G. S. V., and Rennie, T. J. (2003). Microwave finish drying of osmotically dehydrated cranberries. *Drying Technology*, *21*(9), 1797–1810.

Bernstein, A. and Noreña, C. P. Z. (2014). Study of thermodynamic, structural, and quality properties of yacon (*Smallanthus sonchifolius*) during drying. *Food and Bioprocess Technology*, *7*(1), 148–160.

Boekel, V. T. (2008). Kinetic modeling of food quality: A critical review. *Comprehensive Reviews in Food Science and Food Safety*, *7*(1), 144–158.

Botha, G. E., Oliveira, J. C., and Ahrné, L. (2011). Quality optimisation of combined osmotic dehydration and microwave assisted air drying of pineapple using constant power emission. *Food and Bioproducts Processing*, 90(2), 171–179, doi:10.1016/j.fbp.2011.02.006.

Botha, G. E., Oliveira, J. C., and Ahrné, L. (2012). Microwave assisted air drying of osmotically treated pineapple with variable power programmes. *Journal of Food Engineering*, *108*(2), 304–311.

Carmo, J. E. F. D., de Lima, A. G. B., and Silva, C. (2012). Continuous and intermittent drying (tempering) of oblate spheroidal bodies: Modeling and simulation. *International Journal of Food Engineering*, *8*(3), 20.

Chirife, J. and Fontana, A. J. (2008). Introduction: Historical highlights of water activity research. In G. V. Barbosa-Cánovas, A. J. Fontana, S. J. Schmidt and T. P. Labuza (Eds.), *Water Activity in Foods: Fundamentals and Applications*, pp. 3–13. Oxford, UK: Blackwell Publishing Ltd, doi: 10.1002/9780470376454.ch1.

Chong, C. H., Figiel, A., Law, C. L., and WojdyAo, A. (2014). Combined drying of apple cubes by using of heat pump, vacuum-microwave, and intermittent techniques. *Food and Bioprocess Technology*, *7*(4), 975–989.

Chou, S. K., Chua, K. J., Mujumdar, A. S., Hawlader, M. N. A., and Ho, J. C. (2000). On the intermittent drying of an agricultural product. *Food and Bioproducts Processing*, *78*(4), 193–203.

Chua, K. J. and Chou, S. K. (2005). A comparative study between intermittent microwave and infrared drying of bioproducts. *International Journal of Food Science & Technology*, *40*(1), 23–39.

Chua, K. J., Chou, S. K., Ho, J. C., Mujumdar, A. S., and Hawlader, M. N. A. (2000a). Cyclic air temperature drying of guava pieces: Effects on moisture and ascorbic acid contents. *Food and Bioproducts Processing*, *78*(2), 72–78.

Chua, K. J., Mujumdar, A. S., Chou, S. K., Hawlader, M. N. A., and Ho, J. C. (2000b). Convective drying of banana, guava and potato pieces: Effect of cyclical variations of air temperature on drying kinetics and color change. *Drying Technology*, *18*(4), 907–936.

Chua, K. J., Mujumdar, A. S., Hawlader, M. N. A., Chou, S. K., and Ho, J. C. (2001). Batch drying of banana pieces—Effect of stepwise change in drying air temperature on drying kinetics and product colour. *Food Research International*, *34*(8), 721–731.

Cihan, A., Kahveci, K., Hacıhafizoğlu, O., and de Lima, A. G. B. (2008). A diffusion based model for intermittent drying of rough rice. *Heat and Mass Transfer*, *44*(8), 905–911.

Cisse, M., Vaillant, F., Acosta, O., Claudie, D.-M., and Dornier, M. (2009). Thermal degradation kinetics of anthocyanins from blood orange, blackberry, and roselle using the arrhenius, eyring, and ball models. *Journal of Agricultural and Food Chemistry*, *57*(14), 6285–6291.

Dacanal, G. C., Feltre, G., Thomazi, M. G., and Menegalli, F. C. (2016). Effects of pulsating air flow in fluid bed agglomeration of starch particles. *Journal of Food Engineering*, *181*, 67–83.

Da Silva, M. A., Arévalo Pinedo, R., and Kieckbusch, T. G. (2005). Ascorbic acid thermal degradation during hot air drying of Camu-Camu (*Myrciaria dubia*[H.B.K.] McVaugh) slices at different air temperatures. *Drying Technology*, *23*(9–11), 2277–2287.

Demiray, E., Tulek, Y., and Yilmaz, Y. (2013). Degradation kinetics of lycopene, β-carotene and ascorbic acid in tomatoes during hot air drying. *LWT—Food Science and Technology*, *50*(1), 172–176.

Devahastin, S. and Niamnuy, C. (2010). Modelling quality changes of fruits and vegetables during drying: A review. *International Journal of Food Science & Technology*, *45*(9), 1755–1767.

Diop Ndiaye, N., Dhuique-Mayer, C., Cisse, M., and Dornier, M. (2011). Identification and thermal degradation kinetics of chlorophyll pigments and ascorbic acid from ditax nectar (*Detarium senegalense* J.F. Gmel). *Journal of Agricultural and Food Chemistry*, *59*(22), 12018–12027.

Di Scala, K. and Crapiste, G. (2008). Drying kinetics and quality changes during drying of red pepper. *LWT—Food Science and Technology*, *41*(5), 789–795.

Dong, R., Lu, Z., Liu, Z., Nishiyama, Y., and Cao, W. (2009). Moisture distribution in a rice kernel during tempering drying. *Journal of Food Engineering*, *91*(1), 126–132.

Druaux, C. and Voilley, A. (1997). Effect of food composition and microstructure on volatile flavour release. *Trends in Food Science and Technology*, *8*(11), 364–368.

Fathi, M., Mohebbi, M., and Razavi, S. M. A. (2011). Application of image analysis and artificial neural network to predict mass transfer kinetics and color changes of osmotically dehydrated kiwifruit. *Food and Bioprocess Technology*, *4*(8), 1357–1366.

Faulks, R. M. and Southon, S. (2005). Challenges to understanding and measuring carotenoid bioavailability. *Biochimica et Biophysica Acta—Molecular Basis of Disease*, *1740*(2), 95–100.

Fernandes, F. A. N., Rodrigues, S., Law, C. L., and Mujumdar, A. S. (2011). Drying of exotic tropical fruits: A comprehensive review. *Food and Bioprocess Technology*, *4*(2), 163–185.

Frias, J., Peñas, E., Ullate, M., and Vidal-Valverde, C. (2010). Influence of drying by convective air dryer or power ultrasound on the vitamin C and β-carotene content of carrots. *Journal of Agricultural and Food Chemistry*, *58*(19), 10539–10544.

Frías, J. M. and Oliveira, J. C. (2001). Kinetic models of ascorbic acid thermal degradation during hot air drying of maltodextrin solutions. *Journal of Food Engineering*, *47*(4), 255–262.

Fu, M., Qu, Q., Yang, X., and Zhang, X. (2016). Effect of intermittent oven drying on lipid oxidation, fatty acids composition and antioxidant activities of walnut. *LWT—Food Science and Technology*, *65*, 1126–1132.

Gao, Q.-H., Wu, C.-S., Wang, M., Xu, B.-N., and Du, L.-J. (2012). Effect of drying of jujubes (*Ziziphus jujuba* Mill.) on the contents of sugars, organic acids, α-tocopherol, β-carotene, and phenolic compounds. *Journal of Agricultural and Food Chemistry*, *60*(38), 9642.

Górnicki, K. and Kaleta, A. (2007). Drying curve modelling of blanched carrot cubes under natural convection condition. *Journal of Food Engineering*, *82*(2), 160–170.

Goula, A. M. and Adamopoulos, K. G. (2006). Retention of ascorbic acid during drying of tomato halves and tomato pulp. *Drying Technology*, *24*(1), 57–64.

Grosch, W. and Schieberle, P. (2009). *Vitamins*, pp. 403–420. Berlin, Germany: Springer.

Gunasekaran, S. (1999). Pulsed microwave-vacuum drying of food materials. *Drying Technology*, *17*(3), 395–412.

Ho, J. C., Chou, S. K., Mujumdar, A. S., Hawlader, M. N. A., and Chua, K. J. (2001). An optimisation framework for drying of heat-sensitive products. *Applied Thermal Engineering*, *21*(17), 1779–1798.

Holowaty, S. A., Ramallo, L. A., and Schmalko, M. E. (2012). Intermittent drying simulation in a deep bed dryer of yerba maté. *Journal of Food Engineering*, *111*(1), 110–114.

Hui, Y. H. and Barta, J. Z. (2006). *Handbook of Fruits and Fruit Processing*. Ames, IA: Blackwell Pub.

Jangam, S. V. (2011). An overview of recent developments and some R&D challenges related to drying of foods. *Drying Technology*, *29*(12), 1343–1357.

Jay, J. M. (2012). *Modern Food Microbiology*. Springer Science & Business Media, New York, ISBN 0-387-23180-3.

Jiang, L., Zheng, H., and Lu, H. (2014). Use of linear and weibull functions to model ascorbic acid degradation in Chinese winter jujube during postharvest storage in light and dark conditions. *Journal of Food Processing and Preservation*, *38*(3), 856–863.

Joardder, M. U. H., Brown, R. J., Kumar, C., and Karim, M. A. (2015a). Effect of cell wall properties on porosity and shrinkage of dried apple. *International Journal of Food Properties*, *18*(10), 2327–2337.

Joardder, M. U. H., Karim, A., and Kumar, C. (2013). Effect of temperature distribution on predicting quality of microwave dehydrated food. *Journal of Mechanical Engineering and Sciences*, *5*, 562–568.

Joardder, M. U. H., Kumar, C., and Karim, M. A. (2017). Food structure: Its formation and relationships with other properties. *Critical Reviews in Food Science and Nutrition*, *57*(6), 1190–1205.

Joshi, A. P. K., Rupasinghe, H. P. V., and Khanizadeh, S. (2011). Impact of drying processes on bioactive phenolics, vitamin c and antioxidant capacity of red-fleshed apple slices. *Journal of Food Processing and Preservation*, *35*(4), 453–457.

Jumah, R., Al-Kteimat, E., Al-Hamad, A., and Telfah, E. (2007). Constant and intermittent drying characteristics of olive cake. *Drying Technology*, *25*(9), 1421–1426.

Kaminski, W. and Tomczak, E. (2000). Degradation of ascorbic acid in drying process—A comparison of description methods. *Drying Technology*, *18*(3), 777–790.

Karaaslan, M., Yilmaz, F. M., Cesur, Ö., Vardin, H., Ikinci, A., and Dalgiç, A. C. (2014). Drying kinetics and thermal degradation of phenolic compounds and anthocyanins in pomegranate arils dried under vacuum conditions. *International Journal of Food Science & Technology*, *49*(2), 595–605.

Karim, A., Sabah, M., Mohamed, N., Tamara, A., Hanintsoa, F., and Arun, S. M. (2014). Intermittent drying. In A. S. Mujumdar (Ed.), *Handbook of Industrial Drying*, 4th edn., pp. 491–501. New York: CRC Press.

Kaya, A., Aydın, O., and Kolaylı, S. (2010). Effect of different drying conditions on the vitamin C (ascorbic acid) content of Hayward kiwifruits (*Actinidia deliciosa* Planch). *Food and Bioproducts Processing*, *88*(2), 165–173.

Khraisheh, M. A. M., McMinn, W. A. M., and Magee, T. R. A. (2004). Quality and structural changes in starchy foods during microwave and convective drying. *Food Research International*, *37*(5), 497–503.

Kowalski, S. J. and Pawłowski, A. (2011). Energy consumption and quality aspect by intermittent drying. *Chemical Engineering & Processing: Process Intensification*, *50*(4), 384–390.

Kowalski, S. and Szadzińska, J. (2014a). Convective-intermittent drying of cherries preceded by ultrasonic assisted osmotic dehydration. *Chemical Engineering & Processing: Process Intensification*, *82*, 65–70.

Kowalski, S. J. and Szadzinska, J. (2014b). Kinetics and quality aspects of beetroots dried in non-stationary conditions. *Drying Technology*, *32*(11), 1310–1318.

Kowalski, S. J., Szadzińska, J., and Łechtańska, J. (2013). Non-stationary drying of carrot: Effect on product quality. *Journal of Food Engineering*, *118*(4), 393–399.

Kowalski, S. J., Szadzińska, J., and Pawłowski, A. (2015). Ultrasonic-assisted osmotic dehydration of carrot followed by convective drying with continuous and intermittent heating. *Drying Technology*, *33*(13), 1570–1580.

Kumar, C., Joardder, M. U. H., Farrell, T. W., and Karim, A. (2016a). Multiphase porous media model for intermittent microwave convective drying (IMCD) of food. *International Journal of Thermal Science*, *104*, 304–314.

Kumar, C., Joardder, M. U. H., Farrell, T. W., Millar, G. J., and Karim, M. A. (2015). A mathematical model for intermittent microwave convective (IMCD) drying of food materials. *Drying Technology*, *34*(8), 962–973.

Kumar, C., Joardder, M. U. H., Karim, A., Millar, G. J., and Amin, Z. (2014a). Temperature redistribution modeling during intermittent microwave convective heating. *Procedia Engineering*, *90*, 544–549.

Kumar, C., Karim, M. A., and Joardder, M. U. H. (2014b). Intermittent drying of food products: A critical review. *Journal of Food Engineering*, *121*, 48.

Kumar, C., Karim, A., Joardder, M. U. H., and Miller, G. J. (2012a). Modeling heat and mass transfer process during convection drying of fruit. *The Fourth International Conference on Computational Methods*, Gold Coast, Queensland, Australia, November 25–28.

Kumar, C., Karim, A., Saha, S. C., Joardder, M. U. H., Brown, R. J., and Biswas, D. (2012b). Multiphysics modeling of convective drying of food materials. *Proceedings of the Global Engineering, Science and Technology Conference*, Dhaka, Bangladesh, December 28–29.

Kumar, C., Millar, G. J., and Karim, M. A. (2016b). Effective diffusivity and evaporative cooling in convective drying of food material. *Drying Technology*, *33*(2), 227–237.

Lafarge, C., Bard, M. H., Breuvart, A., Doublier, J. L., and Cayot, N. (2008). Influence of the structure of cornstarch dispersions on kinetics of aroma release, *Journal of Food Science*, *73*(2), S104–S109.

Lopez-Velasco, G., Welbaum, G., Boyer, R., Mane, S., and Ponder, M. (2011). Changes in spinach phylloepiphytic bacteria communities following minimal processing and refrigerated storage described using pyrosequencing of 16S rRNA amplicons. *Journal of Applied Microbiology*, *110*(5), 1203–1214.

Madamba, P. S. (2002). The response surface methodology: An application to optimize dehydration operations of selected agricultural crops. *LWT—Food Science and Technology*, *35*(7), 584–592.

Mishkin, M., Saguy, I., and Karel, M. (1984a). A dynamic test for kinetic models of chemical changes during processing: Ascorbic acid degradation in dehydration of potatoes. *Journal of Food Science*, *49*(5), 1267–1270.

Mishkin, M., Saguy, I., and Karel, M. (1984b). Optimization of nutrient retention during processing: Ascorbic acid in potato dehydration. *Journal of Food Science*, *49*(5), 1262–1266.

Mujumdar, A. S. and Jangam, S. (2011). Some innovative drying technologies for dehydration of foods, In *Proceedings of ICEF*, Athens, Greece, 555–556.

Nicoleti, J. F., Silveira, V., Telis-Romero, J., and Telis, V. R. N. (2007). Influence of drying conditions on ascorbic acid during convective drying of whole persimmons. *Drying Technology*, *25*(5), 891–899.

Ong, S. P., Law, C. L., and Hii, C. L. (2012). Optimization of heat pump-assisted intermittent drying. *Drying Technology*, *30*(15), 1676.

Pan, Y. K., Zhao, L. J., Dong, Z. X., Mujumdar, A. S., and Kudra, T. (1999). Intermittent drying of carrot in a vibrated fluid bed: Effect on product quality. *Drying Technology*, *17*(10), 2323–2340.

Pan, Y. K., Zhao, L. J., and Hu, W. B. (1998). The effect of tempering-intermittent drying on quality and energy of plant materials. *Drying Technology*, *17*(9), 1795–1812.

Perera, C. O. (2005). Selected quality attributes of dried foods. *Drying Technology*, *23*(4), 717–730.

Pittia, P. and Antonello, P. (2016). Safety by control of water activity: Drying, smoking, and salt or sugar addition. In V. Prakash, O. Martín-Belloso, L. Keener, S. Astley, S. Braun, H. McMahon, H. Lelieveld (Eds.), *Regulating Safety of Traditional and Ethnic Foods*, pp. 7–28. Waltham, MA: American Press.

Qing-guo, H., Min, Z., Mujumdar, A. S., Wei-hua, D., and Jin-cai, S. (2006). Effects of different drying methods on the quality changes of granular edamame. *Drying Technology*, *24*(8), 1025–1032.

Ramallo, L. A., Lovera, N. N., and Schmalko, M. E. (2010). Effect of the application of intermittent drying on Ilex paraguariensis quality and drying kinetics. *Journal of Food Engineering*, *97*(2), 188–193.

Razavi, M. A., Poreza, H. R., Zenoozian, M. S., Devahastin, S., and Shahidi, F. (2007). Use of artificial neural network and image analysis to predict physical properties of osmotically dehydrated pumpkin. *Drying Technology*, *26*(1), 132–144.

Reyes, A., Mahn, A., Guzmán, C., and Antoniz, D. (2012). Analysis of the drying of broccoli florets in a fluidized pulsed bed. *Drying Technology*, *30*(11), 1368.

Röhr, A., Lüddecke, K., Drusch, S., Müller, M. J., and Alvensleben, R. V. (2005). Food quality and safety—Consumer perception and public health concern. *Food Control*, *16*(8), 649–655.

Sablani, S. (2006). Drying of fruits and vegetables: Retention of nutritional/functional quality. *Drying Technology*, *24*(2), 123–135.

Saxena, A., Maity, T., Raju, P. S., and Bawa, A. S. (2010). Degradation kinetics of colour and total carotenoids in jackfruit (*Artocarpus heterophyllus*) bulb slices during hot air drying, *Food and Bioprocess Technology*, *5*(2), 672–679.

Setiady, D., Tang, J., Younce, F., Swanson, B. A., Rasco, B. A., and Clary, C. D. (2009). Porosity, color, texture, and microscopic structure of russet potatoes dried using microwave vacuum, heated air, and freeze drying. *Applied Engineering in Agriculture*, *25*(5), 719–724. doi:10.13031/2013.28844.

Sikorski, Z. E. (2007). *Chemia _zywnos´ci. Tom 1.* Warszawa, Poland: WNT.

Soysal, Y., Arslan, M., and Keskin, M. (2009a). Intermittent microwave-convective air drying of oregano. *Food Science and Technology International*, *15*(4), 397–406.

Soysal, Y., Ayhan, Z., Eştürk, O., and Arıkan, M. F. (2009b). Intermittent microwave–convective drying of red pepper: Drying kinetics, physical (colour and texture) and sensory quality. *Biosystems Engineering*, *103*(4), 455–463.

Szadzińska, J. (2014). Influence of convective-intermittent drying on the kinetics, energy consumption and quality of green pepper. A research project, No. 32-444/14 DS-BP, Poznań University of Technology, Poznań, Poland.

Telis, V., Telis-Romero, J., and Gabas, A. (2005). Solids rheology for dehydrated food and biological materials. *Drying Technology*, *23*(4), 759–780.

Thomkapanich, O., Suvarnakuta, P., and Devahastin, S. (2007). Study of intermittent low-pressure superheated steam and vacuum drying of a heat-sensitive material. *Drying Technology*, *25*(1), 205–223.

Tokuşoğlu, Ö. and Swanson, B. G. (2014). *Improving Food Quality with Novel Food Processing Technologies*. New York: CRC Press.

Torreggiani, D. (1993). Osmotic dehydration in fruit and vegetable processing. *Food Research International*, *26*(1), 59–68.

Tsotsas, E. and Mujumdar, A. S. (2011). *Modern Drying Technology, Product Quality and Formulation*. Wiley-VCH, Germany.

Uribe, E., Vega-Gálvez, A., Di Scala, K., Oyanadel, R., Saavedra Torrico, J., and Miranda, M. (2011). Characteristics of convective drying of pepino fruit (*Solanum muricatum* Ait.): Application of weibull distribution. *Food and Bioprocess Technology*, *4*(8), 1349–1356.

Vadivambal, R. and Jayas, D. (2010). Non-uniform temperature distribution during microwave heating of food materials—A review. *Food and Bioprocess Technology*, *3*(2), 161–171.

Valdramidis, V., Geeraerd, A., Gaze, J., Kondjoyan, A., Boyd, A., Shaw, H., and Van Impe, J. (2006). Quantitative description of *Listeria monocytogenes* inactivation kinetics with temperature and water activity as the influencing factors; model prediction and methodological validation on dynamic data. *Journal of Food Engineering*, *76*(1), 79–88.

Valdramidis, V. P., Taoukis, P. S., Stoforos, N. G., and Van Impe, J. F. M. (2012). Chapter 14—Modeling the kinetics of microbial and quality attributes of fluid food during novel thermal and non-thermal processes. In P. J. Cullen, B. K. Tiwari, V. P. Valdramidis (Eds.), *Novel Thermal and Non-Thermal Technologies for Fluid Foods*, pp. 433–471. San Diego, CA: Academic Press.

Van Schie, P. and Young, L. (2000). Biodegradation of phenol: Mechanisms and applications. *Bioremediation Journal*, *4*(1), 1–18.

Vega-GÁLvez, A., San MartÍN, R., Sanders, M., Miranda, M., and Lara, E. (2010). Characteristics and mathematical modeling of convective drying of quinoa (*Chenopodium quinoa* willd.): Influence of temperature on the kinetic parameters. *Journal of Food Processing and Preservation*, *34*(6), 945–963.

Vega-Valencia, Y., Cruz y Victoria, M. T., Vizcarra Mendoza, M. G., and Anaya Sosa, I. (2014). Intermittent drying of nopal (*Opuntia ficus indica*) in a fluidized bed pilot dryer adapted with revolving chambers: Intermittent drying of nopal in a fluidized bed pilot dryer, *Journal of Food Process Engineering*, *37*, 211–219.

Zhang, N., Yang, Z., Chen, A. G., and Zhao, S. S. (2014). Effects of intermittent heat treatment on sensory quality and antioxidant enzymes of cucumber. *Scientia Horticulturae*, *170*, 39–44.

Zhao, L. J., Kudra, T., Pan, Y. K., Dong, Z. X., and Mujumdar, A. S. (1999). Intermittent drying of carrot in a vibrated fluid bed: Effect on product quality. *Drying Technology*, *17*(10), 2323–2340.

Zheng, H. and Lu, H. (2011). Use of kinetic, Weibull and PLSR models to predict the retention of ascorbic acid, total phenols and antioxidant activity during storage of pasteurized pineapple juice. *LWT—Food Science and Technology*, *44*(5), 1273–1281.

Zhu, Y., Pan, Z., McHugh, T. H., and Barrett, D. M. (2010). Processing and quality characteristics of apple slices processed under simultaneous infrared dry-blanching and dehydration with intermittent heating. *Journal of Food Engineering*, *97*(1), 8–16.

6 Relationship between Intermittency of Drying, Microstructural Changes, and Food Quality

Mohammad U.H. Joardder,
M.H. Masud, and M. Azharul Karim

CONTENTS

6.1 INTRODUCTION

Fruits and vegetables are important sources of essential dietary nutrients. Since the moisture content of most of the fresh fruits and vegetables varies between 80% and 95%, they are classified as highly perishable commodities (Orsat et al., 2007; Joardder et al., 2013a). Lack of proper processing causes extensive wastage of seasonal fruits in many countries, which is estimated to be 30%–40% in developing countries (Karim and Hawlader, 2005). Drying is an important and the oldest method of food preservation. Many physical and chemical changes occur in food during the drying process. The quality of the dehydrated product is affected by the quality of raw material, method of preparation, processing treatments, and drying conditions (Puranik et al., 2012; Joardder et al., 2015).

Hot air or convection drying is the easiest and oldest method of food drying (Ilknur, 2007; Kumar et al., 2012a,b, 2014a). However, convectional drying with hot air takes a long drying time and has a low energy efficiency (Zhang et al., 2010; Kumar et al., 2015, 2016a). To overcome this difficulty, a combination of convective drying with other techniques has been proven to be a better drying approach. Hybrid drying is one of the best ways to improve product quality attributes and reduce energy consumption (Sagar and Suresh Kumar, 2010). When the microwave or another secondary heating source is combined with hot air drying, not only energy efficiency and product quality are improved but also drying time is reduced significantly (Nishiyama et al., 2006).

However, continuous application of secondary heating, for example, microwave power supply, during the entire drying period may cause quality degradation and higher energy consumption (Soysal et al., 2009). Intermittent use of secondary heating sources with convective drying represents a viable option to overcome the above-mentioned problems. Intermittent drying can effectively reduce the net drying time of the sample, which can minimize process energy and operational cost significantly.The purpose of intermittent drying is to optimize the quality, time, cost, and energy, as shown in Figure 6.1. The moderate drying temperature that occurs during the tempering period of intermittent drying helps enhance the product quality (Kumar et al., 2014b, 2016b,c).

Most of the food materials subjected to a drying process can be treated as the hygroscopic porous amorphous media with the multiphase transport of heat and mass. Chemical composition and physical structure of food materials influence both heat and mass transfer mechanisms (Rosselló, 1992). Since understanding the correlations between physical properties and microstructural changes could help identify optimum drying conditions, several studies have been conducted to investigate such relationship in various food products (Riva et al., 2005; Askari et al., 2009; Yang et al., 2010; Gumeta-Chavez et al., 2011).

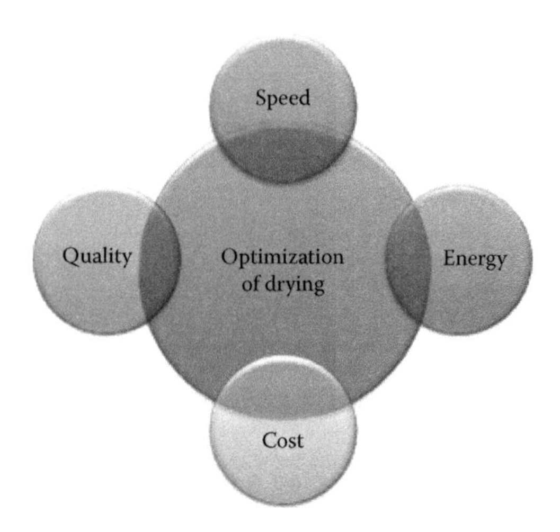

FIGURE 6.1 Optimization of drying process through intermittency.

Food properties, drying methods, and drying conditions influence the microstructural changes of the final dried product (Joardder et al., 2017). Even the same type of raw materials demonstrate different microstructural changes and pore characteristics, depending on the drying method and conditions (Sablani et al., 2007; Joardder et al., 2013c). Therefore, microstructural change is directly dependent on initial water content, temperature, pressure, relative humidity, air velocity, electromagnetic radiation, composition and original microstructure, and the viscoelastic properties of the biomaterials (Saravacos, 1967; Krokida et al., 1997; Guiné, 2006; Joardder et al., 2013b).

Interrelationships between microstructural changes and food quality have been reported in the literature (Joardder et al., 2016). Moreover, a considerable amount of literature has been attempted to study the structure–quality relationships in foods. In particular, it has been demonstrated that many desirable attributes of food, such as texture, color, or flavor, depend on the way foods are structured (Aguilera, 2005). Changes in many physical attributes are eventually due to the modifications in the product microstructure (Ramos et al., 2004; Panyawong and Devahastin, 2007; Witrowa-Rajchert and Rzaca, 2009). Several studies have been conducted to investigate the structure–quality relationship of various food products under different drying process and drying conditions (Achanta and Okos, 1996; Riva et al., 2005; Askari et al., 2009; Yang et al., 2010; Gumeta-Chavez et al., 2011). For example, Servais et al. (2002) found a correlation between sugar particle size in chocolate and its influence on the rheological properties of the chocolate mass.

In this chapter, the relationship between food microstructure and the quality attributes of food is presented.

6.2 GENERAL MECHANISM OF STRUCTURAL CHANGES

Cell collapse and shrinkage are the consequence of microstructural modification over the course of drying. There is a fine difference between collapse and shrinkage; collapse is a process where cellular- or tissue-level structure may break down irreversibly, while the shrinkage refers to a reduction in volume of food material. In most of the cases, the structural change of food materials is the result of a collapse of the cellular wall during the drying period when large amount of water loss takes place (Ramos et al., 2003; Devahastin and Niamnuy, 2010). During the drying process, negative pressure is developed due to the removal of water, which is the major reason behind the volumetric shrinkage, as demonstrated in Figure 6.2. Air inside the pores in food maintains a balanced pressure, but due to the collapse of the pores, negative pressure is developed and as a consequence shrinkage occurs (Pakowski and Adamski, 2012).

The shrinkage can be minimized if sufficient number of pores can be created and maintained during drying. There are several important factors such as porosity of the tissue, intercellular adhesion, and strength of cell walls that influence the nature of cell collapse and shrinkage (Ormerod et al., 2004).

At the beginning of the drying, food materials shrink without cell collapsing. However, cell collapse takes place at the final stage of drying. When the volume of migrated water is exactly equal to the volume reduction of the solid matrix, the deformation phenomenon can be defined as ideal shrinkage.

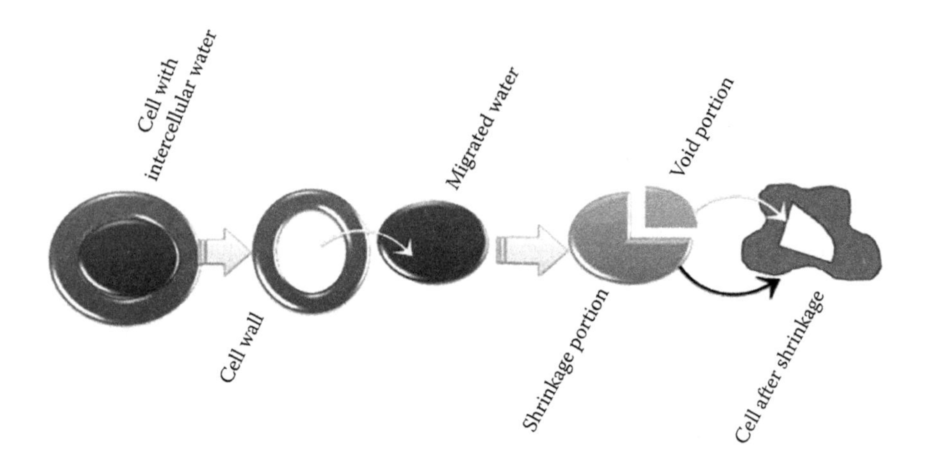

FIGURE 6.2 Cellular shrinkage of food materials during drying.

Process conditions and internal moisture migration mechanisms significantly affect the microstructure of dried foods. Application of different types of intermittent drying results in different internal moisture transfer and heat transfer mechanisms. Internal moisture transfer mechanisms directly or indirectly cause various forms of microstructure changes in the food product. However, it is very difficult to generalize the relationship between the changes in microstructure and particular intermittent drying conditions. There are many ways to attain intermittency by using and combining different drying methods. Describing all these options is beyond the scope of this chapter. In this chapter, the change in microstructure over the course of intermittent microwave-convective drying has been presented to illustrate how intermittency affects the microstructure of food materials.

6.3 CHANGES IN MICROSTRUCTURE DURING INTERMITTENT DRYING

Process conditions such as heating rate, air velocity, and humidity significantly affect the change in food microstructure (Joardder et al., 2013b). In particular, temperature and moisture distributions play an active role in modifying the food microstructure during drying. In general, the regions with high temperature and moisture show less porosity as a consequence of excessive cell rupture. On the contrary, regions with lower temperature show uniform porous microstructure.

6.3.1 Temperature Distribution

Temperature distributions on the surface of the samples captured by thermal imaging camera during convective drying at 70°C, continuous microwave drying (CMD), and intermittent microwave connective drying (IMCD) are presented in Figure 6.3 in order to present how intermittency affect the temperature distribution.

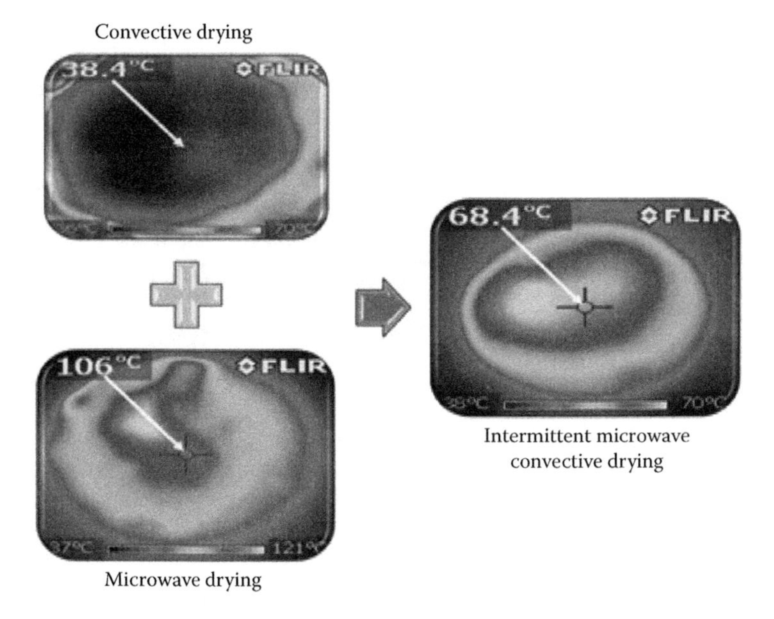

FIGURE 6.3 Temperature distribution in apple slice during three selected drying processes.

The figure shows that in convective drying, higher temperature exists at the edge of the sample and gradually decreases toward the center. Over the drying time, temperature distribution shows the same trend in hot air drying of the sample. Unlike hot air drying, a random temperature distribution has been observed during microwave drying, which is the result of the uneven electromagnetic distribution of microwave power. In microwave drying, higher temperature exists at the core of the sample.

Similar to the continuous microwave heating, IMCD results in higher temperature in the interior of the sample. However, the temperature is redistributed during tempering period and achieves a more uniform pattern in IMCD compared with CMD. Therefore, intermittent application of microwave reduces the impact of nonuniform temperature distribution, resulting in a better overall quality of food materials (Kumar et al., 2014a).

6.3.2 Effect of Temperature Distribution on Structural Modification

Pattern of temperature distribution significantly affects the type of structural modification of food materials over the course of drying. For instance, in hot air drying of fruits and vegetables, tissue is featured by an extensive shrinkage and microstructural changes (Bolin and Huxsoll, 1987; Aguilera and Chiralt, 2003).

Owing to thermal exposure of the surface cells of the sample, there is a significant amount of collapse in these cells. This cell collapse causes case hardening, which can be avoided by balancing the energy flow to the surface and volumetric heating. Application of microwave heating develops higher driving force for moisture flow by increasing vapor pressure difference between the interior and surface of food

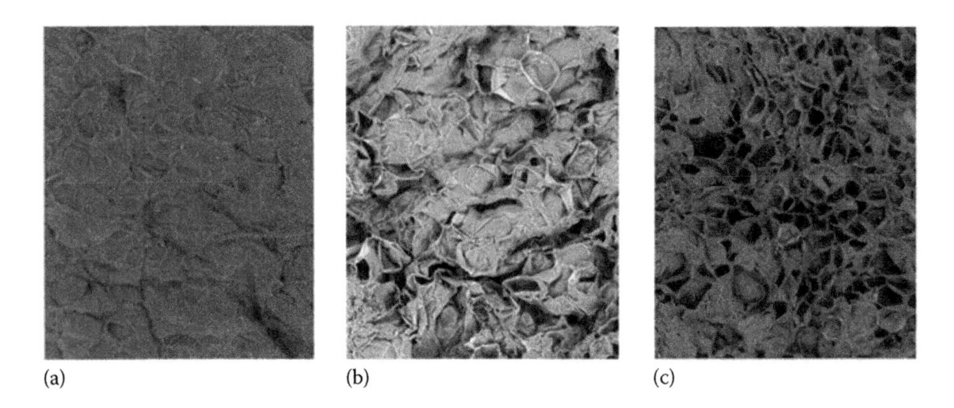

(a) (b) (c)

FIGURE 6.4 Morphology of dried apple after convective, microwave, and intermittent microwave drying: (a) surface pores of convective dried sample (100×), (b) surface pores of CMD sample (100×), and (c) surface pores of IMCD sample (100×).

materials. Sometimes, it may cause puffing within the sample, but eventually, this increase in vapor pressure enhances the porosity of the plant-based food materials (Therdthai and Visalrakkij, 2012).

However, continuous microwave drying may cause overheating, which results in burning of the solid materials, as shown in Figure 6.4. Intermittent application of microwave energy can overcome the above problem. Compared with convective and CMD drying, IMCD provides more uniform heating and moisture distribution, which result in uniform pores in the surface, as shown in Figure 6.4. However, it is reported that the use of microwave energy in the early stage of drying may result in cellular collapse and bulk shrinkage in the final products (Al-Duri, 1992; Zhang et al., 2006). Therefore, some of the studies (Askari et al., 2006; Argyropoulos et al., 2011) recommended microwave application at the finishing stage as it leads to more porous structure than if the microwave is deployed at the earlier stage of drying. However, it is still not conclusive when the microwave should be applied to get the best-finished product.

6.4 RELATIONSHIP BETWEEN INTERMITTENCY, MICROSTRUCTURAL CHANGES, AND QUALITY

Quality attributes can be classified in several ways; however, the most common types are physical and psychical, sensory, chemical, kinetic, and nutritional properties. Properly controlling the process parameters can provide high-quality dried food along with energy savings. However, it is very difficult to determine the single dominating factor that leads to the exact amount of quality deterioration during food drying, as different drying parameters affect various aspects of quality differently during drying, as shown in Figure 6.5. In fact, a combined effect of different process parameters and transport phenomena is the determining factor of quality aspect of dried food.

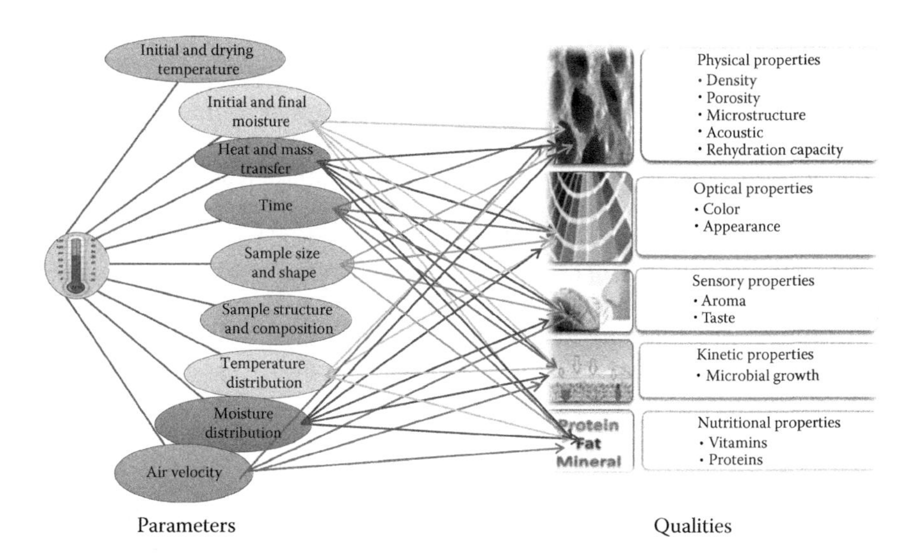

FIGURE 6.5 Interrelationship between process parameters and quality of dried foods.

As transportation of heat and mass through porous biomaterials depends on the pattern of microstructure, process parameters that influence the change in microstructure must be taken into consideration in the design step of the intermittent drying system. Very limited investigation has been conducted in the structural changes that occur in the food materials during intermittent drying, and little research has been done on how these changes affect overall food quality. Aguilera (2005) claimed that changes in different quality attributes of food occur over the course of drying as a result of microstructural changes in food stuff. Knowing the microstructural changes and understanding them are important for minimizing the unexpected changes in quality of foodstuff. An attempt to associate microstructural changes and food properties such as rehydration, mechanical properties, and sensory properties during IMCD has been made in the following sections.

6.4.1 Rehydration

Rehydration is a common practice, as most of the time dried foods are rehydrated before consumption or cooking. Rehydration is an irreversible process, like many other common processes, and it cannot be considered as the opposite to dehydration (Lewicki et al., 1998). There are several factors upon which the rehydration kinetics depend, such as the dimension of capillaries, temperature, porosity, trapped air bubbles, cavities near the surface, the amorphous-crystalline state, and pH of the soaking water and soluble solids (Bai et al., 2001; Weerts et al., 2003). Although all the factors mentioned above affect rehydration, microstructure-related parameters are the dominant ones in dried food materials (Farkas and Singh, 1991; Karathanos et al., 1993). Intermittent drying influences the nature of porosity, and the capability of rehydration depends highly on porosity. Low porosity results in poor rehydration characteristic of a product (Mcminn and Magee, 1976; Mayor and Sereno, 2004).

Moreover, the existence of the trapped air in the pores of the food materials has a direct impact on rehydration capability. The main barrier toward rehydration is the existence of air bubbles in the pores of food materials, which reduces the soaking of rehydrating water. It is worth to mention here that the intermittency of additional heat sources such as microwave may leave large number of air bubbles due to puffing effect.

Therefore, the condition of drying is one of the major factors in which the capability of reconstitution depends. For instance, microwave dried products have a lower capacity to rehydrate than freeze-dried materials due to the unfavorable microstructure caused by microwave drying for effective rehydration (Oikonomopoulou and Krokida, 2013).

6.4.2 Mechanical Properties

The mechanical properties of dried food materials depend on the original composition of the cell wall and solid matrix structure (Gogoi et al., 2000). Based on a study on fresh apples by Vincent (1989), it was found that torsional stiffness ranges from 0.5 to 7 MPa, which is closely related to its porosity range of 0.84 to 0.54. The study also indicated that a linear relation can be established while plotting stiffness as a dependent variable of initial porosity of the sample.

Apart from these factors, the pattern of microstructure significantly affects the hardness of food materials. From several studies, it was found that the decreased pore size improves the hardness of the food materials, and vice versa (Therdthai and Visalrakkij, 2012). Furthermore, it was also found that food materials with larger and higher pores have a solid matrix with a lower strength, which can be ruptured without any difficulty. Therefore, maintaining appropriate intermittency of the secondary energy source is vital as it affects microstructure, which, in turn, influences the mechanical properties of dried food.

6.4.3 Physical Changes

Cracking and case hardening are the common examples of the physical characteristics of food materials, which are greatly influenced by conditions of drying, including the intermittency of the secondary energy source during drying (Kowalski et al., 2010; Kowalski and Pawlowski, 2011a,b). To monitor the micro- and macrocracks produced at the time of drying of clay and wood samples, acoustic emission method was applied by Kowalski and Pawłowski (2011a). It was found that suitable drying condition can reduce the fracture and provide better quality of products. By examining three important factors such as variable humidity, constant temperature, and variable temperature conditions, they compared the intensity of cracks. For two different conditions, they found the best quality samples in the lowest energy consumption situation. Implementing the variable humidity condition, best-quality samples are achieved. The variable air temperature conditions resulted in lowest energy consumption, while the humidity variation strategy consumed more energy. However, this method was not tested for food products yet.

By implementing intermittent drying, some other physical characteristics, such as fissuring of rice kernel (Aquerreta et al., 2007), have been found to increase. Aquerreta et al. (2007) showed that the fissuring of kernels was also reduced by incorporating multistep drying rather than drying in a single step. Although the method of improving physical properties is not described in detail by the authors, the result confirmed better physical properties in intermittent drying. This can be possible only if redistribution of temperature and moisture can be done during tempering period, which finally results in the improvement of internal stresses.

6.4.4 Sensory Quality

There is a great relationship between sensory quality and microstructure of food materials. It is essential to understand the relationship to achieve desired quality of processed food (Langton et al., 1997). Acute taste is experienced while masticating any food with higher pores and larger exposed surface area (Wilson and Brown, 1997).

Literature found the evidence that the kinetics of releasing flavor from food materials can be modified by changing the structure of food matrix and physiochemical properties of food materials (De Roos, 2003; Mezzenga et al., 2005; Laurienzo et al., 2013). A lot of investigations have been carried out to relate the discharge of flavor and fragrance with the solid food matrix (Cayot et al., 2004; Boland et al., 2006; Seuvre et al., 2006; Lafarge et al., 2008).

It can be perceived that foods with similar microstructure have comparable sensorial behavior. Therefore, maintaining fresh food-like structure by means of innovative drying system would deliver the flavor close to fresh foods.

6.4.5 Texture

Microstructural changes significantly affect the texture of food materials (Reeve, 1970; Rahman, 2008). Although the structure-texture relationship is a common phenomenon, there is not sufficient literature available concerning this issue.

Chen and Opara (2013) have examined the relationship between texture and microstructure and found that several factors such as rheology, food structure, and its surface properties affect the food texture (Kravchuk et al., 2012; Stokes et al., 2013).

Owing to diversification of processes and dynamic characteristics of texture evolution during drying, establishment of a relationship between texture evolution and food structure is very difficult (Wilkinson et al., 2000). More research is needed to establish such a relationship.

6.4.6 Color

Color is one of the major quality attributes that influences consumer decision because it is the first impression the user gets about the product. Lengthy exposure to higher drying temperature results in substantial degradation of color (Zhang et al., 2006). Several studies on color changes in intermittent drying process have been performed on different samples like potatoes, bananas (Chua et al., 2000b; Maskan, 2001), and guavas (Chua et al., 2000a). All investigations reported a significantly improved color during intermittent drying than after conventional hot air drying. Moreover, it was reported that different drying conditions influence the color significantly. For example, by using stepwise varying drying condition for banana pieces in two-stage heat pump dryer, Chua et al. (2001) found that there is a great improvement in color of food materials. While performing step-up and step-down intermittency of drying, respectively, in a temperature range between 20°C and 35°C, Chua et al. (2001) reported a reduction in color degradation by 40% and 23%, respectively.

Chua et al. (2000b) pursued the effect of different temperature profiles on color parameters. In that study, changes in different color variations like lightness, redness or greenness, and blueness or yellowness (L, a, b) were investigated. From their investigation, it was found that all the food materials do not maintain a common trend of color change.

During the drying of potatoes and guavas, the lightness was decreased, whereas in banana it was significantly improved. Moreover, changes in color can be reduced by 87%, 75%, and 67%, respectively, for potatoes, guavas, and bananas, if variable temperature drying condition is applied. Owing to the high moisture and low sugar content, the color change in potato samples is lower compared with banana and guava samples. Labuza et al. (1972) found consistency in the color suppression by raising the moisture content.

Taking all these factors into consideration, the intensity of color change may be caused due to the diverse structural and compositional nature of plant-based food materials. Therefore, proper drying conditions can ensure attractive color of the products.

6.4.7 Degradation of Composition

The chemical composition of food materials that undergo different drying conditions varies with the types and intensity of process parameters. For instance, An et al. (2016) found that starch grains in ginger were not well preserved in convective drying due to its dense structure (Figure 6.6). On the contrary, after intermittent microwave-convective drying, the cellular structure was retained along with well-preserved starch grains.

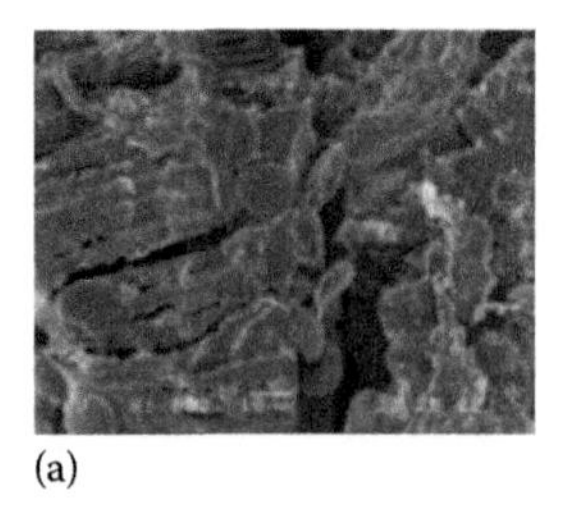

(a)

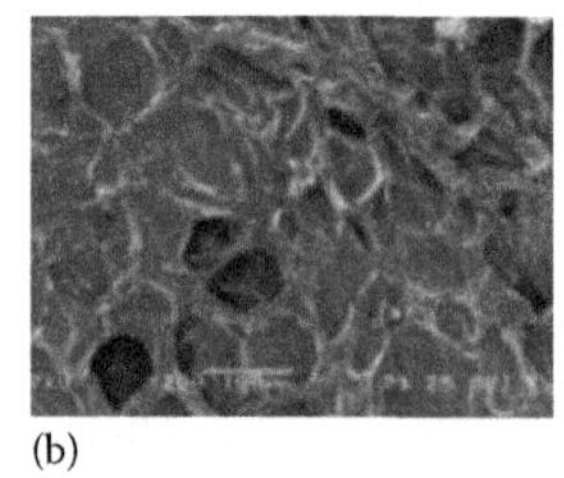

(b)

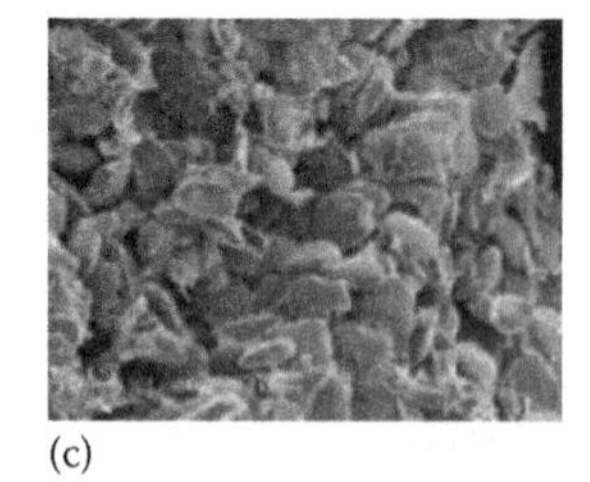

(c)

FIGURE 6.6 Starch grain preservation in ginger tissue after different drying practices: (a) convective dried sample, (b) microwave dried sample, and (c) intermittent microwave convective drying. (Adapted from An, K. et al., *Food Chem.*, 197(Part B), 1292, 2016.)

In general, the migration of volatile and soluble components of food materials depends on microstructure change of food materials. Therefore, appropriate drying conditions are essential for maintaining proper structure and retention of valuable components during the time of drying.

6.5 CONCLUSION

Comprehensive understanding of structural changes over the course of drying is essential for developing more energy-efficient food drying. Maintaining the microstructure and the quality of the fresh foods are of utmost importance for the food industry. Studies show that a relationship between transport process parameters, food microstructure, and food quality has been established. However, more research study will be required to clearly demonstrate such a relationship. Such relationship will help design the right drying process that can ensure microstructural pattern of finished product closer to fresh food structure. The understanding of the physicochemical changes during drying is essential to design optimum drying process for maximizing quality of processed food. Therefore, researchers and scientists from different fields can work together and share their knowledge in order to establish a bridging relationship between process conditions, food structure, and product quality.

REFERENCES

Achanta S and Okos MR. (1996) Predicting the quality of dehydrated foods and biopolymers—Research needs and opportunities. *Drying Technology* 14: 1329–1368.

Aguilera JM. (2005) Why food microstructure? *Journal of Food Engineering* 67: 3–11.

Aguilera JM and Chiralt AFP. (2003) Food dehydration and product structure. *Trends in Food Science & Technology* 14: 432–437.

Al-Duri B. (1992) Comparison of drying kinetics of foods using a fan-assisted convection oven, a microwave oven and a combined microwave/convection oven. *Journal of Food Engineering* 15: 139–155.

An K, Zhao D, Wang Z et al. (2016) Comparison of different drying methods on Chinese ginger (*Zingiber officinale* Roscoe): Changes in volatiles, chemical profile, antioxidant properties, and microstructure. *Food Chemistry* 197(Part B): 1292–1300.

Aquerreta J, Iguaz A, Arroqui C et al. (2007) Effect of high temperature intermittent drying and tempering on rough rice quality. *Journal of Food Engineering* 80: 611–618.

Argyropoulos D, Heindl A, and Müller J. (2011) Assessment of convection, hot-air combined with microwave-vacuum and freeze-drying methods for mushrooms with regard to product quality. *International Journal of Food Science & Technology* 46: 333–342.

Askari GR, Emam-Djomeh Z, and Mousavi SM. (2006) Effects of combined coating and microwave assisted hot-air drying on the texture, microstructure and rehydration characteristics of apple slices. *Food Science and Technology International* 12: 39–46.

Askari GR, Emam-Djomeh Z, and Mousavi SM. (2009) An investigation of the effects of drying methods and conditions on drying characteristics and quality attributes of agricultural products during hot air and hot air/microwave-assisted dehydration. *Drying Technology* 27: 831–841.

Bai Y, Rahman MS, Perera CO et al. (2001) State diagram of apple slices: Glass transition and freezing curves. *Food Research International* 34: 89–95.

Boland AB, Delahunty CM, and van Ruth SM. (2006) Influence of the texture of gelatin gels and pectin gels on strawberry flavour release and perception. *Journal of Food Chemistry* 96: 452–460.

Bolin HR and Huxsoll CC. (1987) Scanning electron microscope/image analyzer determination of dimensional postharvest changes in fruit cells. *Journal of Food Science* 52: 1649–1650.

Cayot N, Pretot F, Doublier JL et al. (2004) Release of isoamyl acetate from starch pastes of various structures: Thermodynamic and kinetic parameters. *Journal of Agriculture of Food Chemistry* 52: 5436–5442.

Chen L and Opara UL. (2013) Approaches to analysis and modeling texture in fresh and processed foods—A review. *Journal of Food Engineering* 119: 497–507.

Chua KJ, Chou SK, Ho JC et al. (2000a) Cyclic air temperature drying of guava pieces: Effects on moisture and ascorbic acid contents. *Food and Bioproducts Processing* 78: 72–78.

Chua KJ, Mujumdar AS, Chou SK et al. (2000b) Convective drying of banana, guava and potato pieces: Effect of cyclical variations of air temperature on drying kinetics and color change. *Drying Technology* 18: 907–936.

Chua KJ, Mujumdar AS, Hawlader MNA et al. (2001) Batch drying of banana pieces—Effect of stepwise change in drying air temperature on drying kinetics and product colour. *Food Research International* 34: 721–731.

De Roos KB. (2003) Effect of texture and microstructure on flavour retention and release. *International Dairy Journal* 13: 593–605.

Devahastin S and Niamnuy C. (2010) Modelling quality changes of fruits and vegetables during drying: A review. *International Journal of Food Science & Technology* 45: 1755–1767.

Farkas BE and Singh RP. (1991) Physical properties of air dried and freeze-dried chicken white meat. *Journal of Food Science* 56: 611–615.

Gogoi BK, Alavi SH, and Rizvi SSH. (2000) Mechanical properties of protein-stabilized starch-based supercritical fluid extrudates. *International Journal of Food Properties* 3: 37–58.

Guiné R de PF. (2006) Influence of drying method on density and porosity of pears. *Food and Bioproducts Processing* 84: 179–185.

Gumeta-Chavez C, Chanona-Perez JJ, Mendoza-Perez JA et al. (2011) Shrinkage and deformation of Agave atrovirens Karw tissue during convective drying: Influence of structural arrangements. *Drying Technology* 29: 612–623.

Ilknur A. (2007) Microwave, air and combined microwave–air-drying parameters of pumpkin slices. *LWT—Food Science and Technology* 40(8): 1445–1451.

Joardder MUH, Karim A, Brown RJ et al. (2016) *Porosity: Establishing the Relationship between Drying Parameters and Dried Food Quality*. Springer, Switzerland.

Joardder MUH, Kumar C, Brown RJ et al. (2015) Effect of cell wall properties on porosity and shrinkage of dried apple. *International Journal of Food Properties* 18: 2327–2337.

Joardder MUH, Kumar C, and Karim MA. (2013a) Better understanding of food material on the basis of water distribution using thermogravimetric analysis. *International Conference on Mechanical, Industrial and Materials Engineering*, Rajshahi, Bangladesh.

Joardder MUH, Kumar C, and Karim MA. (2013b) Effect of moisture and temperature distribution on dried food microstucture and porosity. *Proceedings of From Model Foods to Food Models: The DREAM Project International Conference*, France.

Joardder MUH, Kumar C, and Karim MA. (2013c) Effect of temperature distribution on predicting quality of microwave dehydrated food. *Journal of Mechanical Engineering and Sciences* 5: 562–568.

Joardder MUH, Kumar C, and Karim MA. (2017) Food structure: Its formation and relationships with other properties. *Critical Reviews in Food Science and Nutrition* 57: 1190–1205.

Karathanos V, Anglea S, and Karel M. (1993) Collapse of structure during drying of celery. *Drying Technology* 11: 1005–1023.
Karim MA and Hawlader MNA. (2005) Mathematical modelling and experimental investigation of tropical fruits drying. *International Journal of Heat and Mass Transfer* 48: 4914–4925.
Kowalski SJ, Musielak G, and Banaszak J. (2010) Heat and mass transfer during microwave-convective drying. *AIChE Journal* 56: 24–35.
Kowalski SJ and Pawłowski A. (2011a) Energy consumption and quality aspect by intermittent drying. *Chemical Engineering and Processing: Process Intensification* 50: 384–390.
Kowalski SJ and Pawłowski A. (2011b) Intermittent drying: Energy expenditure and product quality. *Chemical Engineering & Technology* 34: 1123–1129.
Kravchuk O, Torley P, and Stokes JR. (2012) *Food Texture Is Only Partly Rheology*. Oxford, U.K.: Wiley-Blackwell.
Krokida MK, Zogzas NP, and Maroulis ZB. (1997) Modelling shrinkage and porosity during vacuum dehydration. *International Journal of Food Science & Technology* 32: 445–458.
Kumar C, Joardder MUH, Farrell TW, and Karim A. (2016a) Multiphase porous media model for intermittent microwave convective drying (IMCD) of food. *International Journal of Thermal Science* 104: 304–314.
Kumar C, Joardder MUH, Farrell TW, Millar GJ, and Karim MA. (2015) A mathematical model for intermittent microwave convective (IMCD) drying of food materials. *Drying Technology* 34(8): 962–973.
Kumar C, Joardder MUH, Karim A, Millar GJ, and Amin Z. (2014a) Temperature redistribution modeling during intermittent microwave convective heating. *Procedia Engineering* 90: 544–549.
Kumar C, Karim MA, and Joardder MUH. (2014b) Intermittent drying of food products: A critical review. *Journal of Food Engineering* 121: 48–57.
Kumar C, Karim MA, Joardder MUH, and Miller GJ. (2012a) Modeling heat and mass transfer process during convection drying of fruit. *The Fourth International Conference on Computational Methods*, Gold Coast, Queensland, Australia, November 25–28, 2012.
Kumar C, Karim A, Saha SC, Joardder MUH, Brown RJ, and Biswas D. (2012b) Multiphysics modeling of convective drying of food materials. *Proceedings of the Global Engineering, Science and Technology Conference*, Dhaka, Bangladesh, December 28–29, 2012.
Kumar C, Millar GJ, and Karim MA. (2016b) Effective diffusivity and evaporative cooling in convective drying of food material. *Drying Technology* 33(2): 227–237.
Kumar C, Saha SC, Sauret E, Karim A, and Gu YT. (2016c) Mathematical modeling of heat and mass transfer during intermittent microwave-convective drying (IMCD) of food materials. *Australasian Heat and Mass Transfer Conference 2016*, Brisbane, Queensland, Australia.
Labuza TP, McNally L, Gallagher D, and Hawkes J. (1972) Stability of intermediate moisture foods. I. Lipid oxidation. *Journal of Food Science* 37: 154–159.
Lafarge C, Bard MH, Breuvart A et al. (2008) Influence of the structure of cornstarch dispersions on kinetics of aroma release. *Journal of Food Science* 73: S104–S109.
Langton M, Åström A, and Anne-Marie H. (1997) Influence of the microstructure on the sensory quality of whey protein gels. *Food Hydrocolloids* 11: 217–230.
Laurienzo P, Cammarota G, Di Stasio M et al. (2013) Microstructure and olfactory quality of apples de-hydrated by innovative technologies. *Journal of Food Engineering* 116: 689–694.
Lewicki PP, Pomaranska-Lazuka W, Witrowa-Rajchert D et al. (1998) Effect of mode of drying on storage stability of colour of dried onion. *Journal of Food and Nutrition Sciences* 7(48): 701–706.
Maskan M. (2001) Kinetics of colour change of kiwifruits during hot air and microwave drying. *Journal of Food Engineering* 48: 169–175.

Mayor L and Sereno AM. (2004) Modelling shrinkage during convective drying of food materials: A review. *Journal of Food Engineering* 61: 373–386.

Mcminn WAM and Magee TRA. (1976) Physical characteristics of dehydrated potatoes—Part II. *Journal of Food Engineering* 33: 49–55.

Mezzenga R, Schurtenberger P, Burbidge A et al. (2005) Understanding foods as soft materials. *Nature Materials* 4: 729–740.

Nishiyama Y, Cao W, and Li B. (2006) Grain intermittent drying characteristics analyzed by a simplified model. *Journal of Food Engineering* 76: 272–279.

Oikonomopoulou VP and Krokida MK. (2013) Novel aspects of formation of food structure during drying. *Drying Technology* 31: 990–1007.

Ormerod AP, Ralfs JD, Jackson R et al. (2004) The influence of tissue porosity on the material properties of model plant tissues. *Journal of Materials Science* 39: 529–538.

Orsat V, Yang W, Changrue V et al. (2007) Microwave-assisted drying of biomaterials. *Food and Bioproducts Processing* 85: 255–263.

Pakowski Z and Adamski R. (2012) Formation of underpressure in an apple cylinder during convective drying. *Drying Technology* 30: 1238–1246.

Panyawong S and Devahastin S. (2007) Determination of deformation of a food product undergoing different drying methods and conditions via evolution of a shape factor. *Journal of Food Engineering* 78: 151–161.

Puranik V, Srivastava P, Mishra V, and Saxena DC. (2012) Effect of different drying techniques on the quality of garlic: A comparative study. *American Journal of Food Technology* 7(5): 311–319.

Rahman MS. (2008) Dehydration and microstructure. In C. Ratti (Ed.), *Advances in Food Dehydration*. CRC Press, London, pp. 97–122.

Ramos IN, Brandão TRS, and Silva CLM. (2003) Structural changes during air drying of fruits and vegetables. *Food Science and Technology International* 9: 201–206.

Ramos IN, Silva CLM, Sereno AM et al. (2004) Quantification of microstructural changes during first stage air drying of grape tissue. *Journal of Food Engineering* 62: 159–164.

Reeve RM. (1970) Relationships of histological structure to texture of fresh and processed fruits and vegetables. *Journal of Texture Studies* 1: 247–284.

Riva M, Campolongo S, Leva AA et al. (2005) Structure–property relationships in osmo-air-dehydrated apricot cubes. *Food Research International* 38: 533–542.

Rosselló C. (1992) Simple mathematical model to predict the drying rates of potatoes. *Journal of Agricultural and Food Chemistry* 40: 2374–2378.

Sablani SS, Rahman MS, Al-Kuseibi MK et al. (2007) Influence of shelf temperature on pore formation in garlic during freeze–drying. *Journal of Food Engineering* 80: 68–79.

Sagar VR and Suresh Kumar P. (2010) Recent advances in drying and dehydration of fruits and vegetables: A review. *Journal of Food Science & Technology* 47: 15–26.

Saravacos GD. (1967) Effect of the drying method on the water sorption of dehydrated apple and potato. *Journal of Food Science* 32: 81–84.

Servais C, Jones R, and Roberts I. (2002) The influence of particle size distribution on the processing of food. *Journal of Food Engineering* 51: 201–208.

Seuvre AM, Philippe E, Rochard S et al. (2006) Retention of aroma compounds in food matrices of similar rheological behavior and different compositions. *Journal of Food Chemistry* 96: 104–114.

Soysal Y, Ayhan Z, Eştürk O et al. (2009) Intermittent microwave–convective drying of red pepper: Drying kinetics, physical (colour and texture) and sensory quality. *Biosystems Engineering* 103: 455–463.

Stokes JR, Boehm MW, and Baier SK. (2013) Oral processing, texture and mouthfeel: From rheology to tribology and beyond. *Current Opinion in Colloid & Interface Science* 18: 349–359.

Therdthai N and Visalrakkij T. (2012) Effect of osmotic dehydration on dielectric properties, microwave vacuum drying kinetics and quality of mangosteen. *International Journal of Food Science & Technology* 47: 2606–2612.

Vincent JFV. (1989) Relationship between density and stiffness of apple flesh. *Journal of the Science of Food and Agriculture* 47: 443–462.

Weerts AH, Lian G, and Martin D. (2003) Modeling rehydration of porous biomaterials: Anisotropy effects. *Journal of Food Science* 68: 937–942.

Wilkinson C, Dijksterhuis GB, and Minekus M. (2000) From food structure to texture. *Trends in Food Science & Technology* 11: 442–450.

Wilson CE and Brown WE. (1997) Influence of food matrix structure and oral breakdown during mastication on temporal perception of flavor. *Journal of Sensory Studies* 12: 69–86.

Witrowa-Rajchert D and Rzaca M. (2009) Effect of drying method on the microstructure and physical properties of dried apples. *Drying Technology* 27: 903–909.

Yang J, Di Q, Jiang Q et al. (2010) Application of pore size analyzers in study of Chinese angelica slices drying. *Drying Technology* 28: 214–221.

Zhang M, Jiang H, and Lim R-X. (2010) Recent developments in microwave-assisted drying of vegetables, fruits, and aquatic products—Drying kinetics and quality considerations. *Drying Technology* 28: 1307–1316.

Zhang M, Tang J, Mujumdar AS et al. (2006) Trends in microwave-related drying of fruits and vegetables. *Trends in Food Science & Technology* 17: 524–534.

7 Microwave-Assisted Pulsed Fluidized and Spouted Bed Drying

Yuchuan Wang, Arun S. Mujumdar, and Min Zhang

CONTENTS

7.1 INTRODUCTION

Fluidized and spouted bed dryers have been studied extensively in the literature and have also found numerous industrial applications to dry particulate materials in continuous as well as batch models. More recently, pulsing of the spouting air has been studied to enhance hydrodynamic mixing of the particles yielding better heat and mass transfer between the gas and the particles. In most instances, the heat transfer is by pure convection between the gas and particles. Addition of conduction heating enhances both the drying rate and the energy efficiency of the dryers. Another option is supplement heating by generating microwave field within the drying chamber. Volumetric heating of the wet particles enhances the drying rate without excessive heating of the particles, which can also improve product quality with lower energy consumption and smaller carbon footprint.

Intermittent drying involves time-varying operations to match energy input levels to the requirement of the drying rate characteristic of the materials being dried. In principle, if done correctly, the product quality is enhanced by avoiding overheating while allowing a tempering period for the internal moisture to migrate to the surface for removal during the period when heat is supplied and the vapor is removed by convection or by applying vacuum. Note that all operating conditions are subject to periodic pulsation, for example, air velocity, temperature, humidity, as well as heat input by conduction, radiation, or volumetric heat by microwave or radio frequency (RF) fields. It has been shown by numerous experimental studies that intermittent drying is advantageous for batch drying of heat-sensitive materials with significant internal moisture.

In this chapter, we consider microwave-assisted drying in fluidized and spouted beds. Experimental results are presented for drying of various agricultural and food materials carried out in laboratory-scale equipment as well as in large-scale equipment. Effects of parameters such as fluidizing/spouting air velocity, temperature, as well as microwave power level (W/g of material) are examined. Note that the design of pulsed bed is identical to that for conventional convection dryers as discussed in many standard references, for example, Law and Mujumdar (2014).

7.2 ADVANTAGES AND DRAWBACKS OF MICROWAVE-ASSISTED PULSED FLUIDIZED AND SPOUTED BED DRYERS

Many researchers have shown that fluidized and spouted bed dryers have some distinct advantages including high thermal and mass transfer efficiency, high rate of moisture removal, easy material transport inside the dryer, uniform drying, ease of control, and low maintenance cost. However, there are a few limitations that exist in the conventional fluidized and spouted bed dryers, such as high pressure drop, high air and energy consumption, poor drying quality of some particulate products, nonuniform product quality in certain types of fluidized and spouted bed dryers, erosion of pipes and vessels, entrainment of fine particles, attrition or pulverization of particles, agglomeration of fine particles, and so on (Law and Mujumdar, 2006, 2014).

Pulsating fluidized and spouted bed dryers can be used in processing many products such as vegetables, fruits, edible mushrooms, aquatic products, and snacks (Law and Mujumdar, 2006, 2014). Compared to conventional fluidized and spouted bed dryers, pulsating fluidized and spouted bed dryers save airflow and energy by pulsating the flow resulting in energy cost saving and enhanced drying performance without affecting the product quality and process performance or additional capital costs, as well as reduce problem of mechanical damage to the particles due to continuous vigorous particle–particle collision as well as attrition-induced dusting (Law and Mujumdar, 2006).

Microwave-assisted conventional hot air fluidized and spouted bed dryers have been reported in both archival and patent literature. Compared to conventional fluidized and spouted bed dryers, microwave-assisted fluidized and spouted bed dryers can enhance thermal efficiency, shorten drying time, and, sometimes, improve product quality, as well as extend the size range of particulate products to be dried because of rapid volumetric heating of microwave heating. But, nonuniformity in drying of particulate products exists in microwave-assisted fluidized and spouted bed dryers caused by an uneven spatial distribution of the electromagnetic field inside the drying cavity (Li et al., 2011; Vadivambal and Jayas, 2010).

Pulsed fluidized and spouted bed microwave vacuum drying and pulsed fluidized and spouted bed microwave freeze drying are two kinds of innovative drying technology that have emerged in recent years. They combine the advantages of pulsed fluidized and spouted bed drying, microwave vacuum drying, and microwave freeze drying. This has solved the drying uniformity issue of the dried products existing in both microwave vacuum drying and microwave freeze drying. Microwave vacuum drying and microwave freeze drying with pulsed fluidized and spouted bed model can improve product quality such as flexibility, color, flavor, nutritional value, microbial stability, enzyme inactivation, rehydration capacity, crispiness, and fresh-like appearance.

7.3 PULSED FLUIDIZED AND SPOUTED BED MICROWAVE DRYING SYSTEM

7.3.1 Pulsed Fluidized and Spouted Bed Air Drying (PF/SBAD)

The design for pulsed fluidized and spouted bed dryers using hot air as heat source is similar to that of the conventional fluidized and spouted bed dryers while the gas control system is different. Pulse gas generation can be achieved in several ways. For example, pulsating gas can be generated by a pulsed electrical valve, which is installed in the gas inlet duct. Hence, the pulsating airflow can be controlled by a manual or automatic regulating valve that can be rotated at a specific angular velocity (Lu et al., 2014; Rosa et al., 2013); alternatively, the pulsating gas can be generated by a rotating hollow cylinder with two slots on the opposite sides of its lateral surface. The rotating velocity can be controlled, yielding a rectangular pulse that sweeps across the base of the column in a single direction (Reyes et al., 2006). These two types of pulsating beds spouting are suitable for drying beds with small diameter, while others are based on the relocation of the gas stream to the bottom of the bed. Relocation of the gas stream can be achieved by: (1) the use of a perforated rotating disc located below the circular bed; (2) the use of a rotating slotted horizontal cylinder that directs the gas stream to different sections of the base of a circular bed; and (3) the use of a rotating distributor valve, which makes the gas stream sweep across the specified chambers of a rectangular bed (Law and Mujumdar, 2006). Several varieties of mechanical design are possible.

7.3.2 Pulsed Fluidized and Spouted Bed Microwave Air Drying (PF/SBMAD)

Pulsed fluidized and spouted bed microwave-assisted hot air drying system is more complex in structure. Spouted bed drying chamber in microwave field needs to be designed and selected carefully. In addition, wet particle characteristic must be taken into consideration, for example, moisture content, size, shape, sugar contents, and stickiness, and so on. In general, microwave-assisted pulsed spouted bed hot air drying system consists of the following six basic systems (Lu et al., 2014; Rosa et al., 2013; Wang et al., 2013c): fluidized or spouted bed drying chamber, microwave heating system, hot air heating system, pulsed gas supply system, collector system for dried particle products, and appropriate control system.

Two methods are generally used to design a fluidized and spouted bed hot air dryer. The conventional fluidized and spouted bed hot air dryer is an integrated fluidized and spouted bed. Microwave heating cavity is designed as the spouted or fluidized bed drying chamber. This type of microwave-assisted fluidized and spouted bed dryer is simple in structure with low manufacturing costs. For the all-in-one design of the spouted bed dryer, some wet and fine particle materials with high moisture content and high sugar content are difficult to be spouted and can stick inside walls of the microwave cavity, especially the particle products that remain in the microwave source feed inlet. On the one hand, a large "hole" formed in the center of wet particles inside fluidized and spouted bed dryer at the beginning of drying can result in spouting deterioration. Also, these drawbacks of sticky wall and residual inside surface of microwave heating cavity may cause partial scorching of the product due to poor drying uniformity. Another design of the fluidized and spouted bed hot air dryer involves separated fluidized and spouted bed. Microwave heating system and spouted bed drying chamber are designed and manufactured separately. Such fluidized or spouted bed dryer is made of glass and can reduce the amount of wet particles sticking to the chamber wall. In addition, this structure can avoid accumulation of the feed in the microwave power inlet and protect the magnetron from overheating.

For hot air heating systems, air is heated electrically in the experimental fluidized and spouted bed apparatus and by high pressure steam using heat exchangers in industrial equipment. Hot air is injected into the fluidized or spouted bed chamber through a perforated distributor and airflow can be adjusted using an inventor adjustable valve.

For pulsed gas supply system, pulse pipe is fixed at the center of the distributor, and the pulsed gas is supplied by a high pressure vessel connected to an air compressor. At initial spouted drying stages, using high pressure pulse gas wet particles may

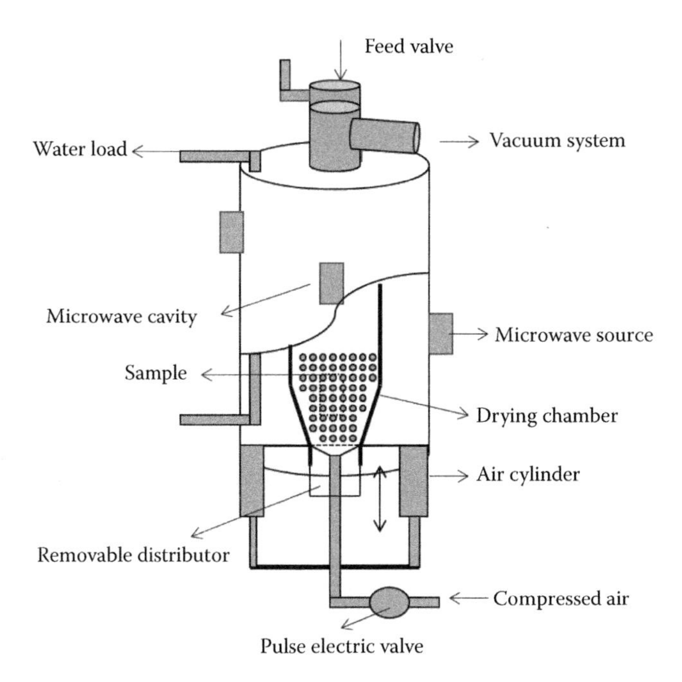

FIGURE 7.1 Schematic diagram of dryer for PSMVD and PSMFD.

cause a spatial movement like a fountain and achieve heating uniformity. As the moisture content drops, high pressure pulse gas flow may be adjusted. The use of lower air velocity only to carry away the water vapor in a pulsed bed makes it possible to reduce the consumption of air and save effective energy costs.

For the microwave heating system, the energy required for drying particle products is mainly supplied by microwaves during pulse spouted bed drying. Each magnetron's power output can be regulated by the combination of a microwave power controller and a water load. Besides regulation of the microwave power, a water load system is often necessary to prevent the magnetron from overheating.

For the collector system of dried particle products, a moving distributor together with pulse gas pipe has to be designed. This system is different from conventional fluidized and spouted bed dryer with some advantages such as simple structure, same residence time in dryer, and ease of cleaning. In industrial pulse spouted bed microwave/hot air drying system, multifluidized or spouted bed drying chambers are commonly designed and located in a microwave heating cavity. See Figures 7.1 and 7.2.

7.3.3 Pulsed Fluidized and Spouted Bed Microwave Vacuum Drying (PF/SBMVD)

The design of pulsed spouted bed microwave vacuum dryer follows the same design principle as the conventional spouted bed. In order to maintain normal vacuum pressure during drying, the pulse spouting period must be taken into consideration.

FIGURE 7.2 A pulsed spouted bed microwave hot air dryer producing 300 T dried vegetables per year. (From Hai Tong Group, Ningbo, China.)

The pulsed spouted bed microwave vacuum drying system consists of the following six basic subsystems (Wang et al., 2013c) (see Figures 7.1 and 7.3): spouted bed dryer, microwave heating system, pulsed gas supply system, water load system, vacuum system, and control system. The initial vacuum spouted bed was made of stainless steel with a few magnetrons fixed around the vacuum spouted bed surface. Vacuum seal between magnetron and spouted bed is obtained by a Teflon plate. In the newly developed apparatus, the microwave heating system and vacuum

FIGURE 7.3 A pilot-scale pulsed spouted bed microwave vacuum/freeze drying chamber. (From Jiangnan University, Wuxi, China.)

spouted bed are designed separately to avoid aerial discharge. The microwave feed position is at high microwave power intensity and simultaneously reduces vacuum leakage point. The microwave heating system is a cylindrical multimode microwave cavity (stainless steel); microwave generators are distributed symmetrically along the microwave cavity height. The spouted bed drying chamber is made of Teflon or glass and fixed inside the microwave cavity. Pulsed gas spouted system is equipped with a set of adjustable gas flow and distributor unit. The distributor is made of Teflon and can move up and down by a set of air cylinders. The spouted height of particle when spouting can be controlled by adjusting the gas pressure and flow rate. Gas used in this system may be inert gas (nitrogen) or air. The gas flow can be adjusted using a regulating valve. Each magnetron's power output can be adjusted by a combination of microwave power controller and water load. Besides assisting regulation of microwave power, the water load system prevents the magnetron from overheating. The vacuum system is equipped with a cooler, a water-ring vacuum pump, and a vacuum vessel from which a quick vacuum pressure recovery can be attained after pulse spouting. The fiber-optic temperature probe is fixed at the bottom of the distributor and measures the product temperature in real time during drying.

In the pulsed spouted bed microwave vacuum drying system, the wet particles can be put into the vacuum spouted bed chamber from a three-way valve fixed on the top of the vacuum spouted bed chamber. After drying, the distributor is separated from the inlet of the vacuum spouted bed chamber by the air cylinder, and the dried particles are taken out of the outlet of the spouted bed chamber. During drying, samples are spouted in specified time intervals by allowing air/gas to flow periodically into the vacuum drying chamber by use of an electromagnetic valve according to a preset time. The airflow rate was regulated by a manual flux adjusting valve, ensuring that the pressure fluctuated within preset values.

7.3.4 Pulsed Fluidized and Spouted Bed Microwave Freeze Drying (PF/SBMFD)

Some drawbacks of the conventional microwave freeze drying prevent its extensive use for commercial purposes, for example, corona discharge and nonuniform drying. Many experimental researches prove that the quality uniformity of microwave freeze-dried products is influenced by many factors such as vacuum cavity design, product attributes, and their spatial location in the microwave cavity. Based on the design concept of pulsed spouted bed dryer and combining regulation and control of microwave power and product dielectric attributes, newly developed pulse spouted bed microwave freeze drying system can overcome these drawbacks. Taking maintaining normal freeze drying vacuum pressure into consideration when instantaneous pulse spouting occurs, microwave power controller must be designed to avoid melting of ice in frozen products during drying period. Hence, unlike conventional tray-type microwave freeze dryer, pulsed spouted bed microwave freeze dryer can make drying particles with equal probability of microwave power exposure resulting from their spatial movement in the spouting chamber.

The pulsed spouted bed microwave freeze-drying system consists of the following seven basic subsystems (Wang et al., 2013b) (see Figures 7.1 and 7.3): spouted bed drying system, microwave heating system, pulsed gas supply system, water load system, vacuum system, refrigeration system, and control system. Unlike ordinary tray-type microwave freeze dryer, the microwave heating cavity and spouted bed drying chamber are designed separately to avoid corona discharge at the microwave feed position at high microwave power intensity and simultaneously minimize vacuum loss due to leakage. The spouted bed dryer made of glass or Teflon is used as the freeze drying chamber, and a cylindrical cavity made of stainless steel is used as the microwave heating cavity. The spouted bed is fixed vertically inside the microwave heating cavity. Several microwave generators (magnetrons) are distributed symmetrically along the helix in the microwave cavity. The angle and distance between adjacent microwave generators are 90° and 50 cm, respectively. Pulsed gas spouted system is equipped with a set of adjustable gas flow and a distributor unit. The distributor is made of Teflon and can be moved up and down using an air cylinder. The spouted height of particles can be controlled by adjusting the gas pressure and gas flow rate. Gas used in this system may be from inert gas (nitrogen gas source) or depurative air, and gas flow can be adjusted using regulating valve. Each magnetron's power output could be regulated by combination of microwave power controller and water load. Besides assisting regulation of microwave power, the water load system is added to prevent the magnetron from overheating. It was done by a cooling/heating circulating water unit. The vacuum system is equipped with an oil-sealed vacuum pump and a vacuum vessel from which quick vacuum pressure recovery can be attained after pulse spouting. The pressure in the drying chamber reaches 133 Pa from the atmospheric pressure in less than 1 min and can be regulated by a vacuum controller. The refrigeration system consists of a set of air-cooling refrigeration compressor unit and a vapor condenser (evaporating temperature −50°C to −30°C). A fiber-optic temperature probe is fixed at the bottom of the distributor and measures the particles' temperature in real time during drying.

In the pulsed spouted bed microwave freeze-drying system, frozen particles can be put into the freeze-drying chamber from a inlet valve fixed on the top of the freeze-drying chamber. When freeze-drying is completed, the dried particles can be taken out from the outlet valve fixed on the bottom of the freeze-drying chamber. During drying, the samples are spouted by allowing air/gas to flow periodically into the chamber by use of an electromagnetic valve according to preset time. The airflow rate is regulated by a manual adjusting valve, ensuring that the pressure fluctuated within the preset values. When the sample is spouted, microwave power automatically stops, and restarts when the pressure returns to its set value to prevent formation of ice crystals in the samples being dried from melting due to pressure fluctuation.

7.4 CHARACTERISTICS OF PULSED FLUIDIZED AND SPOUTED BED MICROWAVE DRYING

7.4.1 PF/SBMAD

Compared to conventional spouted and fluidized bed drying, microwave-assisted spouted and fluidized bed drying can greatly enhance drying rate and reduce energy consumption. The design and operating conditions such as air temperature, airflow rate, relative humidity, initial moisture content, microwave power level, and pulsation frequency affect the performance of the drying for foods. Table 7.1 shows drying time using pulsed fluidized and spouted bed microwave hot air drying at different drying conditions.

7.4.2 PS/FBMVD

In comparison with conventional microwave vacuum drying of a product with stationary or rotating turntable modes, microwave vacuum drying with pulsed spouted or fluidized bed showed significant decrease in drying time at the same microwave power level. This is because throughout the drying process no vapor condensation occurred on the chamber wall and sample dish surface. Although condensed water on the glass turntable surface increases turntable temperature and accelerates drying process, it may cause more non-uniform temperature distribution of dried products and more microwave energy consumption. Increasing microwave power density reduces the drying time of PSBMVD and MVD. However, over the critical value of microwave power level, there is no obvious difference of drying time exists in both drying systems, resulting in undesired dried product quality. Table 7.2 shows the drying time of some fruits and vegetables using PSBMVD, MVD, and VD.

7.4.3 PS/FBMFD

Similar to PSBMVD, microwave freeze drying with pulsed spouted/fluidized bed can reduce the required drying time at same microwave power level compared to that of microwave freeze drying in a tray. This difference is related to the design of the microwave freeze dryer. It can be explained by the fact that after PSBMVD for about 2 h, ice crystals on sample's surface are completely removed due to sublimation and shrinkage of the drying sample volume, especially those attached to the drying chamber wall. This results in the falling of drying product particles in the vertical drying chamber, exposing the drying products containing more ice crystals to the inner surface of the drying chamber. Moreover, the pulsed spouting during PSBMFD leads to equal distribution of microwave energy within the particles due to particle movement, and ensures that the drying sample achieves uniform heating and maintains a constant drying rate before the falling rate period starts. Table 7.2 shows the drying time of some fruits and vegetables dried using PSBMFD, MFD, and FD.

7.5 ENERGY ASPECTS OF PULSED FLUIDIZED AND SPOUTED BED MICROWAVE DRYING

Specific energy consumption can be expressed as the amount of energy required per unit moisture evaporated or per unit of dried products. In general, the specific energy consumption is closely related to drying method and the scale of drying equipment. It is also a more complex function of inlet hot air temperature, air velocity, moisture

TABLE 7.1
Drying Time and Energy Consumption for Selected Products Using Spouted and Fluidized Beds Microwave Hot Air Drying

Material	Size	Weight	Drying Method	Apparatus Scale	Initial/Final MC (%)	Drying Condition	Drying Time (min)	Energy Consumption	References
Apple	12.7×9.5×6.4 mm	40 g	SBMAD	Lab	22.8/5.5 (w.b.)	3.7 W/g, 70°C, 1.9 m/s	22.5		Feng and Tang (1998)
						6.1 W/g, 70°C, 1.9 m/s	9.5		
			SBAD			70°C, 1.9 m/s	147		
White kernel	Whole	300 g	FBMAD	Lab	60/12 (d.b.)	40°C, 5 m/s, 500 W	86		Kaensup et al. (1998)
						60°C, 5 m/s, 500 W	25		
						90°C, 5 m/s, 500 W	13		
			FBAD			40°C, 5 m/s	145		
						60°C, 5 m/s	29.7		
						90°C, 5 m/s	17.5		
Peppercorn	Diameter: 0.6 mm	300 g	FBMAD	Lab	300/12 (d.b.)	2 m/s, 50°C, 800 W	50		Kaensup and Wongwises (2004)
						2 m/s, 70°C, 80 W	20		
						2 m/s, 90°C, 800 W	15		
			FBAD			3.5 m/s, 50°C	460		
						3.5 m/s, 70°C	170		
						3.5 m/s, 90°C	70		

(Continued)

TABLE 7.1 (*Continued*)
Drying Time and Energy Consumption for Selected Products Using Spouted and Fluidized Beds Microwave Hot Air Drying

Material	Size	Weight	Drying Method	Apparatus Scale	Initial/Final MC (%)	Drying Condition	Drying Time (min)	Energy Consumption	References
Carrot	6×6× 6 mm	1800 g	FBMAD	Pilot	90.5/12 (w.b.)	0.73 W/g, 52°C–48°C	80	0.08 kg/MJ	Stanisławski (2005)
					89.3/8.4 (w.b.)	1.2 W/g, 65°C–42°C	65	0.141 kg/MJ	
					88.8/8.4 (w.b.)	1.2 W/g, 78°C–36°C	55	0.162 kg/MJ	
			FBAD		91/11.7 (w.b.)	54°C–52°C	166	0.055 kg/MJ	
					90.1/9.1 (w.b.)	70°C–68°C	150	0.077 kg/MJ	
Turnip seed	Diameter: 0.6 mm	150 g	FBMAD	Lab	54/10 (d.b.)	50°C, 0.55 m/s, RH8%, 150 W	13		Reyes et al. (2006)
						50°C, 0.55 m/s, RH30%, 150 W	16.3		
						50°C, 0.55 m/s, RH8%, 300 W	12.4		
						50°C, 0.55 m/s, RH30%, 300 W	12.3		
			PFBMAD			50°C, 0.55 m/s, RH8%, 150 W	10.3		
						50°C, 0.55 m/s, RH30%, 150 W	10.5		
						50°C, 0.55 m/s, RH8%, 300 W	8.6		
						50°C, 0.55 m/s, RH30%, 300 W	9		
Parboiled wheat kernel	Whole	200 g	SBMAD	Lab	–/12 (w.b.)	50°C, 1 m/s, 288 W	80		Kahyaoglu et al. (2012)
						50°C, 1 m/s, 624 W	35		
						70°C, 1 m/s, 288 W	60		
						70°C, 1 m/s, 624 W	25		
			SBAD			50°C, 1 m/s	270		
						70°C, 1 m/s	165		
						90°C, 1 m/s	105		

(*Continued*)

TABLE 7.1 ***(Continued)***
Drying Time and Energy Consumption for Selected Products Using Spouted and Fluidized Beds Microwave Hot Air Drying

Material	Size	Weight	Drying Method	Apparatus Scale	Initial/Final MC (%)	Drying Condition	Drying Time (min)	Energy Consumption	References
Shelled corn	Whole	100 g	PSBMAD	Lab	26/12.5 (d.b.)	40°C, 0.0108 m^3/s, 540 W	25		Momenzadeh et al. (2011)
						50°C, 0.0108 m^3/s, 540 W	15		
						60°C, 0.0108 m^3/s, 540 W	10		
			PSBAD			40°C, 0.0108 m^3/s	65		
						50°C, 0.0108 m^3/s	80		
						60°C, 0.0108 m^3/s	65		
Burdock root	5×5× 5 mm	40 g	PSBAD	Lab	82/8 (w.b.)	1 W/g, 50°C, 1.5 m/s	50		Lu et al. (2014)
						2 W/g, 50°C, 1.5 m/s	40		
						3 W/g, 50°C, 1.5 m/s	40		
Soy bean	Whole	14 kg	PSBMAD	Large	51.5/9 (w.b.)	24 kW	25	0.525 kgce/kg	Zhang et al. (2016)
			AD	Large	53.4/10.7 (w.b.)	90°C	108	1.267 kgce/kg	

SBMAD, spouted bed microwave hot air drying; SBAD, spouted bed hot air drying; FBMAD, fluidized bed microwave hot air drying; FBAD, fluidized bed hot air drying; PFBMAD, pulsed fluidized bed microwave hot air drying; PSBMAD, pulsed spouted bed microwave hot air drying.

TABLE 7.2
Drying Time and Energy Consumption for Dried Products Using Pulsed Spouted Bed Microwave Vacuum/Freeze Drying

Material	Size	Weight	Drying Method	Apparatus Scale	Initial/Final MC (%)	Drying Condition	Drying Time (min)	Energy Consumption	References
Stem lettuce	Diameter: 12 mm Thickness: 5 mm	200 g	PSBMVD	Lab	96/6.5 (w.b.)	2.4 W/g, 7 kPa, 3 s/2 s	60		Wang et al. (2013c)
			MVD			2.4 W/g, 7 kPa	120		
			VD			7 kPa, 60°C	270		
Apple	Cube: 5 mm	200 g	PSBMVD	Lab	87.5/6.2 (w.b.)	2.4 W/g, 7.5 kPa, 8 s/2 s	80		Mothibe et al. (2014)
			MVD			2.4 W/g, 7.5 kPa	75		
			VD			7.5 kPa, 70°C	400		
Restructured chips	Diameter: 0 mm Height: 300 mm	300 g	PSBMVD	Pilot	85/6 (w.b.)	2.0 W/g, 7.5 kPa, 88 s/2 s	49		Liu et al. (2015)
						2.6 W/g, 7.5 kPa, 88 s/2 s	43		
						3.0 W/g, 7.5 kPa, 88 s/2 s	38		
			MVD	Lab		2.0 W/g, 7.5 kPa	49		
						3.0 W/g, 7.5 kPa	38		
			VD	Lab		70°C, 7.5 kPa	140		
Okra	Whole	300 g	PSBMVD	Pilot	91/6 (w.b.)	500 W, 7.5 kPa, 8 s/2 s	90		Huang and Zhang (2016)
			MVD	Lab		500 W, 7.5 kPa	120		
			FD	Lab		50°C, 80 Pa	720		
Stem lettuce	Diameter:12 mm Thickness: 5 mm	200 g	PSBMVD	Lab	96/6.5 (w.b.)	2.4 W/g, 7 kPa, 3 s/2 s	60	52.2 kWh/kg	Wang (2013)
			VD	Lab	96/6.5 (w.b.)	60°C, 7 kPa, single layer	240	121.2 kWh/kg	
					96/6.5 (w.b.)	60°C, 7 kPa, multilayer	900	454.3 kWh/kg	

(*Continued*)

TABLE 7.2 (*Continued*)
Drying Time and Energy Consumption for Dried Products Using Pulsed Spouted Bed Microwave Vacuum/Freeze Drying

Material	Size	Weight	Drying Method	Apparatus Scale	Initial/Final MC (%)	Drying Condition	Drying Time (min)	Energy Consumption	References
Pea	Whole	5 kg	PSBMVD	Pilot	88/6.5 (w.b.)	2.4 W/g, 7 kPa, 60 s/2 s	62	10.24 kWh/kg	Lu et al.
			MVD		88/6.5 (w.b.)	7 kPa, 70°C	420	73.09 kWh/kg	(2015)
Stem lettuce	Diameter: 12 mm	200 g	PSBMFD	Lab	96/3.5 (w.b.)	2 W/g, 80 kPa, 600 s/2 s	240		Wang et al.
	Thickness: 5 mm		MFD	Lab		2 W/g, 80 kPa	300		(2013)
UDDEW	300×200×10 mm	200 g	PSBMFD	Lab	93/3.5 (w.b.)	2 W/g, 80 kPa, 600 s/2 s	240		Wang et al.
NUDDEW	300×200×10 mm	200 g	PSBMFD	Lab	93/3.5 (w.b.)	2 W/g, 80 kPa, 600 s/2 s	270		(2013)
FDEW	300×200×10 mm	200 g	PSBMFD	Lab	93/3.5 (w.b.)	2 W/g, 80 kPa, 600 s/2 s	300		
Banana	Thickness: 0.8 mm	200 g	PSBMFD	Lab	3.5/3.5 (d.b.)	2 W/g, 80 kPa, 600 s/2 s	360		Jiang et al.
			MFD	Lab		2 W/g, 80 kPa	360		(2014)
			FD	Lab		80 kPa	660		

PSBMVD, pulsed spouted bed microwave vacuum drying; MVD, microwave vacuum drying; VD, vacuum drying; PSBMFD, pulsed spouted bed microwave freeze drying; MFD, microwave freeze drying; FD, pulsed fluidized bed hot air freeze; FDEW, fresh duck egg white; UDDEW, desalted duck egg white with ultrasound pretreatment; NUDDEW, desalted duck egg white without ultrasound pretreatment.

content, pulsating frequency, operating pressure, applied microwave power level, as well as drying stages. Tables 7.1 and 7.2 list a few data of the energy consumption of some fruits and vegetables to be dried using microwave-assisted spouted and fluidized bed dryers. High moisture content results in high energy consumption. The increase of microwave power at constant inlet air temperature results in the decrease of energy consumption. During pulsed spouted and fluidized bed microwave hot air drying, at same air velocity and microwave power level, the specific energy consumption decreases with increasing air temperature. The specific energy consumption increases with increasing air velocity at same air temperature and microwave power level. For a pulsed spouted/fluidized bed dryer, several experimental investigations (Rosa et al., 2013; Stanisławski, 2005) have shown that the specific energy consumption using PS/FBMAD is lower than that using PS/FBAD, and under vacuum condition the specific energy consumption using PS/FBMVD is also much lower than that using PS/FBVD.

7.6 QUALITY CHARACTERISTICS OF PULSED FLUIDIZED AND SPOUTED BED MICROWAVE DRYING

During conventional fluidized and spouted bed drying process, higher temperature and longer drying time may cause serious damage to the quality attributes of the products such as color degeneration, reduction in bulk density and rehydration capacity, and loss in characteristic flavor and nutrients in dried products.

In addition, case-hardening is a common problem in dried products using conventional fluidized and spouted bed dryer due to rapid drying. The reason of this phenomenon is that during drying, if the rate of water evaporation is lower than the rate of water movement to the product surface, the outer skin of the products being dried will be firstly dried resulting in case-hardening.

Pulsed fluidized and spouted bed drying has no distinct difference in dried product quality compared with fluidized and spouted bed drying without pulsating model because pulsed fluidized and spouted model mainly reduces hot air usage leading to lower energy consumption, but hardly impact drying temperature and drying time of the products. But, uniform drying of the particular products may be obtained resulting from movement of the particular products in conventional fluidized and spouted bed dryer with and without pulsating model.

Compared with conventional fluidized and spouted bed drying using hot air, microwave-assisted fluidized and spouted bed drying could greatly reduce the drying time of biological materials without damaging the quality attributes of the finished products due to the high thermal efficiency of microwave drying. Moreover, microwave-assisted fluidized and spouted bed drying could also inhibit occurrence of high surface temperatures, continuation of product respiration, lowered product temperatures when combined with vacuum drying, and reduction in the loss of water-soluble components. In addition, combined microwave fluidized and spouted bed drying could eliminate the problem of case-hardening due to lower surface temperatures and higher rate of moisture removal from the products being dried. A significant distinction may exist in dried product quality using fluidized and spouted bed drying with and without pulsating model depending on pulsating parameters and the structure of the drying bed. An optimal design of pulsed fluidized and spouted bed drying system can reduce the damage to the quality attributes of the dried products due to more drying uniformity of the products being dried compared to continuous fluidized and spouted bed drying system. Non-uniformity especially in the moisture content of the dried particular products can cause effect on product safety from microorganism in the stages of transporting distribution and storage.

7.6.1 Optical Properties

Table 7.3 shows color of dried products using pulsed spouted and fluidized bed dryers with and without microwave. It can be seen that under atmospheric drying condition, the color of dried products using SBMAD shows a slight further darkening in comparison with that of dried products using SBAD. This means that color degradation of the product caused by SBAD drying was slightly more than that by SBMAD drying. The lower color degradation of dried products with SBMAD is due to the substantial reduction in drying time. SBAD drying also can obtain acceptable color because of higher heat and mass transfer that facilitate a higher drying rate. High drying temperature and long drying time would accelerate the development of browning during apple drying. The lightness L^* of dried apple dices using SBMAD is slightly higher than using SBAD. This contributes to lower drying time and uniform drying temperature caused by using SBMAD.

Under vacuum condition, regardless of adopting PSBMVD or PSBMFD, the color of dried products is better than that using SBMAD and SBAD. For microwave vacuum drying, the pulsed spouted bed method retains more color due to the shorter drying time and lower drying temperature. Even after rehydration, pulsed spouted mode products exhibit lower color variation compared to those of the rotating turntable mode. Overheating is the most common problem of MVD, which can lead to heat spots and, furthermore, cause local charring during drying. However, PSBMVD can solve the problem of local overheating by ensuring the samples' spatial movement during drying. So, the mean values L^*, a^*, and b^* of samples dried by PSBMVD are better than MVD. However, microwave power density, pulsating frequency, and operating pressure recovery time are important drying parameters in order to obtain better color quality. Higher microwave power density and pulsating frequency as well as long operating pressure recovery time can cause color degradation. For microwave freeze drying, the pulsed spouted bed method also reduces color degradation due to the shorter drying time and high drying uniformity. Even after rehydration, pulsed spouted mode products exhibit lower color variation compared to those of the static mode. Similar to PSBMVD, microwave power density and control, pulsating frequency, and operating pressure recovery time are important drying parameters to PSBMFD in order to obtain better color quality. Higher microwave power density and pulsating frequency as well as long operating pressure recovery time will result in melting of ice crystals and cause color degradation during PSBMFD.

TABLE 7.3
Color of Dried Products Using Pulsed Spouted and Fluidized Bed Dryers with and without Microwave

Material	Size	Weight	Drying Method	Drying Condition	Lightness L^*	Redness a^*	Yellowness b^*	
Parboiled wheat bulgur	Whole	200 g	SBMAD	50°C, 288 W	40.9	9.7	39.5	Kahyaoglu et al. (2010)
				50°C, 624 W	42.3	8.9	40.1	
				70°C, 288 W	41.7	9.4	40.2	
				70°C, 624 W	42.8	13.1	40.7	
			SBAD	50°C	41.8	10.1	40.3	
				70°C	40.8	9.9	40.0	
				90°C	40.3	13.2	39.2	
Stem lettuce	Diameter: 12 mm Thickness: 5 mm	200 g	PSBMVD	2.4 W/g, 7 kPa, 3 s/2 s	6.34	–6.34	13.35	Wang et al. (2013)
			MVD	2.4 W/g, 7 kPa	9.21	–2.89	2.93	
Apple	5×5×5 mm	200 g	PSBMVD	2.4 W/g, 7 kPa, 3 s/2 s	78.02	–2.01	43.13	Mothibe et al. (2014)
			MVD	2.4 W/g, 7.5 kPa	58.91	7.73	26.54	
			VD	7.5 kPa, 70°C	68.96	4.20	35.68	
			Fresh		76.09	–3.13	18.20	
Restructured chips	Diameter: 40 mm Height: 300 mm	300 g	PSBMVD	2.2 W/g, 88 s/2 s	71.69	–2.69	17.69	Liu et al. (2015)
				2.6 W/g, 88 s/2 s	74.10	–2.75	16.32	
				3.0 W/g, 88 s/2 s	73.68	–2.93	17.05	
			MVD	2.2 W/g, 7.5 kPa	73.16	–2.73	17.23	
				2.6 W/g, 7.5 kPa	75.73	–2.89	16.12	
				3.0 W/g, 7.5 kPa	74.84	–3.13	18.32	
			VD	60°C, 7.5 kPa	65.21	–1.816	21.12	
			Fresh		78.31	–3.515	13.23	

(*Continued*)

TABLE 7.3 (*Continued*)
Color of Dried Products Using Pulsed Spouted and Fluidized Bed Dryers with and without Microwave

Material	Size	Weight	Drying Method	Drying Condition	Lightness L^*	Redness a^*	Yellowness b^*	
Puffed salted duck egg white/starch products	8×8×3 mm	300 g	PSBMVD	0.4 kW, 140 min, 1.34 kW, 15 min	77.71	1.96	18.49	Wang et al. (2016)
				0.4 kW, 140 min, 2.01 kW, 15 min	74.65	2.65	21.70	
				0.4 kW, 140 min, 2.68 kW, 15 min	65.64	8.49	28.70	
Okra	Whole	300 g	PSBMVD	500 W	69.08	−11.32	24.5	Huang and Zhang (2016)
			MVD	500 W	63.64	−8.66	22.53	
			FD		71.46	−11.51	23.3	
Stem lettuce	Diameter: 12 mm Thickness: 5 mm	200 g	PSBMFD	2 W/g, 80 kPa, 600 s/2 s	30.18	−4.89	22.02	Wang et al. (2013)
			MFD	2 W/g, 80 kPa	31.19	−3.91	22.72	
			FD	2 W/g, 80 kPa	33.78	−3.24	19.86	
Banana	Thickness: 0.8 mm	200 g	PSBMFD	2 W/g, 10 min/2 s, 80 Pa	58.41	2.29	17.73	Jiang et al. (2014)
			MFD	2 W/g, 80 Pa	55.70	−0.58	12.42	
			FD	50°C	67.09	1.09	22.43	
			Fresh		65.78	−1.92	17.79	

SBMAD, spouted bed microwave hot air drying; SBAD, spouted bed hot air drying; PSBMVD, pulsed spouted bed microwave vacuum drying; MVD, microwave vacuum drying; VD, vacuum drying; PSBMFD, pulsed spouted bed microwave freeze drying; MFD, microwave freeze drying; FD, pulsed fluidized bed hot air freeze.

7.6.2 Sensory Characteristics

Some researches (Huang and Zhang, 2016; Liu et al., 2015) show that the retention of volatile components responsible for flavor is more in hot air microwave drying compared to conventional hot air drying alone. For the example of burdock cubes (Lu et al., 2014) dried using hot air drying and MPSBD, it is observed that the cubes dried using MPSBD are richer in flavor compounds compared to those dried using hot air. Microwave drying can induce a few of flavor compounds including elemene in ginseng, methyl salicylate in Keemun black tea, isobutyl phthalate in fruit, as well as 3-methyl butanal, phenyl acetaldehyde, hexanal, and decanal. Hence, MPSBD can be used to improve the flavor matrix of dried products.

In recent years, electronic nose device is used to evaluate odor of dried products with different methods. For example of dried restructured chips using VD, MVD and PSBMVD (Liu et al., 2016), electronic nose device can distinguish the differences in the odors of the dried restructured chips with these three drying methods; the odors of dried samples using VD are different from those using PSBMVD and MVD, and the odor differences of dried samples using MVD and PSBMVD are not significant, and browning reaction in VD is more serious than in MVD and PSMVD.

7.6.3 Rehydration Characteristics

Many investigations have proved that none of the dried products regain their initial moisture. Irreversible physical and chemical changes might have occurred during drying. This might be due to the fact that a higher proportion of pores are made as a result of the cellular and structural disruption that take place during drying. This results in reduced hydrophilic properties and inability to imbibe sufficient water, leaving pores unfilled. In theory, high drying temperature during hot-air drying can cause damage to cells and gel structures in food, resulting in the migration of soluble solids to the surface to form a crust and then leading to a relatively closed surface structure.

Microwave drying facilitates a predominant vapor migration from the interior of the material as compared to a more predominant transfer of soluble solids during conventional drying. This difference in vapor and soluble solids transfer, combined with high internal pressure, is likely to result in a more porous structure compared with conventional hot-air dried products. Such enhanced porous structure forms a higher rehydration capacity in dried products using SBMAD.

For vacuum drying and freeze drying, low drying temperature and long drying time can cause damage to cells and gel structures in food. No distinct crust is formed in the surface of products dried by conventional vacuum drying and freeze-drying. Moreover, conventional heating method is milder and slower in heat transfer as compared to microwave heating. So, physical and chemical changes in the structure and composition of dried products using conventional heating are lower than that using microwave heating. These results show that microwave energy might cause greater changes in structure and composition of dried materials during MVD and MFD compared to FD. The rehydration capacities for dried products with VD and FD are higher than those with MVD and MFD, respectively. In comparison with MVD with static and rotating turntable types, PSBMVD can offer higher rehydration capacity of dried products. Similarly, PSBMFD can obtain higher rehydration capacity of dried products compared to MFD with static and tray types. This shows that the pulsed spouting has no significant influence on the rehydration capacity of the dried samples, although vacuum changed periodically.

7.6.4 Textural Properties

For vacuum drying, compared to VD and MVD, PSBMVD can obtain higher surface hardness due to intercollision between the products being dried at high pulsating frequency. This indicates that products dried by PSBMVD have the highest elastic behavior after rehydration, and the sample dried by MVD is the softest. For freeze drying, different from vacuum drying, measured hardness value of the samples dried by FD is higher than that of the samples dried by both MFD methods, especially the drying of some food containing high ions. After rehydration the samples dried by PSBMFD have higher firmness values. This reason is not caused by pulsating spouted model because of low pulsating frequency used during PSBMFD. This may be explained by the fact that more ions and soluble components in the samples might have transferred from the inner layer to the surface of the dried samples during MFD process as a result of an entrainment effect at a relatively high drying rate during this process. FD samples are harder because all ions and soluble components remain in the original positions in the dried portions of the samples due to a relatively low drying rate.

7.6.5 Nutritional Characteristics

For fruits, vegetables, and edible mushrooms, ascorbic acid, flavonoids, phenolic, chlorophyll, and total sugar are often used to evaluate nutritional loss during the drying. These nutritional ingredients could be damaged by light, oxygen, heat, moisture, or radiation. The drying method has important impact on the degradation of these nutritional ingredients. Different drying conditions and equipment performance cause different kinds of damage to the nutritional ingredients in dried products. In general, higher drying temperature, long drying time, and nonuniform drying due to microwave field are the main reason resulting in nutritional loss in the final product. Compared to FD and MFD, PSBMFD can retain higher ascorbic acid content. Shorter drying time than FD and better uniformity of temperature distribution than in MFD are the main reason for better retention by PSBMFD. Compared with MVD, PSBMVD not only provides better extrinsic features but also better nutritional attributes. Under vacuum drying condition, the ascorbic acid content in PSBMVD samples is also relatively higher than that in MVD samples. For total flavonoids and total phenolic content, compared with PSBMVD and MVD, VD has greater loss of flavonoids and phenolics due to the longer drying time. However, at the same microwave power level, flavonoids and phenolics contents of samples dried by PSBMVD are slightly lower than those of MVD products. This indicates that long drying time and pulsating air have important influence on flavonoids and phenolics under vacuum condition. This perhaps can be explained by the release of air carrying oxygen when spouting during PSBMVD. For chlorophyll content, the difference between the chlorophyll content value of MVD and PSBMVD products is very significant; using PSBMVD can obtain better chlorophyll retention than using MVD (Huang and Zhang, 2016). The problem of local overheating during MVD causes a serious loss of chlorophyll. For total sugar content, it is reported that the PSBMVD product has higher total sugar content than intermediate-infrared dried and hot-air dried products; the hot-air dried product is the lowest (Qi et al., 2014).

7.7 DRYING UNIFORMITY IN PULSED FLUIDIZED AND SPOUTED BED MICROWAVE DRYING

Drying uniformity of dried samples is evaluated using standard deviation (SD) or relative standard deviation (RSD) by some researchers. The smaller the RSD or SD, the more uniform is the drying of the sample. A few of researchers use qualified rate as evaluation method of drying uniformity of dried product particles. Qualified rate reflects the degree of carbonization of dried product, which is defined as the ratio of qualified product particles and the total product particles. In practice, drying uniformity (DU) of dried products is well defined by the following equation (Wang, 2013):

$$DU = (1 - RSD) * 100\%$$

RSD is the ratio of standard deviation to mean measurement value of temperature, moisture content, color, and shrinkage. In general, temperature, moisture content, color, and shrinkage are often used as evaluating objects of drying uniformity.

For microwave drying under atmospheric condition, the drying uniformity can be improved by combining microwave drying with spouted bed drying. This contribute SBMAD can provide pneumatic agitation to help avoid uneven microwave heating because of increasing particle–air heat and mass transfer by high air velocity and effective mixing.

For microwave drying under vacuum condition, the drying uniformity can be also well enhanced by hybridizing pulsed spouted bed drying and microwave vacuum drying and microwave freeze drying, respectively. After this novel drying technology has emerged in the recent years, some investigators have studied its drying uniformity using some foods as experimental materials such vegetables and fruits. For instance, in stem lettuce cubes drying, using PSBMVD and PSBMFD can offer significant improvement to the drying uniformity in the product temperature, moisture content, color, and shrinkage ratio compared to conventional tray or turntable model microwave vacuum drying and microwave freeze drying.

From Table 7.4 it can be observed that the measured drying uniformity value in the product temperature, moisture content, color, and shrinkage ratio of the samples being dried by FD is the best, followed by that for PSBMFD samples, while that for MFD samples is the poorest. This indicates that the pulse-spouted mode can effectively improve uniformity in color temperature, moisture content, color, and shrinkage ratio distribution within microwave freeze-dried sample as compared to static mode. This may be also related to the design of microwave freeze dryer, which involves rational microwave power supply system as well as the water load for its ability to adjust microwaves absorbed. Both PSBMVD and PSBMFD can also be

TABLE 7.4
Drying Degree of Uniformity of Dried Products Using Pulsed Spouted and Fluidized Bed Microwave Drying

Material	Size	Weight	Apparatus Scale	Drying Method	Drying Condition	Degree of Uniformity				
						T (°C)	MC (%)	SR (%)	Color	
Carrot	5 × 5 × 5 mm	14 kg	Large	PSBMAD	24 kW, 3 s/2 s		9987	87.93	85.2	Cao et al.
Corn	Whole	14 kg	Large	PSBMAD	24 kW, 3 s/2 s		99.9	89.9	85.1	(2016)
Soy bean	Whole	14 kg	Large	PSBMAD	24 kW, 3 s/2 s		99.9	90.9	86.1	
Stem lettuce	Diameter: 12 mm	200 g	Lab	PSBMVD	2.4 W/g, 7 kPa, 3 s/2 s	±2.1	97.8	90.4	91.4	Wang et al.
	Thickness: 5 mm			MVD	2.4 W/g, 7 kPa	±20.8	8.1	15.8	15.3	(2013)
Pea	Whole	1.2 kg	Pilot	PSBMVD	5 kW, 3 s/2 s		93.3	98.1	97.0	Lu et al. (2015)
Soy bean	Whole	1.2 kg	Pilot	PSBMVD	5 kW, 3 s/ 2 s		92.5	96.9	98.0	
Potato	Diameter: 12 mm Thickness: 5 mm	1.2 kg	Pilot	PSBMVD	5 kW, 3 s/2 s		95.0	92.8	97.2	
Restructured chips	Diameter: 40 mm	300 g	Pilot	PSBMVD	88 s/2 s, 2.8 W/g, 7.5 kPa		97.8			Liu et al. (2015)
	Height: 300 mm			MVD	2.6 W/g, 7.5 kPa		91.2			
Okra	Whole	300 g	Pilot	PSBMVD	500 W, 7.5 kPa		56	94.1	96.9	Huang and
				MVD	500 W, 7.5 kPa		71.4	94.3	98.5	Zhang (2016)
Stem lettuce	Diameter: 12 mm	200 g	Lab	PSBMFD	600 s/2 s, 2 W/g, 80 Pa	2–5	91.1	91.7	92.5	Wang et al.
	Thickness: 5 mm			MFD	2 W/g, 80 Pa	0–40	79.4	81	90.4	(2013)
				FD	50°C, 80 Pa			88	95.7	
Banana	Thickness: 0.8 mm	200 g	Lab	PSBMFD	600 s/2 s, 2 W/g, 80 Pa	5–12	98			Jiang et al.
				MFD	2 W/g, 80 Pa	8–27	93			(2014)
				FD	50°C, 80 Pa	4–13	97.1			

PSBMAD, pulsed spouted bed microwave hot air drying; PSBMVD, pulsed spouted bed microwave vacuum drying; MVD, microwave vacuum drying; PSBMFD, pulsed spouted bed microwave freeze drying; MFD, microwave freeze drying; FD, freeze drying.

suitable to drying of long cylindrical foods such as okra and long rectangular foods such as carrot strip, and get uniform product quality without breakage. In addition, pretreatment of materials before drying, such as apple pretreated by ultrasonic wave and stem lettuce by microwave, can improve drying uniformity of the materials in PSBMFD.

7.8 CLOSING REMARKS

Based on the recent experimental studies carried out in senior author's laboratory, it is concluded that application of microwave field and pulsation to fluidized or spouted bed can intensify drying kinetics as well as enhance dried product quality with improved energy efficiency as compared to conventional fluidized and spouted beds. Both atmospheric pressure and vacuum operation yield similar enhancement in drying rate, efficiency, as well as quality of the products. There is additional capital cost due to the incorporation of magnetron and application of vacuum, although the running cost may be low due to shortened drying time. Optimization of the operating parameters using experimental and modeling is needed in future R&D to make these novel dryers commercially attractive. A few industrial-scale units are operating successfully in China.

ACKNOWLEDGMENT

The authors would like thank 863-Hi-Tech Research and Development Program of China for supporting the research under contract no. 2011AA100802.

REFERENCES

Cao, Y. F., Liu, C. G., and Xue, G. F. 2016. Examining report of microwave pulsed spouted dryer (MSD-200) from Quality Examination Center for Mechanical Industry Food, Beijing, China.

Feng, H. and Tang, J. 1998. Microwave finish drying of diced apples in a spouted bed. *Journal of Food Science* 63(4): 679–683.

Huang, J. P. and Zhang, M. 2016. Effect of three drying methods on the drying characteristics and quality of okra. *Drying Technology* 34(8): 900–911.

Jiang, H., Zhang, M., Mujumdar, A. S., and Lim, R. X. 2014. Comparison of drying characteristic and uniformity of banana cubes dried by pulse-spouted microwave vacuum drying, freeze drying and microwave freeze drying. *Journal of the Science of Food and Agriculture* 94(9): 1827–1834.

Kaensup, W. and Wongwises, S. 2004. Combined microwave/fluidized bed drying of fresh peppercorns. *Drying Technology* 22(4): 779–794.

Kaensup, W., Wongwises, S., and Chutima, S. 1998. Drying of pepper seeds using a compared microwave/fluidized bed dryer. *Drying Technology* 16(3–5): 853–862.

Kahyaoglu, L.N., Sahin, S., and Sumnu, G. 2010. Physical properties of parboiled wheat and bulgur produced using spouted bed and microwave assisted spouted bed drying. *Journal of Food Engineering* 98(2): 159–169.

Kahyaoglu, L. N., Sahin, S., and Sumnu, G. 2012. Spouted bed and microwave-assisted spouted bed drying of parboiled wheat. *Food and Bioproducts Processing* 90(2): 301–308.

Law, C. L. and Mujumdar, A.S. 2006. Fluidized bed dryers, in *Handbook of Industrial Drying*, 3rd ed. Mujumdar, A.S. ed., Marcel Dekker, New York, Chapter 8.

Law, C. L. and Mujumdar, A.S. 2014. Fluidized bed dryers, in *Handbook of Industrial Drying*, 4th ed. Mujumdar, A.S. ed., Marcel Dekker, New York, Chapter 8.

Li, Z., Wang, R., and Kudra, T. 2011. Uniformity issue in microwave drying. *Drying Technology* 29(6): 652–660.

Liu, Z., Zhang, M., and Wang, Y. C. 2016. Drying of restructured chips made from the old stalks of *Asparagus officinalis*: Impact of different drying methods. *Journal of the Science of Food and Agriculture*, 96(8): 2815–2824.

Lu, J. F., Song, X. N., and Zhang, Y. N. 2015. Examining report of pulsed spouted microwave vacuum dryer (PSMVD-5) from Quality Examination Center for Mechanical Industry Food Mechanical Product, Nan Jiang, China.

Lu, Y., Zhang, M., Sun, J. C., Cheng, X. F., and Adhikari, B. 2014. Drying of burdock root cubes using a microwave-assisted pulsed spouted bed dryer and quality evaluation of the dried cubes. *Drying Technology* 32(15): 1785–1790.

Momenzadeh, L., Zomorodian, A., and Mowla, D. 2011. Experimental and theoretical investigation of shelled corn drying in a microwave-assisted fluidized bed dryer using Artificial Neural Network. *Food and Bioproducts Processing* 89(1): 15–21.

Mothibe, K. J., Zhang, M., Mujumdar, A. S., Wang, Y. C., and Cheng, X. F. 2014. Microwave-assisted pulse-spouted vacuum drying of apple cubes. *Drying Technology* 32(15): 1762–1768.

Qi, L. L., Zhang, M., Mujumdar, A. S., Meng, X. Y., and Chen, H. Z. 2014. Comparison of drying characteristics and quality of shiitake mushrooms (*Lentinus edodes*) using different drying methods. *Drying Technology* 32(15): 1751–1761.

Reyes, A., Campos, C., and Vega, R. 2006. Drying of turnip seeds with microwaves in fixed and pulsed fluidized beds. *Drying Technology* 24(11): 1469–1480.

Rosa, G. S., Marsaioli, A. Jr., and Rocha, S. C. S. 2013. Energy analysis of poly-hydroxybutirate (PHB) drying using a combined microwave/rotating pulsed fluidized bed (MW/RPFB) dryer. *Drying Technology* 31(7): 795–801.

Stanisławski, J. 2005. Drying of diced carrot in a combined microwave-fluidized bed dryer. *Drying Technology* 23(8): 1711–1721.

Vadivambal, R. and Jayas, D. 2010. Non-uniform temperature distribution during microwave heating of food materials—A review. *Food and Bioprocess Technology* 3(2): 161–171.

Wang, T., Zhang, M., Fang, Z. X., and Liu, Y. P. 2016. Effect of processing parameters on the pulsed-spouted microwave vacuum drying of puffed salted duck egg white/starch products. *Drying Technology* 34(2): 206–214.

Wang, Y. C. 2013. Studies on mechanism and technology of negative pressure drying assisted by microwave for stem lettuce cubes with efficiency, energy-saving and uniformity (PhD), School of Food Science and Technology, Jiang Nang University, WuXi, China.

Wang, Y. C., Zhang, M., Adhikari, B., Mujumdar, A. S., and Zhou, B. 2013a. The application of ultrasound pretreatment and pulse-spouted bed microwave freeze drying to produce desalted duck egg white powders. *Drying Technology* 31(15): 1826–1836.

Wang, Y. C., Zhang, M., Mujumdar, A. S., and Mothibe, K. J. 2013b. Microwave-assisted pulse-spouted bed freeze-drying of stem lettuce slices—Effect on product quality. *Food and Bioprocess Technology* 6(12): 3530–3543.

Wang, Y. C., Zhang, M., Mujumdar, A. S., Mothibe, K. J., and Roknul Azam, S. M. 2013c. Study of drying uniformity in pulsed spouted microwave–vacuum drying of stem lettuce slices with regard to product quality. *Drying Technology* 31(1): 91–101.

Zheng, S., Xu X., and Cheng, J. L. 2016. Examining report of microwave pulsed spouted dryer (MSD-200) from Ning Bo Energy Examination Company, Ning Bo, China.

8 Mathematical Modelling of Intermittent Drying

Chandan Kumar, Aditya Putranto, Md. Mahiuddin, S.C. Saha, Y.T. Gu, and M. Azharul Karim

CONTENTS

8.1 INTRODUCTION

Drying has several advantages compared to other food processing methods. It reduces moisture level, decreases weight and thus reduces packaging cost and increases shelf life. However, the product quality attributes such as colour, texture, aroma and nutritional values may degrade during drying. Moreover, high energy consumption remains another great concern for drying operations. Intermittent drying is one of the methods to reduce energy consumption and quality degradations (Chua et al. 2003). Intermittent drying can be achieved by several techniques including controlled or intermittent supply of thermal energy during drying that varies periodically (Kowalski and Pawlowski 2011b, Kumar et al. 2014b). A general classification of intermittent drying is presented by Kumar et al. 2014b).

Intermittent drying has been considered as one of the most energy-efficient drying processes (Chua et al. 2002c, 2003, Kowalski and Pawłowski 2011a). Various studies showed that energy requirement for intermittent drying was lower than continuous convection drying in different products such as yerba mate (Ramallo et al. 2010), squash slice (Pan et al. 1998), grain (Jumah 1995), kaolin (Kowalski and Pawlowski 2011a,b) and *Ganoderma tsugae* (Chin and Law 2010). On the other hand, the quality of dried food is another important issue in food drying. Intermittent drying can prevent case-hardening (Zeki 2009) and can provide better-quality dried food (Joardder et al. 2015a,b, Kowalski and Pawłowski 2011a). In intermittent drying, determination of optimum tempering time and process parameters is essential to obtain the best possible outcome. Moreover, the most favourable intermittent drying will take place when energy supplied to the dryer matches with the energy requirement for the moisture removal at any time of drying. Temperature and moisture distribution inside the sample can help in determining the right energy requirement at the specific time of drying. Mathematical modelling is necessary for evaluating the effect of process parameters and investigating the heat and mass transfer during the intermittent drying process. Therefore, this chapter presents the mathematical models for intermittent drying obtained by varying the convective energy. Moreover, the mathematical modelling of intermittent drying obtained by varying alternative source of energy such as microwave and infrared are discussed. The reaction engineering approaches (REA) for intermittent drying are reviewed and presented. Finally, the solution techniques for solving the modelling equations are presented.

8.2 MATHEMATICAL MODELLING OF INTERMITTENT CONVECTIVE DRYING

This section discusses the modelling of intermittent drying. The categories of modelling approaches for intermittent drying are shown in Figure 8.1. The modelling approaches of various intermittent drying based on Figure 8.1 are discussed. Modelling of intermittent drying achieved by various heat sources is discussed in Section 8.6. Then, the reaction engineering approaches are discussed in Section 8.7.

8.3 EMPIRICAL MODEL

Empirical models are simple to apply and often used to describe drying curve. Table 8.1 shows the list of the empirical models used to describe the kinetics of intermittent drying. Among them, the Page and Exponential models are most commonly used.

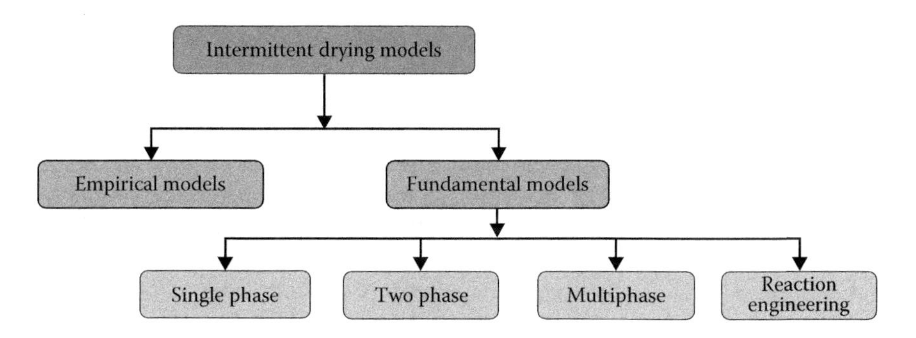

FIGURE 8.1 Modelling approach of intermittent drying.

TABLE 8.1
Empirical Models Used in Drying

Model	Name of the Model	References
$MR = \exp(-kt)$	Newton	Sutar and Thorat (2011) and Baini and Langrish (2007)
$MR = \exp(-kt^n)$	Page	
$MR = \exp(-(kt)^n)$	Modified page	
$MR = a\exp(-kt)$	Henderson and Pabis	
$MR = a\exp(-kt) + c$	Logarithmic	
$MR = a\exp(-k_0 t) + a\exp(-k_1 t)$	Two-term	
$MR = 1 + at + bt^2$	Wand and Singh	
$MR = a\exp(-kt) + (1-a)\exp(-kbt)$	Approximation of diffusion	
$MR = a\exp(-kt) + (1-a)\exp(-gt)$	Verma	
$MR = a\exp(-kt) + b\exp(-gt) + c\exp(-ht)$	Modified Henderson and Pabis	
$MR = a\exp(-kt) + (1-a)\exp(-kat)$	Two-term exponential	
$MR = \exp\left(-k\left(\frac{t}{L^2}\right)^n\right)$	Modified page equation—II	
$MR = 1 - \frac{t}{a+bt}$	Peleg model	da Silva et al. (2015)
$MR = \exp\left(-at - b\sqrt{t}\right)$	Silva et al.	da Silva et al. (2015)

Here, M_t is moisture content at time t, M_e is equilibrium moisture content and M_0 is initial moisture content, the moisture ratio, $MR = (M_t - M_e)/(M_0 - M_e)$, and a, b, c are model constants (dimensionless) and k, g, h are the drying constants (s^{-1}). These models are originally derived from Newton's law of cooling and Fick's law of diffusions (Erbay and Icier 2010). Regression analysis is often used to find constant and fit the drying curve.

da Silva et al. (2015) used empirical models to describe continuous and intermittent drying processes. They found that the Peleg model best described the intermittent drying of bananas (with a R^2 value of 0.9992–0.9995), whereas the continuous drying was best described by the Page model ($R^2 = 0.9995$). However, the Page model was most commonly applied for intermittent drying, employed by Holowaty et al. (2012), Ramallo et al. (2010) and Zhu et al. (2010), to predict moisture variation during intermittent drying. Zhu et al. (2010) regressed the empirical coefficients (k and n) of the Page model with processing variables (slice thickness and surface temperature).

Foroughi-Dahr et al. (2015) dried Iranian rough rice by intermittent drying with different drying air temperatures of 40°C, 50°C and 60°C and air velocities of 2.2 and 3.2 m s^{-1}. They applied 11 empirical models for intermittent drying of rough rice and performed statistical analysis of the models. The Midilli and two-term models were found to be the most appropriate for the first and second drying stages, respectively. They expressed constants and coefficients of these two models as a polynomial function of the temperature and velocity of drying air. They compared the predicted moisture contents with the experimental moisture ratio values from all the experiments and found a perfect agreement between the calculated and experimental data, further indicating the suitability of Midilli and two-term models in describing the drying behavior of rough rice.

8.3.1 Artificial Neural Network (ANN) Model

Apart from those empirical models, an artificial neural network (ANN)-based model can be used to simulate intermittent drying. ANN models are more efficient to model a complex and ill-defined process like drying (Jumah and Mujumdar 2005). The main advantage of the ANN model is that the user can learn from successive experiments and, in this way, improve the model (Chen et al. 2001). ANN consists of three layers: an input layer, one or more hidden layers and an output layer, as shown in Figure 8.2. Each unit of the layers is called node. The input layer receives information from the outside world, such as process parameters in case of drying, for example drying air temperature, drying air velocity and air relative humidity, and then transmits it to the hidden layers for processing. The output layer is mainly the prediction layer which

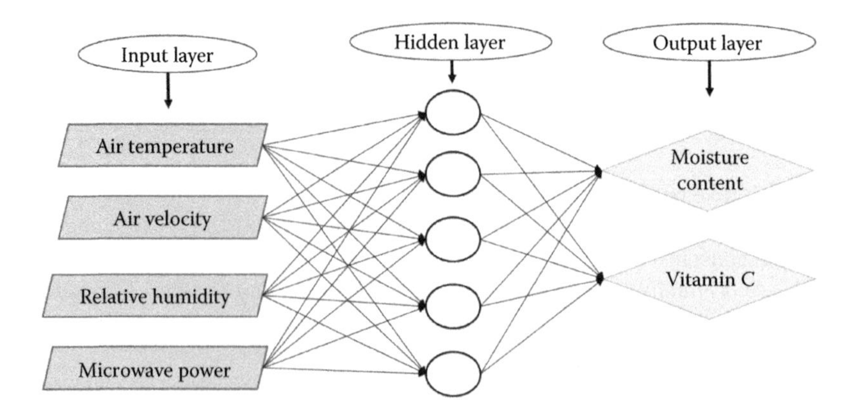

FIGURE 8.2 Schematic representation showing the structure of the artificial neural network models which have four input nodes, five nodes in hidden layer and two output nodes.

predicts the desired parameters based on the input layer and processing in the hidden layer (Chen et al. 2001). The network can be trained with the iterative process by adjusting the network connection weight in the hidden layer in response to process parameters for better prediction. A sufficiently trained network is expected to produce outputs that are satisfactorily close to actual outputs.

However, all the empirical models are only applicable in the range of experimental parameters; also they are not able to capture the physical during drying. In contrast to empirical model, the fundamental models can capture the physical during drying (Kumar et al. 2014a,b).

8.4 SINGLE-PHASE MODEL

The single-phase models consider only one phase in the food domain, that is, liquid water diffusion through solid media. These are often called diffusion-based models and are very popular because of their simplicity and good predictive capability (Arballo et al. 2010, Kumar et al. 2012b, Perussello et al. 2014). Therefore, the majority of the intermittent drying models are single phase.

The single-phase model considers only diffusive water transport as shown in Equation 8.1:

$$\frac{\partial c_w}{\partial t} + \nabla \cdot \left(-D_{eff} \nabla c_w\right) + u \cdot \nabla c_w = 0 \tag{8.1}$$

where

c_w is the moisture concentration (mol m^{-3})
D_{eff} is the effective diffusion coefficient (m^2 s^{-1})
u is the convective flow (m s^{-1})

The effective diffusion coefficient, D_{eff}, in those models has to be determined experimentally. Thus, this model can be called a semi-empirical model.

The heat transfer equation for single phase can be written as:

$$\rho c_p \frac{\partial T}{\partial t} + \rho c_p u \cdot \nabla T = \nabla \cdot \left(k \nabla T\right) + Q_e \tag{8.2}$$

where

T is the temperature at time t (K)
ρ is density (kg m^{-3})
c_p is the specific heat of material (J kg^{-1} K^{-1})
k is thermal conductivity (W m^{-2} K^{-1})
Q_e is internal heat source or sink (J)

The heat source term is zero for convection drying; however, when electromagnetic heating such as microwave is involved, then it should be added to the heat transfer equation. The heat and mass transfer coefficient calculation is crucial for the modelling of drying. The majority of the drying research calculated the heat and mass transfer coefficient from the well-established correlations of Nusselt, $Nu = h_T L/k$, and Sherwood number, $Sh = h_m L/D$, as described by Kumar et al. (2016b). These empirical relationships have been used in drying by many other researchers (Golestani et al. 2013, Karim and Hawlader 2005, Montanuci et al. 2014, Perussello et al. 2014).

A single-phase moisture transport model was developed by Nishiyama et al. (2006). They presented the tempering effects of intermittent drying of grain at different tempering temperatures using the sphere drying model. Simulation results agreed well with the experimental results. The range of standard error was between 0.1% and 0.47% db. Figure 8.3 shows the average moisture contents (AMC) and surface moisture contents (SMC) obtained by experiment and model. It can be seen that after the tempering period, the SMC increases, which helps to increase drying rate in the following drying time. The drying rate was improved after the tempering period, demonstrating the benefit of allowing moisture redistribution. Tempering temperature had a significant effect on required tempering time. Long-grain rough rice required more tempering than short-grain rough rice.

A model that considers the partial pressure of vapour was developed by Kowalski and Pawłowski (2010). In their model, partial pressure is considered as a function of drying air temperature and tempering temperature for continuous and intermittent drying, respectively. They used finite difference method to solve the coupled system

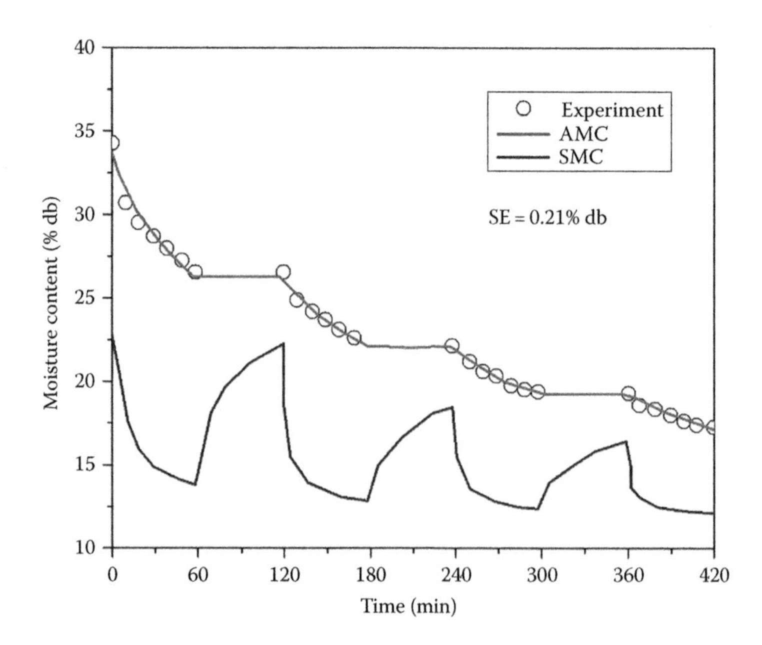

FIGURE 8.3 The drying curve for intermittent drying. AMC is the calculated average moisture content, SMC is the calculated surface moisture content. (From Nishiyama, Y. et al., *J. Food Eng.*, 76(3), 272, 2006.)

of equations. The variable air temperature in their intermittent model was incorporated by the following equation:

$$T_a(t) = \frac{T_{\max} + T_{\min}}{2} + \frac{T_{\max} - T_{\min}}{2} \cos\left[\frac{2\pi}{\lambda}(t - t_{ch})\right] \tag{8.3}$$

where

T_{min} and T_{max} denote minimum and maximum of the variable air temperatures
t is the period of temperature variation
t_{ch} denotes the time at which the change of drying parameters begins

In numerical calculation, t_{ch} was about 0.5–1.0 h earlier than critical time t_{cr}, which was read each time from the experimental drying curves.

Figure 8.4 illustrates the temperature of dried product obtained by numerical calculations when the air temperature varies periodically between 65°C and 43°C. The presented model showed very good agreement with experimental data.

Váquiro et al. (2009) modelled and optimized intermittent drying of mango taking enthalpy as an objective function, using COMSOL Multiphysics. They neglected the shrinkage and considered the effective diffusivity as a function of both moisture and temperature. They have measured heat capacity using differential scanning calorimetry (DSC) and used a predictive model similar to Floury et al. (2008) to

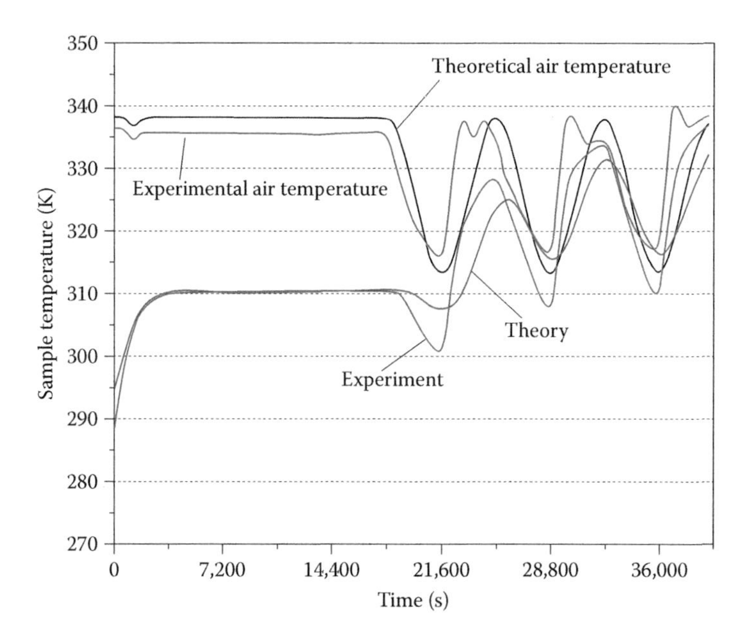

FIGURE 8.4 Periodic change in temperature during intermittent drying for periodically changed drying air temperature between 65°C and 43°C. (From Kowalski, S.J. and Pawlowski, A., *Chem. Eng. Technol.*, 34(7), 1123, 2011b.)

estimate the thermal conductivity of mango. The sorption isotherm was determined from the water activity and moisture content. Heat and mass transfer coefficient were calculated from Nusselt and Sherwood number, respectively, by using empirical correlations.

8.5 TWO-PHASE MODEL

Two-phase models consider two species, namely water vapour and liquid water, for mass transfer during the modelling process. The general governing equations for the double-phase model are presented in Equations 8.4 and 8.5.

$$\text{Liquid phase of water: } \frac{\partial c_l}{\partial t} + \nabla\cdot\left(-D_l \nabla c_l\right) + u_l \cdot \nabla c_l = -R \tag{8.4}$$

$$\text{Vapor phase of water: } \frac{\partial c_v}{\partial t} + \nabla\cdot\left(-D_v \nabla c_v\right) + u_v \cdot \nabla c_v = R \tag{8.5}$$

Here

- c_l and c_v are the concentration of liquid water and water vapour
- D_l and D_v are the diffusion coefficient for liquid water and water vapour, respectively
- u_l and u_v are the convective flow of liquid water and vapour

The R is the production or consumption of species (i.e. evaporation or condensation).

The heat transfer equation for double-phase model (shown in Equation 8.6) is similar to the single-phase model as shown in Equation 8.2; however, it will include a heat source/sink term, I, due to the latent heat of evaporation:

$$\rho c_p \frac{\partial T}{dt} + \rho c_p u \cdot \nabla T = \nabla\cdot\left(k \nabla T\right) + Q_e + I \tag{8.6}$$

The source or sink term can be calculated by

$$I = -h_{fg} R \tag{8.7}$$

Here, h_{fg} is latent heat of evaporation.

There are some models that consider double phase of intermittent drying developed by Ho et al. (2002), Chua et al. (2002a,b) and Chou et al. (2000). The drying rate curves for banana sample of intermittent drying obtained by step-up and step-down temperature variation are shown in Figure 8.5 (Chua et al. 2001). In both cases, the tempering period allowed the moisture migration and higher drying rate in subsequent drying period. The step-up profile showed two drying peaks due to slow initial drying with lower temperature, followed by 5°C increment in drying temperature.

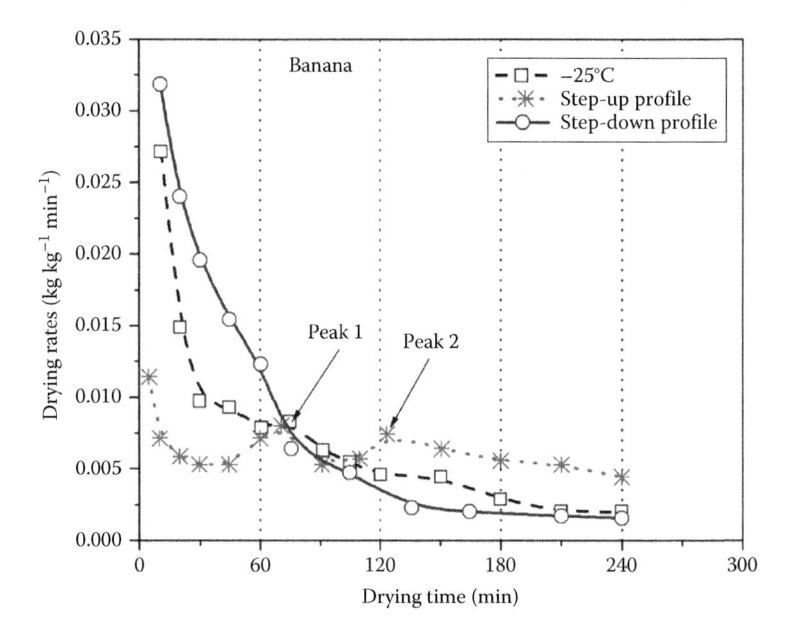

FIGURE 8.5 Drying rate of banana and guava samples versus drying time for step-wise profiles. (From Chua, K.J. et al., *Dry. Technol.*, 19(8), 1949, 2001.)

8.6 MULTIPHASE MODEL

Multiphase models consider more than two phases in food materials such as water, air, water vapour and solid matrix. Multiphase models are considered to be more comprehensive and realistic because they include diffusion and pressure-driven transport mechanisms for various phases. The single-phase and double-phase models only consider lumped diffusion and often ignore the pressure-driven flow or convective flow. The assumption of lumping all the water transport as diffusion cannot be justified in all situations, particularly during microwave drying where pressure-driven flow plays a significant role. In these cases, multiphase models are more applicable. The multiphase models are applied to a wide range of food processes such as frying (Bansal et al. 2014, Ni and Datta 1999), microwave heating (Chen et al. 2014, Rakesh et al. 2010), puffing (Rakesh and Datta 2013), baking (Zhang et al. 2005) and meat cooking (Dhall and Datta 2011). However, application of these models in the intermittent convective drying of food materials is limited. Recently, Kumar et al. (2016a) applied multiphase model for intermittent microwave convective drying (IMCD). The details of multiphase modelling applied to IMCD can be found at Kumar et al. (2016a).

8.7 MODELLING INTERMITTENT DRYING WITH EXTERNAL HEAT SOURCE

This section presents the modelling of intermittent drying assisted by an external heat source such as microwave and infrared. The governing equations are the same for these models except an additional heat generation due to the external heat sources. Therefore, this section discusses the calculation of heat generation due to these external sources.

8.7.1 Microwave Heat Source

The microwave is the most common energy source in intermittent drying application. The combination of intermittent microwave with convective drying was modelled both in single phase (Kumar et al. 2015) and multiphase (Kumar et al. 2016a). Lambert's law for microwave energy distribution in food products during drying is the most common for modelling intermittent microwave-assisted drying (Abbasi Souraki and Mowla 2008, Arballo et al. 2012, Hemis et al. 2012, Mihoubi and Bellagi 2009, Salagnac et al. 2004). This law has been used to calculate microwave energy absorption inside the food samples. It considers exponential attenuation of microwave absorption within the product, as expressed by the following equation:

$$P_{mic} = P_0 \exp^{-2\alpha(h-z)} \tag{8.8}$$

Here

P_0 is the incident power at the surface (W)
α is the attenuation constant (m^{-1})
h is the thickness of the sample (m)
$(h - z)$ represents the distance from sample surface (m)

The measurement of P_0 via experiments was discussed by Kumar et al. (2015).

The attenuation constant, α, is given by

$$\alpha = \frac{2\pi}{\lambda}\sqrt{\varepsilon'\left[\frac{\sqrt{\left(1+\left(\varepsilon''/\varepsilon'\right)^2\right)}-1}{2}\right]}, \tag{8.9}$$

where

λ is the wavelength of the microwave in free space (λ = 12.24 cm at 2450 MHz and air temperature 202°C)
ε' and ε'' are the dielectric constant and dielectric loss, respectively

However, Maxwell's equation provided a more accurate solution for microwave propagation in samples (Chandrasekaran et al. 2012), which can provide three-dimensional distribution of electromagnetic field. Maxwell's equations that govern the propagation and microwave heating of material are given as follows:

$$\nabla \times E = \frac{\partial B}{\partial t} \tag{8.10}$$

$$\nabla \times H = \frac{\partial D}{\partial t} + J_c \tag{8.11}$$

$$\nabla \cdot B = 0 \tag{8.12}$$

$$\nabla \cdot D = \rho \tag{8.13}$$

where

E is the electric field distribution (V m^{-1})
D is the electric flux density distribution (C m^{-2})
H is the magnetic field distribution (A m^{-1})
B is the magnetic flux density distribution (W m^{-2})
J_c is the current density (A m^{-2})
ρ is the volume charge density (C m^{-3})

Alternatively, for a known frequency, frequency domain harmonic Maxwell's equations can be solved to find electric field:

$$\nabla \times \left(\frac{1}{\mu'} \nabla \times \vec{E} \right) - \frac{\omega^2}{c} (\varepsilon' - i\varepsilon'') \vec{E} = 0 \tag{8.14}$$

Here

$\vec{E}$ is the electric field strength (V m^{-1})
f is the microwave frequency (Hz)
c is the speed of light (m s^{-1})
ε', ε'', μ are the dielectric constant, dielectric loss factor and electromagnetic permeability of the material, respectively

Then, the volumetric heat generation due to microwave can be calculated by Equation 8.15, which needs to be incorporated into the energy equation while modelling intermittent microwave-assisted drying:

$$Q_m = \pi f \varepsilon_0 \varepsilon'' \left| \vec{E} \right|^2 \tag{8.15}$$

The dielectric constant and dielectric loss of the material are the most important parameters in microwave heating and drying applications because these properties define how materials interact with electromagnetic energy (Sosa-Morales et al. 2010). The evaluation of dielectric properties is critical in modelling and process development (Ikediala et al. 2000). Dielectric properties of materials define how much microwave energy will be converted to heat (Chandrasekaran et al. 2013). Recently, several papers presented multiphase models for intermittent microwave–convective drying (Kumar et al. 2015, 2016a). They have validated their model with experimental data and found a quite satisfactory match with the experimental result (with $R^2 = 0.99359$). Another advantage of the multiphase model is that the vapour and water flux can separately be quantified using the model. A more detailed description of water and vapour flux during intermittent microwave convective drying was given by Kumar et al. (2016b). They have found that the temperature on the surface and centre of the sample fluctuated due to intermittent microwave heat source. Thus, the model can be used to provide the effect of intermittency on the temperature of the product, which could be used to improve quality of the product.

Yang and Gunasekaran developed an implicit finite-difference model considering cylindrical geometry (Gunasekaran and Yang 2007, Yang and Gunasekaran 2001). They used Lambert's law for microwave power calculation. For a cylindrical sample in a uniform plane wave field, the total power (P_{total}) and the incident power (P_o) are expressed as the volume integral of the $P(x)$ function as shown in Equations 8.16 and 8.17.

$$P_{total} = \int_V P(x)dV = \int_0^Z \int_0^{2\pi} \int_0^R P_o e^{-2\beta x}\, dx\,du\,dz \qquad (8.16)$$

where, Z and R are the height and radius of the test sample. Integrating between limits and solving for P_o gives

$$P_o = \frac{\beta \rho V C_p \Delta T_{av}}{\pi Z t\left(1 - e^{2\beta R}\right)} \qquad (8.17)$$

8.7.2 Infrared (IR) Heat Source

Limited work has been done on modelling of intermittent infrared (IR) drying. Zhu and Pan (2009) developed a simultaneous infrared dry-blanching and dehydration process for producing high-quality blanched and partially dehydrated apple slices with three different thicknesses, 5, 9 and 13 mm, which were heated using infrared for up to 10 min at 4000 W m^{-2} IR intensity. In their model development, the apple slices were assumed as one-dimensional with heat and mass transfer only occurring in one direction. The governing equation for one-dimensional heat transfer from a cylinder in the z-direction is given in Equation 8.18:

$$\rho(T)\cdot c_p(T)\cdot\frac{\partial T}{\partial t} = \frac{\partial}{\partial z}\left[k(T)\cdot\frac{\partial T}{\partial t}\right] + q_{incident}e^{\left[-S\cdot(H-z)\right]} + \lambda(T)\cdot\rho_s(M)\cdot\frac{\partial M}{\partial t} \qquad (8.18)$$

Here

- k is the temperature-dependent thermal conductivity
- c_p is the specific heat capacity
- ρ is density
- λ is the latent heat of vaporization

The coefficient of dissipation of radiation (S) for far-IR radiation was determined from the literature as 1200 (m^{-1}) (Ginzburg 1969), where a smaller dissipation coefficient means a slower decay of the IR radiation and thus a larger penetration depth.

Shrinkage was considered only in the z-direction. The vapour transfer or vaporization was the dominant transfer mechanism of water from the interior to the surface of the apple slice. The predicted temperature and moisture profiles and inactivation rate of enzymes were found to be in good agreement with the experimental data. This indicated that the process of simultaneous dry blanching and dehydration of apple slices under IR heating can be predicted with the developed models.

Infrared induce a superficial heating for opaque material, then the heat transmit by conduction (Salagnac et al. 2004). Salagnac et al. (2004) calculated the absorbed heat flux due to IR by multiplying absorption coefficient with incident power.

8.8 REACTION ENGINEERING MODEL

8.8.1 Review of Reaction Engineering Approach

The general reaction engineering approach (REA) is an application of chemical reaction engineering principles to model drying kinetics, which was first reported in 1996–1997 (Chen and Xie 1997). Generally, the drying rate of a material can be expressed as:

$$m_s \frac{d\bar{X}}{dt} = -h_m A\left(\rho_{v,s} - \rho_{v,b}\right) \tag{8.19}$$

Equation 8.19 is a basic mass transfer equation where $\bar{X}$ is the average moisture content, m_s is the mass of dried sample (kg), A is the surface area (m^2), h_m is the mass transfer coefficient (m s^{-1}), $\rho_{v,s}$ is the surface water vapour concentration (kg m^{-3}) and $\rho_{v,b}$ is the water vapour concentration in drying medium (kg m^{-3}).

The mass transfer coefficient (h_m) is determined based on the established Sherwood number correlations for the geometry and flow condition of concern or established experimentally for the specific drying conditions involved. The surface vapour concentration ($\rho_{v,s}$) is then scaled against saturated vapour concentration ($\rho_{v,sat}$) using the following equation:

$$\rho_{v,s} = \exp\left(\frac{-\Delta E_v}{RT_s}\right)\rho_{v,sat}\left(T_s\right) \tag{8.20}$$

where ΔE_v represents the additional difficulty to remove moisture from the material beyond the free water effect (J mol^{-1}). This ΔE_v is moisture content (X) dependent. T_s is the surface temperature of the material being dried (K) and $\rho_{v,sat}$ is the saturated water vapour concentration (kg m^{-3}).

The mass balance equation (Equation 8.19) can then be expressed as (Chen and Putranto 2013):

$$m_s \frac{d\bar{X}}{dt} = -h_m A\left[\exp\left(\frac{-\Delta E_v}{RT_s}\right)\rho_{v,sat}\left(T_s\right) - \rho_{v,b}\right] \tag{8.21}$$

From Equation 8.21 it can be observed that the lumped REA is expressed in the first-order ordinary differential equation with respect to time.

The activation energy (ΔE_v) is determined experimentally by placing the parameters required for Equation 8.21 in its rearranged form:

$$\Delta E_v = -RT_s \ln \left[\frac{-m_s \dfrac{d\bar{X}}{dt} \dfrac{1}{h_m A} + \rho_{v,b}}{\rho_{v,sat}} \right] \tag{8.22}$$

where dX/dt, average moisture content, surface area and temperature is experimentally determined. The dependence of activation energy on moisture content on a dry basis (X) can be normalized as:

$$\frac{\Delta E_v}{\Delta E_{v,b}} = f\left(\bar{X} - \bar{X}_b\right) \tag{8.23}$$

where

f is a function of water content difference

$\Delta E_{v,b}$ is the 'equilibrium' activation energy representing the maximum ΔE_v under the relative humidity and temperature of the drying air

$$\Delta E_{v,b} = -RT_b \ln\left(RH_b\right) \tag{8.24}$$

where

RH_b is the relative humidity of drying air

T_b is the drying air temperature (K)

In order to generate the relative activation energy ($\Delta E_v/\Delta E_{v,b}$) shown by Equation 8.23, the activation energy (ΔE_v) is evaluated by Equation 8.22 from one accurate drying experiment. The activation energy is divided by the equilibrium activation energy ($\Delta E_{v,b}$) indicated by Equation 8.24 to yield the relative activation energy during drying. For similar drying condition and initial water content, it is possible to obtain the necessary REA parameters (apart from the equilibrium isotherm), expressed in the relative activation energy ($\Delta E_v/\Delta E_{v,b}$) as indicated in Equation 8.23, in one accurate drying experiment. This gives a tremendous advantage in applying the REA in the industrial setting. The relative activation energy ($\Delta E_v/\Delta E_{v,b}$) generated can then be implemented to other drying conditions provided same material and similar initial moisture content since the relative activation energy would collapse to the similar profiles (Putranto and Chen 2014a,b, 2015, 2016a,b, Putranto et al. 2015).

The reaction engineering approach in its lumped format is further called the lumped reaction engineering approach (L-REA). It is used to project the global drying kinetics and average moisture content (Putranto and Chen 2014a,b). The reaction engineering approach is also applicable to model the local evaporation/condensation rate at micro-level by combining with a set of equations of conservation of heat and mass transfer to yield the spatial reaction engineering approach (S-REA) (Putranto and Chen 2015, 2016a,b, Putranto et al. 2015). Discussions on the L-REA are presented in Section 8.8.2, while those on the S-REA are shown in Section 8.8.3.

8.8.2 Mathematical Modelling for Intermittent Drying Using the Lumped Reaction Engineering Approach (L-REA)

As an example of the application of the REA for time-varying drying, the REA was implemented to describe the intermittent drying of mango tissues (Váquiro et al. 2009), whose experimental details are reviewed briefly here for better understanding of the modelling framework. The samples of mango tissues were cubes with initial side-length of 2.5 cm. The initial moisture content and temperature were 9.3 kg kg^{-1} and 10.8°C, respectively. The laboratory drier was implemented for drying of the samples. For monitoring the drying kinetics, the weight of the sample and the centre temperature were recorded. The intermittency of temperature was created by heating and resting period listed in Table 8.2. The resting period was allowed by holding the samples at ambient temperature of 27°C ± 1.6°C and relative humidity of 60%.

By implementing the reaction engineering approach (REA), the mass balance can be expressed as (Chen and Putranto 2013):

$$m_s \frac{dX}{dt} = -h_m A \left[\exp\left(\frac{-\Delta E_v}{RT_s} \right) \rho_{v,sat}\left(T_s\right) - \rho_{v,b} \right] \tag{8.25}$$

The relative activation energy derived from continuous convective drying of mango tissues at drying air temperature of 55°C can be written as (Putranto et al. 2011):

$$\frac{\Delta E_v}{\Delta E_{v,b}} = -9.92 \times 10^{-4} \left(X - X_b\right)^3 + 9.74 \times 10^{-3} \left(X - X_b\right)^2 - 0.101\left(X - X_b\right) + 1.053 \tag{8.26}$$

TABLE 8.2
Scheme of Intermittent Drying of Mango Tissues

Drying Air Temperature (°C)	Period of First Heating (s)	Period of Resting (at 27°C ± 1.6) (s)	Period of Second Heating (s)
45	31,440	14,400	19,800
55	20,760	14,400	6,240

Source: Váquiro, H.A. et al., *Chem. Eng. Res. Des.*, 87(7), 885, 2009.

The heat balance can be written as:

$$\frac{d\left(mC_pT_{avg}\right)}{dt} \approx hA\left(T_b - T_s\right) + m_s\frac{dX}{dt}\Delta H_v \tag{8.27}$$

For modelling the intermittent drying, the relative activation energy shown in Equation 8.26 was implemented and combined with the equilibrium activation energy evaluated according to corresponding drying air temperature and humidity. The heat balance also implements these conditions in each drying period. In order to yield the profiles of moisture content and temperature during drying, Equations 8.25 and 8.27 are solved simultaneously in conjunction with the relative activation energy shown in Equation 8.26.

8.8.2.1 Results of Modelling Using the Lumped Reaction Engineering Approach (L-REA)

The results of modelling of intermittent drying of mango tissues are shown in Figures 8.6 and 8.7. Figure 8.6 shows that at drying air temperature of 45°C, the results of modelling using the REA matched well with the experimental data (R^2 of 0.997). A good agreement between the predicted and experimental data is observed. The REA described well the profiles of moisture content during drying at drying air temperature of 45°C. For the drying air temperature of 55°C, a close agreement

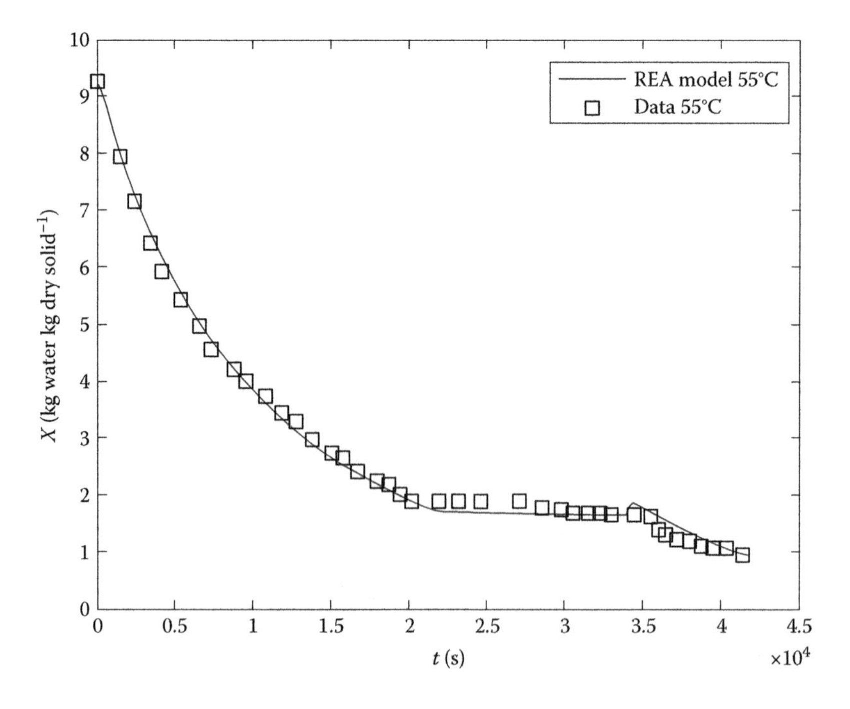

FIGURE 8.6 Profiles of moisture content during intermittent drying of mango tissues at drying air temperature of 55°C modelled using the lumped reaction engineering approach (L-REA).

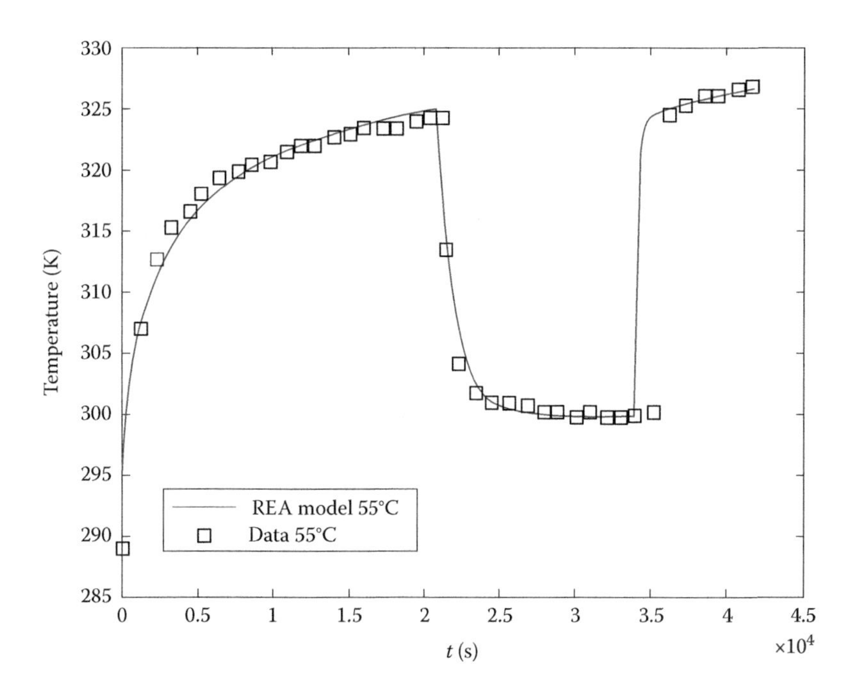

FIGURE 8.7 Profiles of temperature during intermittent drying of mango tissues at drying air temperature of 55°C modelled using the lumped reaction engineering approach (L-REA).

between the predicted and experimental data is indicated by R^2 of 0.998. The REA is accurate to represent the profiles of moisture content during intermittent drying of mango tissues. The moisture reduction during resting period is also described accurately by the REA.

The temperature profiles resulting from the REA are shown in Figure 8.7. The REA estimates well the temperature profiles during intermittent drying at drying air temperature of 45°C. The temperature profiles during first heating, resting and second heating period were well represented by the REA. In conjunction with the close agreement of profiles of moisture content as shown in Figure 8.6, the REA described well the profiles of temperature during intermittent drying at drying air temperature of 55°C. The applicability of the REA in yielding the profiles of temperature is indicated by R^2 higher than 0.998.

The applicability of the REA in modelling the intermittent drying could be widely acceptable because of the flexibility of the relative activation energy. The combination of the relative activation energy and equilibrium activation energy results in unique relationships of the activation energy. The effects of changing environmental conditions may be captured well by the equilibrium activation energy and reflected on the material structure of mango tissues by the relative activation energy. Unlike the characteristic drying rate approach (CDRC), which was shown as not able to describe time-varying drying condition (Baini and Langrish 2007), the REA is able to describe well the cyclic drying.

It is interesting that the relative activation energy established from continuous convective drying was able to describe the intermittent drying. This further highlights the simplicity of the REA in generating the drying parameters. Previously, it was emphasized that only one accurate drying run was necessary to generate the drying parameters since different drying conditions yielded similar profiles of activation energy (Chen and Putranto 2013). Here, the generic application of the relative activation energy is widened. This simplicity is also combined with the ease in mathematical modelling. The REA is represented in the format of ordinary differential equation which is accurate to yield the average moisture content and temperature, favourable for quick-decision making in industry.

8.8.3 Mathematical Modelling for Intermittent Drying Using the Spatial Reaction Engineering Approach (S-REA)

For the intermittent drying of mango tissues presented in Section 8.8.2, the spatial reaction engineering (S-REA) is set up. The S-REA is essentially a set of combination of equations of conservation of heat and mass transfer with the REA in which the REA is used to describe the local evaporation/condensation rates. The REA is a non-equilibrium multiphase model to describe the local moisture content, concentration of water vapour and temperature. In S-REA, the concentration of water vapour in void spaces of porous materials and the moisture content in the solid matrix are considered to be not in equilibrium. The REA is employed to represent the evaporation/condensation rate between the water in the solid matrix and in the vacant pores inside the materials.

For the intermittent drying, the mass balance of water in the liquid phase (liquid water) is written as:

$$\frac{\partial(C_s X)}{\partial t}=\frac{\partial}{\partial x}\left(D_w\frac{\partial(C_s X)}{\partial x}\right)+\frac{\partial}{\partial y}\left(D_w\frac{\partial(C_s X)}{\partial y}\right)+\frac{\partial}{\partial z}\left(D_w\frac{\partial(C_s X)}{\partial z}\right)-\dot{I} \tag{8.28}$$

where

X is the concentration of liquid water (kg H_2O kg dry solids^{-1})
D_w is the capillary water diffusivity (m^2 s^{-1})
C_s is the solids concentration (kg dry solids m^{-3}) which can change if the structure is shrinking
$\dot{I}$ is the evaporation or condensation rate (kg H_2O m^{-3} s^{-1})
$\dot{I}$ is > 0 when evaporation occurs locally

The mass balance of water vapour is expressed as:

$$\frac{\partial C_v}{\partial t}=\frac{\partial}{\partial x}\left(D_v\frac{\partial C_v}{\partial x}\right)+\frac{\partial}{\partial y}\left(D_v\frac{\partial C_v}{\partial y}\right)+\frac{\partial}{\partial z}\left(D_v\frac{\partial C_v}{\partial z}\right)+\dot{I} \tag{8.29}$$

where

D_v is the effective water vapour diffusivity (m^2 s^{-1})
C_v is the concentration of water vapour inside the pore (kg m^{-3})

The energy balance can be expressed as:

$$\rho C_p \frac{\partial T}{\partial t} = \frac{\partial}{\partial x}\left(k \frac{\partial T}{\partial x}\right) + \frac{\partial}{\partial y}\left(k \frac{\partial T}{\partial y}\right) + \frac{\partial}{\partial z}\left(k \frac{\partial T}{\partial z}\right) + \dot{I}\Delta H_v \tag{8.30}$$

where

T is the sample temperature (K)
k is the thermal conductivity (W m^{-1} K^{-1})
ΔH_v is the vaporization heat of water (J kg^{-1})
ρ is the sample density (kg m^{-3})
C_p is the sample heat capacity (J kg^{-1} K^{-1})

The initial and boundary conditions for Equations 8.28 through 8.30 are:

$$t = 0, \quad X = X_o, \quad C_v = C_{vo}, \quad T = T_o \tag{8.31}$$

$$x = 0, \quad \frac{dX}{dx} = 0, \quad \frac{dC_v}{dx} = 0, \quad \frac{dT}{dx} = 0 \tag{8.32}$$

For convective boundary for liquid water transfer at $x = L$,

$$-C_s D_w \frac{dX}{dx} = h_m \varepsilon_w \left(\frac{C_{v,s}}{\varepsilon} - \rho_{v,b}\right) \tag{8.33}$$

For convective boundary for water vapour transfer,

$$-D_v \frac{dC_v}{dx} = h_m \varepsilon_v \left(\frac{C_{v,s}}{\varepsilon} - \rho_{v,b}\right) \tag{8.34}$$

For convective boundary for heat transfer,

$$k \frac{dT}{dx} = h\left(T_b - T\right) \tag{8.35}$$

The local evaporation rate within the solid structure ($\dot{I}$), shown in Equations 8.28 through 8.30, is described as:

$$\dot{I} = h_{m_{in}} A_{in} \left(C_{v,s} - C_v\right) \tag{8.36}$$

where

h_{min} is the internal mass transfer coefficient (m s^{-1})
A_{in} is the total internal surface area per unit volume available for phase change (m^2 m^{-3})

By implementing the REA, internal-surface water vapour concentration can be written as:

$$C_{v,s} = \exp\left(\frac{-\Delta E_v}{RT}\right) C_{v,sat} \tag{8.37}$$

where

$C_{v,s}$ is the internal-solid-surface water vapour concentration (kg m^{-3})
$C_{v,sat}$ is the internal-saturated water vapour concentration (kg m^{-3})

The relative activation energy shown in Equation 8.26 is implemented here as the 'local' relative activation energy by replacing the average moisture content in Equation 8.26 with the local moisture content. For incorporating the intermittency during drying, the equilibrium activation energy is evaluated according to the corresponding humidity and drying air temperature in each drying period. Similarly, the boundary conditions (Equations 8.33 through 8.35) apply the external drying conditions in each drying period. In order to yield the profiles of moisture content and the concentration of water vapour and temperature, Equations 8.28 through 8.30 are solved simultaneously in conjunction with the initial and boundary conditions shown in Equations 8.31 through 8.35.

8.8.3.1 Results of Modelling Using the Spatial Reaction Engineering Approach (S-REA)

For modelling using the S-REA, the relative activation energy used to describe the overall drying kinetics (shown in Equation 8.31) is still applicable. However, the average moisture content is substituted by the local moisture content. Therefore, the 'global' relative activation energy becomes the 'local' activation energy. This way allows the REA to describe the local evaporation/condensation rate at the micro- or structural level. Previously, it has been shown that the REA is applicable to model the drying kinetics of various sizes of samples (Putranto and Chen 2014b). The 'local' relative activation energy is essentially extrapolation of the relative activation energy to describe the evaporation/condensation rate at structural level (Putranto et al. 2015).

Figure 8.8 shows the spatial profiles of moisture content during intermittent drying at drying air temperature of 45°C. As drying progresses, the moisture content of the samples reduces. During drying, the moisture content at the outer part of the samples was lower than that at the inner part, which indicates that moisture migrates outwards during drying. This was reasonable since the surface water vapour concentration is lower than the water vapour concentration inside the void spaces at the core of samples. Initially, the gradient of moisture content was relatively large but it decreased during drying, in agreement with the reduction of moisture content. During resting period, the gradient of moisture content was relatively small, which may also be attributed to the relatively low sample temperature which lowered the capillary diffusivity. Although the sample temperature was relatively high during the second heating period, the gradient was basically low due to the depletion of moisture content inside the samples. At the end of second heating period, there was no

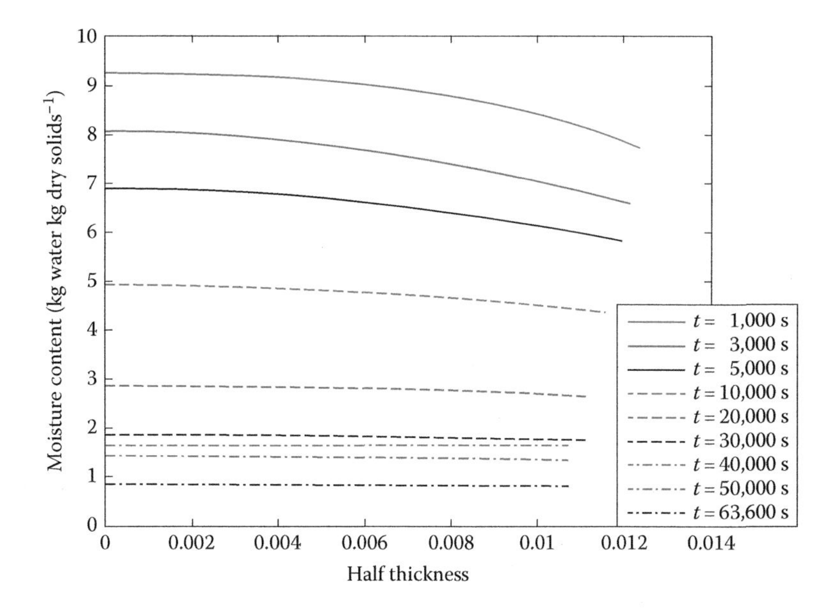

FIGURE 8.8 Spatiotemporal profiles of moisture content during intermittent drying of mango tissues at drying air temperature of 45°C modelled using the spatial reaction engineering approach (S-REA).

noticeable difference in the moisture content of the samples, which indicated that the equilibrium condition was approached.

The spatial profiles of concentration of water vapour during intermittent drying are shown in Figure 8.9. In agreement with the reduction of moisture content during drying, an increase of concentration of water vapour during drying was observed. The concentration of water vapour at the outer part was higher than that at the core of the samples, which could be because of the higher temperature at the outer part. This was also in line with the lower moisture content at the outer part. Initially, the gradient is relatively large but it reduced as drying progressed. Due to the relatively low temperature during resting period, the gradient was relatively small. The low gradient was also observed during the second heating period, which was in agreement with the spatial profiles of moisture content shown in Figure 8.8. At the end of drying, the relatively uniform concentration of water vapour was observed, which indicated the attainment of equilibrium.

It has been shown here that the S-REA is applicable to model the intermittent drying. While the L-REA is effective in describing the overall drying kinetics, the interacting effects of moisture content, temperature and concentration of water vapour as affected by material structure are well represented by the S-REA. It seems that the S-REA is the first mathematical model that describes the spatial profiles of concentration of water vapour during intermittent drying. This capability is essentially due to the accuracy of the REA to serve as the local evaporation/condensation rate. The REA has successfully linked the mass balance of liquid water and water vapour as shown in Equations 8.36 and 8.37. This further indicates that the 'local' relative

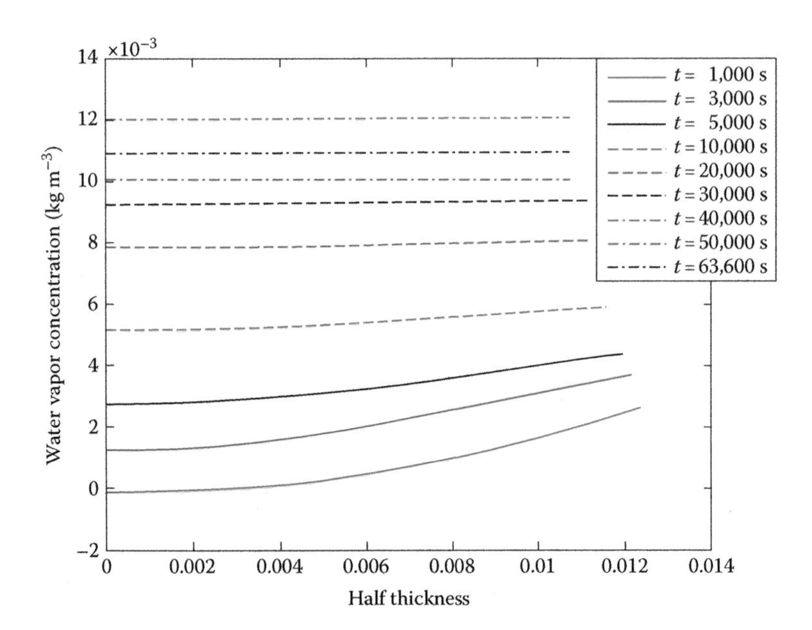

FIGURE 8.9 Spatiotemporal profiles of concentration of water vapour during intermittent drying of mango tissues at drying air temperature of 45°C modelled using the spatial reaction engineering approach (S-REA).

activation energy is applicable to model the local evaporation/condensation rate not only during continuous convective drying but also during intermittent drying. It is accurate to model these rates during the heating and resting period. This applicability may be because of the flexibility of the relative activation energy. Combined with the equilibrium activation energy, it yields unique relationship of activation energy that describes the structural changes of materials at micro-level as affected by the external drying air conditions. The S-REA is readily implemented to describe the intermittent drying under time-varying humidity, infrared heating and microwave intensity.

8.9 SOLUTION TECHNIQUE

This section describes different solution techniques for solving the governing equations of the model presented earlier. The choice of solution technique depends on computational resources, expected accuracy, flexibility and experience of researchers.

8.9.1 Coding

A coding system is required to solve the governing equations along with their relevant boundary and initial conditions. Although the coding is initially time-consuming, it provides huge advantage during its application because the code can be modified easily to be implemented for various applications easily. Coding to solve the governing equation in food processing models has been used widely (Boukadida and Ben Nasrallah 1995, Farkas et al. 1996, Mercier et al. 2014, Nasrallah and Perre 1988, Ni and Datta 1999, Yamsaengsung and Moreira 2002, Zhang et al. 2005). Various programming languages such as FORTRAN, C++, VISUAL BASIC, MATLAB®, and JAVA can be used in coding system. One of the major disadvantages of coding is that it is difficult for a researcher to understand and use a code that has been developed by another researcher.

8.9.2 Specialized Computer Program

There are several specialized programs for simulating transport in porous media. PORE-FLOW is a computer program for CFD application developed at the University of Wisconsin-Milwaukee using the FE/CV filling algorithm (PORE-FLOW 2016). This software has been used for solving a model of the capillary suction-driven liquid flow in swelling porous media behind a clear liquid-front (Masoodi et al. 2012). TOUGH2 is the basic simulator for non-isothermal multiphase flow in fractured porous media (LBNL 2012). The TOUGH2 simulator was developed for problems involving actively heat-driven flow that occur during the drying of the porous material. Solving the pressure gradient term using the conservation equation can be easily solved by using TOUGH2 coding software (Pruess 2004). Finally, LSODES is an excellent coding software that is based on FORTRAN. It is easily implementable and less time-consuming and can be successfully used for drying (Khan et al. 2016a, Stanish et al. 1986).

8.9.3 Commercial CFD-Based Software

There are many CFD-based commercial software that can be used for solving heat and mass transfer process in porous media. Among them COMSOL Multiphysics (Comsol Inc, www.comsol.com), FLUENT (Fluent Inc, www.ansys.com), ANSYS (Ansys Inc, www.ansys.com), CFX(Ansys Inc, www.ansys.com), STAR-CD (CD Adapco Group, www.cd-adapco.com), PHOENICS(CHAM Ltd, www.cham.co.uk) and ADINA (ADINA Inc, www.adina.com) are popular. COMSOL Multiphysics is a multi-disciplinary, advanced simulation software for modelling and simulating physics-based problems. This software is very user-friendly and can easily account for coupled or multiphysics phenomena involving various physics such as electrical, mechanical, fluid flow and chemical applications. As COMSOL can easily implement material properties and input parameters as a function of independent variables, it is very useful in drying application because the material properties change with temperature and moisture content during drying (Khan et al. 2016b). Recently, many researchers have used this software to efficiently implement their model in drying (Curcio 2010, Dhall et al. 2012, Khan et al. 2016b, Kumar 2015, Kumar et al. 2012a, 2015, 2016a,c, Perussello et al. 2014, Zhang and Kong 2012). Kumar et al. (2015) used COMSOL Multiphysics for simulating convection drying and intermittent microwave convective drying (IMCD). Their solution procedure using COMSOL Multiphysics is shown in Figure 8.10.

ANSYS FLUENT is a multi-purpose finite-volume-based menu-structured CFD software which contains physical models for a wide range of applications including

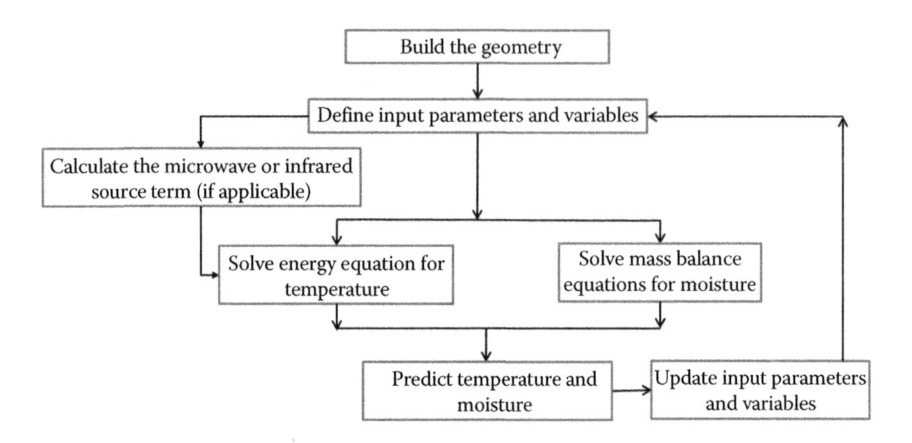

FIGURE 8.10 Solution procedure using commercial software.

turbulent flows, heat transfer, reacting flows, chemical mixing and multiphase flows. The fluent software also has been used for designing and predicting heat and moisture transfer during food drying (Darabi et al. 2015, Erriguible et al. 2007, Wang et al. 2008).

8.10 CONCLUSION

Intermittent drying of food material offers lower energy consumption and better product quality without increasing the capital and operating cost. The tempering period during intermittent drying helps the moisture to diffuse toward the surface and thus increases the drying rate in the subsequent drying period. Mathematical modelling can help to determine the suitable operating conditions of intermittent drying such as optimum tempering period, temperature variation, and so on, in order to reduce drying time and energy consumption while improving product quality. This chapter presented the approaches for modelling intermittent drying which includes empirical models, single-phase, two-phase, and multiphase models and reaction engineering approaches (REA). Although empirical and single-phase models are widely used, the multiphase models are more fundamental and can capture the real physics more accurately. Both lumped reaction engineering approach (L-REA) and spatial reaction engineering approach (S-REA) for modelling time-varying drying condition were presented along with their important results. The governing equations of the model can be solved by coding, special coding software and commercial software which were discussed.

REFERENCES

Abbasi Souraki, B. and D. Mowla. 2008. Experimental and theoretical investigation of drying behaviour of garlic in an inert medium fluidized bed assisted by microwave. *Journal of Food Engineering* 88(4):438–449.

Arballo, J. R., L. A. Campañone, and R. H. Mascheroni. 2010. Modeling of microwave drying of fruits. *Drying Technology* 28(10):1178–1184.

Arballo, J. R., L. A. Campanone, and R. H. Mascheroni. 2012. Modeling of microwave drying of fruits. Part II: Effect of osmotic pretreatment on the microwave dehydration process. *Drying Technology* 30(4):404–415.

Baini, R. and T. A. G. Langrish. 2007. Choosing an appropriate drying model for intermittent and continuous drying of bananas. *Journal of Food Engineering* 79(1):330–343.

Bansal, H. S., P. S. Takhar, and J. Maneerote. 2014. Modeling multiscale transport mechanisms, phase changes and thermomechanics during frying. *Food Research International* 62:709–717.

Boukadida, N. and S. B. Nasrallah. 1995. Two dimensional heat and mass transfer during convective drying of porous media. *Drying Technology* 13(3):661–694.

Chandrasekaran, S., S. Ramanathan, and T. Basak. 2012. Microwave material processing—A review. *AIChE Journal* 58(2):330–363.

Chandrasekaran, S., S. Ramanathan, and T. Basak. 2013. Microwave food processing—A review. *Food Research International* 52(1):243–261.

Chen, C. R., H. S. Ramaswamy, and I. Alli. 2001. Prediction of quality changes during osmo-convective drying of blueberries using neural network models for process optimization. *Drying Technology* 19(3–4):507–523.

Chen, J., K. Pitchai, S. Birla, M. Negahban, D. Jones, and J. Subbiah. 2014. Heat and mass transport during microwave heating of mashed potato in domestic oven—Model development, validation, and sensitivity analysis. *Journal of Food Science* 79(10):E1991–E2004.

Chen, X. D. and A. Putranto. 2013. *Modeling Drying Processes: A Reaction Engineering Approach*. Cambridge University Press, Cambridge, UK.

Chen, X. D. and G. Z. Xie. 1997. Fingerprints of the drying behaviour of particulate or thin layer food materials established using a reaction engineering model. *Food and Bioproducts Processing* 75(4):213–222.

Chin, S. K. and C. L. Law. 2010. Product quality and drying characteristics of intermittent heat pump drying of *Ganoderma tsugae* Murrill. *Drying Technology* 28(12):1457–1465.

Chou, S. K., K. J. Chua, A. S. Mujumdar, M. N. A. Hawlader, and J. C. Ho. 2000. On the intermittent drying of an agricultural product. *Food and Bioproducts Processing* 78(4):193–203.

Chua, K. J., S. K. Chou, M. N. A. Hawlader, A. S. Mujumdar, and J. C. Ho. 2002a. Modeling the moisture and temperature distribution within an agricultural product undergoing time-varying drying schemes. *Biosystems Engineering* 81(1):99–111.

Chua, K. J., S. K. Chou, M. N. A. Hawlader, A. S. Mujumdar, and J. C. Ho. 2002b. PH—Postharvest technology: Modeling the moisture and temperature distribution within an agricultural product undergoing time-varying drying schemes. *Biosystems Engineering* 81(1):99–111.

Chua, K. J., M. N. A. Hawlader, S. K. Chou, and J. C. Ho. 2002c. On the study of time-varying temperature drying—Effect on drying kinetics and product quality. *Drying Technology* 20(8):1559–1577.

Chua, K. J., A. S. Mujumdar, and S. K. Chou. 2003. Intermittent drying of bioproducts—An overview. *Bioresource Technology* 90(3):285–295.

Chua, K. J., A. S. Mujumdar, M. N. A. Hawlader, S. K. Chou, and J. C. Ho. 2001. Convective drying of agricultural products: Effect of continuous and stepwise change in drying air temperature. *Drying Technology* 19(8):1949–1960.

Curcio, S. 2010. A multiphase model to analyze transport phenomena in food drying processes. *Drying Technology* 28(6):773–785.

Darabi, H., A. Zomorodian, M. H. Akbari, and A. N. Lorestani. 2015. Design a cabinet dryer with two geometric configurations using CFD. *Journal of Food Science and Technology* 52(1):359–366.

da Silva, W. P., A. F. Rodrigues, C. M. D. P. S. Silva, D. S. de Castro, and J. P. Gomes. 2015. Comparison between continuous and intermittent drying of whole bananas using empirical and diffusion models to describe the processes. *Journal of Food Engineering* 166:230–236.

Dhall, A. and A. K. Datta. 2011. Transport in deformable food materials: A poromechanics approach. *Chemical Engineering Science* 66(24):6482–6497.

Dhall, A., G. Squier, M. Geremew, W. A. Wood, J. George, and A. K. Datta. 2012. Modeling of multiphase transport during drying of honeycomb ceramic substrates. *Drying Technology* 30(6):607–618.

Erbay, Z., and F. Icier. 2010. A review of thin layer drying of foods: Theory, modeling, and experimental results. *Critical Reviews in Food Science and Nutrition* 50(5):441–464.

Erriguible, A., P. Bernada, F. Couture, and M.-A. Roques. 2007. Simulation of vacuum drying by coupling models. *Chemical Engineering and Processing: Process Intensification* 46(12):1274–1285.

Farkas, B. E., R. P. Singh, and T. R. Rumsey. 1996. Modeling heat and mass transfer in immersion frying. I. Model development. *Journal of Food Engineering* 29(2):211–226.

Floury, J., A. Le Bail, and Q. T. Pham. 2008. A three-dimensional numerical simulation of the osmotic dehydration of mango and effect of freezing on the mass transfer rates. *Journal of Food Engineering* 85(1):1–11.

Foroughi-Dahr, M., M. Golmohammadi, R. Pourjamshidian, M. Rajabi-Hamaneh, and S. J. Hashemi. 2015. On the characteristics of thin-layer drying models for intermittent drying of rough rice. *Chemical Engineering Communications* 202(8):1024–1035.

Ginzburg, A. S. 1969. *Application of Infra-Red Radiation in Food Processing*. Leonard Hill Books.

Golestani, R., A. Raisi, and A. Aroujalian. 2013. Mathematical modeling on air drying of apples considering shrinkage and variable diffusion coefficient. *Drying Technology* 31(1):40–51.

Gunasekaran, S. and H.-W. Yang. 2007. Optimization of pulsed microwave heating. *Journal of Food Engineering* 78(4):1457–1462.

Hemis, M., R. Choudhary, and D. G. Watson. 2012. A coupled mathematical model for simultaneous microwave and convective drying of wheat seeds. *Biosystems Engineering* 112(3):202–209.

Ho, J. C., S. K. Chou, K. J. Chua, A. S. Mujumdar, and M. N. A. Hawlader. 2002. Analytical study of cyclic temperature drying: Effect on drying kinetics and product quality. *Journal of Food Engineering* 51(1):65–75.

Holowaty, S. A., L. A. Ramallo, and M. E. Schmalko. 2012. Intermittent drying simulation in a deep bed dryer of yerba maté. *Journal of Food Engineering* 111(1):110–114.

Ikediala, J. N., J. Tang, S. R. Drake, and L. G. Neven. 2000. Dielectric properties of apple cultivars and codling moth larvae. *Transactions of the ASAE—American Society of Agricultural Engineers* 43(5):1175–1184.

Joardder, M. U. H., C. Kumar, R. J. Brown et al. 2015a. Effect of cell wall properties on porosity and shrinkage of dried apple. *International Journal of Food Properties* 18(10):2327–2337.

Joardder, M. U. H., C. Kumar, and M. A. Karim. 2015b. Food structure: Its formation and relationships with other properties. *Critical Reviews in Food Science and Nutrition* 57(6): 1190–1205.

Jumah, R. and A. S. Mujumdar. 2005. Modeling intermittent drying using an adaptive neuro-fuzzy inference system. *Drying Technology* 23(5):1075–1092.

Jumah, R. Y. 1995. Flow and drying characteristics of a rotating jet spouted bed. PhD NN08118, McGill University, Montreal, Quebec, Canada.

Karim, M. A. and M. N. A. Hawlader. 2005. Mathematical modeling and experimental investigation of tropical fruits drying. *International Journal of Heat and Mass Transfer* 48(23–24):4914–4925.

Khan, M. I. H., M. U. H. Joardder, C. Kumar, and M. A. Karim. 2016a. Multiphase porous media modeling: A novel approach to predicting food processing performance. *Critical Reviews in Food Science and Nutrition*, 1–19, doi: 10.1080/10408398.2016.1197881.

Khan, M. I. H., C. Kumar, M. U. H. Joardder, and M. A. Karim. 2016b. Determination of appropriate effective diffusivity for different food materials. *Drying Technology* 35(3): 335–346, doi: 10.1080/07373937.2016.1170700.

Kowalski, S. J. and A. Pawłowski. 2010. Modeling of kinetics in stationary and intermittent drying. *Drying Technology* 28(8):1023–1031.

Kowalski, S. J. and A. Pawłowski. 2011a. Energy consumption and quality aspect by intermittent drying. *Chemical Engineering and Processing: Process Intensification* 50(4):384–390.

Kowalski, S. J. and A. Pawłowski. 2011b. Intermittent drying: Energy expenditure and product quality. *Chemical Engineering & Technology* 34(7):1123–1129.

Kumar, C. 2015. Modeling intermittent microwave convective drying (IMCD) of food materials. PhD, Chemistry Physics and Mechanical Engineering, Queensland University of Technology, Brisbane, Queensland, Australia.

Kumar, C., M. U. H. Joardder, T. W. Farrell, and A. Karim. 2016a. Multiphase porous media model for intermittent microwave convective drying (IMCD) of food. *International Journal of Thermal Science* 104:304–314.

Kumar, C., M. U. H. Joardder, T. W. Farrell, G. J. Millar, and M. A. Karim. 2015. A mathematical model for intermittent microwave convective (IMCD) drying of food materials. *Drying Technology* 34(8):962–973.

Kumar, C., M. U. H. Joardder, A. Karim, G. J. Millar, and Z. Amin. 2014a. Temperature redistribution modeling during intermittent microwave convective heating. *Procedia Engineering* 90:544–549.

Kumar, C., A. Karim, M. U. H. Joardder, and G. J. Miller. 2012a. Modeling heat and mass transfer process during convection drying of fruit. *The Fourth International Conference on Computational Methods*, Gold Coast, Queensland, Australia, November 25–28.

Kumar, C., A. Karim, S. C. Saha, M. U. H. Joardder, R. J. Brown, and D. Biswas. 2012b. Multiphysics modeling of convective drying of food materials. *Proceedings of the Global Engineering, Science and Technology Conference*, Dhaka, Bangladesh, December 28–29.

Kumar, C., M. A. Karim, and M. U. H. Joardder. 2014b. Intermittent drying of food products: A critical review. *Journal of Food Engineering* 121(0):48–57.

Kumar, C., G. J. Millar, and M. A. Karim. 2016b. Effective diffusivity and evaporative cooling in convective drying of food material. *Drying Technology* 33(2):227–237.

Kumar, C., S. C. Saha, E. Sauret, A. Karim, and Y. T. Gu. 2016c. Mathematical modeling of heat and mass transfer during intermittent microwave-convective drying (IMCD) of food materials. *Australasian Heat and Mass Transfer Conference 2016*, Brisbane, Queensland, Australia.

Lawrence Berkeley National Laboratory (LBNL). 2012. TOUGH: Suite of simulators for non-isothermal multiphase flow and transport in fractured porous media. http://esd.lbl.gov/research/projects/tough/software/ (accessed June 02, 2017).

Masoodi, R., H. Tan, and K. M. Pillai. 2012. Numerical simulation of liquid absorption in paper-like swelling porous media. *AIChE Journal* 58(8):2536–2544.

Mercier, S., B. Marcos, C. Moresoli, M. Mondor, and S. Villeneuve. 2014. Modeling of internal moisture transport during durum wheat pasta drying. *Journal of Food Engineering* 124:19–27.

Mihoubi, D. and A. Bellagi. 2009. Drying-induced stresses during convective and combined microwave and convective drying of saturated porous media. *Drying Technology* 27(7–8):851–856.

Montanuci, F. D., C. A. Perussello, L. M. de Matos Jorge, and R. M. M. Jorge. 2014. Experimental analysis and finite element simulation of the hydration process of barley grains. *Journal of Food Engineering* 131:44–49.

Nasrallah, S. B. and P. Perre. 1988. Detailed study of a model of heat and mass transfer during convective drying of porous media. *International Journal of Heat and Mass Transfer* 31(5):957–967.

Ni, H. and A. K. Datta. 1999. Moisture, oil and energy transport during deep-fat frying of food materials. *Food and Bioproducts Processing* 77(3):194–204.

Nishiyama, Y., W. Cao, and B. Li. 2006. Grain intermittent drying characteristics analyzed by a simplified model. *Journal of Food Engineering* 76(3):272–279.

Pan, Y. K., L. J. Zhao, and W. B. Hu. 1998. The effect of tempering-intermittent drying on quality and energy of plant materials. *Drying Technology* 17(9):1795–1812.

Perussello, C. A., C. Kumar, F. de Castilhos, and M. A. Karim. 2014. Heat and mass transfer modeling of the osmo-convective drying of yacon roots (*Smallanthus sonchifolius*). *Applied Thermal Engineering* 63(1):23–32.

PORE-FLOW. 2016. A computer program developed at University of Wisconsin-Milwaukee to model fluid motion in porous media using the FE/CV filling algorithm.

Pruess, K. 2004. The TOUGH codes—A family of simulation tools for multiphase flow and transport processes in permeable media. *Vadose Zone Journal* 3(3):738–746.

Putranto, A. and X. D. Chen. 2013. Spatial reaction engineering approach as an alternative for nonequilibrium multiphase mass-transfer model for drying of food and biological materials. *AIChE Journal* 59(1):55–67.

Putranto, A. and X. D. Chen. 2014a. Modeling of water vapor sorption process by employing the reaction engineering approach (REA). *Separation and Purification Technology* 122:456–461.

Putranto, A. and X. D. Chen. 2014b. Examining the suitability of the reaction engineering approach (REA) to modeling local evaporation/condensation rates of materials with various thicknesses. *Drying Technology* 32(2):208–221.

Putranto, A. and X. D. Chen. 2015. An assessment on modeling drying processes: Equilibrium multiphase model and the spatial reaction engineering approach (S-REA). *Chemical Engineering Research and Design* 94:660–672.

Putranto, A. and X. D. Chen. 2016a. Drying of a system of multiple solvents: Modeling by the reaction engineering approach. *AIChE Journal* 62: 2144–2153.

Putranto, A. and X. D. Chen. 2016b. Microwave drying at various conditions modeled using the reaction engineering approach (REA). *Drying Technology* 34: 2144–2153.

Putranto, A., X. D. Chen, and P. A. Webley. 2011. Modeling of drying of food materials with thickness of several centimeters by the reaction engineering approach (REA). *Drying Technology* 29(8):961–973.

Putranto, A., X. D. Chen, and W. Zhou. 2015. Bread baking and its color kinetics modeled by the spatial reaction engineering approach (S-REA). *Food Research International* 71:58–67.

Rakesh, V. and A. K. Datta. 2013. Transport in deformable hygroscopic porous media during microwave puffing. *AIChE Journal* 59(1):33–45.

Rakesh, V., Y. Seo, A. K. Datta, K. L. McCarthy, and M. J. McCarthy. 2010. Heat transfer during microwave combination heating: Computational modeling and MRI experiments. *AIChE Journal* 56(9):2468–2478.

Ramallo, L. A., N. N. Lovera, and M. E. Schmalko. 2010. Effect of the application of intermittent drying on Ilex paraguariensis quality and drying kinetics. *Journal of Food Engineering* 97(2):188–193.

Salagnac, P., P. Glouannec, and D. Lecharpentier. 2004. Numerical modeling of heat and mass transfer in porous medium during combined hot air, infrared and microwaves drying. *International Journal of Heat and Mass Transfer* 47(19–20):4479–4489.

Sosa-Morales, M. E., L. Valerio-Junco, A. López-Malo, and H. S. García. 2010. Dielectric properties of foods: Reported data in the 21st Century and their potential applications. *LWT—Food Science and Technology* 43(8):1169–1179.

Stanish, M. A., G. S. Schajer, and F. Kayihan. 1986. A mathematical model of drying for hygroscopic porous media. *AIChE Journal* 32(8):1301–1311.

Sutar, P. P. and B. N. Thorat. 2011. Drying of roots. In Jangam, S. V., Law, C. L., and Mujumdar, A. S., eds., *Drying of Foods, Vegetables and Fruits*, Vol. 2, pp. 43–74.

Váquiro, H. A., G. Clemente, J. V. García-Pérez, A. Mulet, and J. Bon. 2009. Enthalpy-driven optimization of intermittent drying of *Mangifera indica* L. *Chemical Engineering Research and Design* 87(7):885–898.

Wang, H. G., W. Q. Yang, P. Senior, R. S. Raghavan, and S. R. Duncan. 2008. Investigation of batch fluidized-bed drying by mathematical modeling, CFD simulation and ECT measurement. *AIChE Journal* 54(2):427–444.

Yamsaengsung, R. and R. G. Moreira. 2002. Modeling the transport phenomena and structural changes during deep fat frying. Part II: Model solution & validation. *Journal of Food Engineering* 53(1):11–25.

Yang, H. W. and S. Gunasekaran. 2001. Temperature profiles in a cylindrical model food during pulsed microwave heating. *Journal of Food Science* 66(7):998–1004.

Zeki, B. 2009. Chapter 22—Dehydration. *Food Process Engineering and Technology*, pp. 459–510. San Diego, CA: Academic Press.

Zhang, J., A. K. Datta, and S. Mukherjee. 2005. Transport processes and large deformation during baking of bread. *AIChE Journal* 51(9):2569–2580.

Zhang, Z. and N. Kong. 2012. Nonequilibrium thermal dynamic modeling of porous medium vacuum drying process. *Mathematical Problems in Engineering* 2012:1–22.

Zhu, Y. and Z. Pan. 2009. Processing and quality characteristics of apple slices under simultaneous infrared dry-blanching and dehydration with continuous heating. *Journal of Food Engineering* 90(4):441–452.

Zhu, Y., Z. Pan, T. H. McHugh, and D. M. Barrett. 2010. Processing and quality characteristics of apple slices processed under simultaneous infrared dry-blanching and dehydration with intermittent heating. *Journal of Food Engineering* 97(1):8–16.

9 Cellular Level Water Distribution and Its Investigation Techniques

Md. Imran H. Khan, R. Mark Wellard, Md. Mahiuddin and M. Azharul Karim

CONTENTS

9.1 INTRODUCTION

Food materials, specifically plant-based foods, are complex in nature. They have hygroscopic and porous properties and contain about 80%–90% water (Khan et al. 2017). Food security is a major concern in large parts of the world with about one-third of the global food production, or 1.3 billion tons, lost annually due to a lack of proper processing (Gustavsson et al. 2011; Kumar et al. 2012b). It is estimated that the per capita food waste by consumers in Europe and North America is 95–115 kg per year (Gustavsson et al. 2011). Food waste means not only the loss of food for nutrition but also a waste of the resources used in the production of that food such as land, water, energy and labour inputs (Khan et al. 2016a). Producing food that will not be consumed leads to unnecessary CO_2 emissions which are a significant contributor to today's global warming problem. Therefore, the importance of proper food processing must be emphasized to help reduce this massive loss, promote food security, reduce global warming and combat hunger.

Drying is a method of food preservation that inhibits the growth of bacteria, yeasts and mould through the removal of water (Kumar et al. 2015, 2016a). Dried foods have gained commercial importance, and their growth on a commercial scale has become an important sector of the agricultural industry (Karim and Hawlader 2005; Kumar et al. 2014a). The main advantage of drying is that it is an excellent way to preserve foods that can add variety to meals and provide delicious, nutritious snacks (Kumar et al. 2016b). One of the biggest advantages of dried foods is that they take much less storage space than canned or frozen foods (Kumar et al. 2012a). Another advantage is the minimal energy that is required for long-term storage. Conventional drying uses convection or hot air drying, which is the easiest way of drying (Kumar et al. 2014b). However, compared to other food storage methods (freezing, canning, etc.), long processing times, low energy efficiency and poor quality are factors associated with convective drying (Chou and Chua 2001). These problems can be overcome by intermittent drying. Intermittent application of heat energy has proven to be an alternative method to avoid uneven heating, and to improve product quality and energy utilization by allowing redistribution of temperature and moisture profiles within the product (Gunasekaran 1999, Kumar et al. 2016c). To maintain a better quality of the dried product, an understanding of actual heat and mass transfer is crucial during the intermittent drying process. Heat and mass transfer during drying in food tissue depends on the mechanical properties at different levels of structure: the cellular level (i.e. the architecture of the tissue cells and their interaction) and the organ level (i.e. the arrangement of cells into tissues and their chemical and physical interactions) (Ilker and Szczesniak 1990; Jackman and Stanley 1995; Waldron et al. 1997). Therefore, an appropriate understanding of cellular level water distribution of the raw food material and its evolution during processing is crucial in order to understand and accurately describe dehydration processes during intermittent drying. Therefore, the one aim of this chapter is to discuss the cellular structure of food tissue, cellular level water distribution and evaluation technique, and the analysis of cellular level water migration mechanism(s) during drying.

9.2 MICROSTRUCTURE OF FOOD TISSUE

9.2.1 Cellular Structure

Food materials are a composite of different elements with cellular tissue. In plant-based food material, cells are the smallest structural units that are capable of functioning independently. According to biological analysis, a cell of a plant-based food material is composed of protoplasm surrounded by the plasma membrane (plasma lemma) and the cell wall. The protoplasm is an aqueous colloidal complex of proteins and other organic and inorganic substances. It comprises the living nucleus and the cytoplasm (or ground substance). The nucleus, cytoplasm and plasma membrane of a

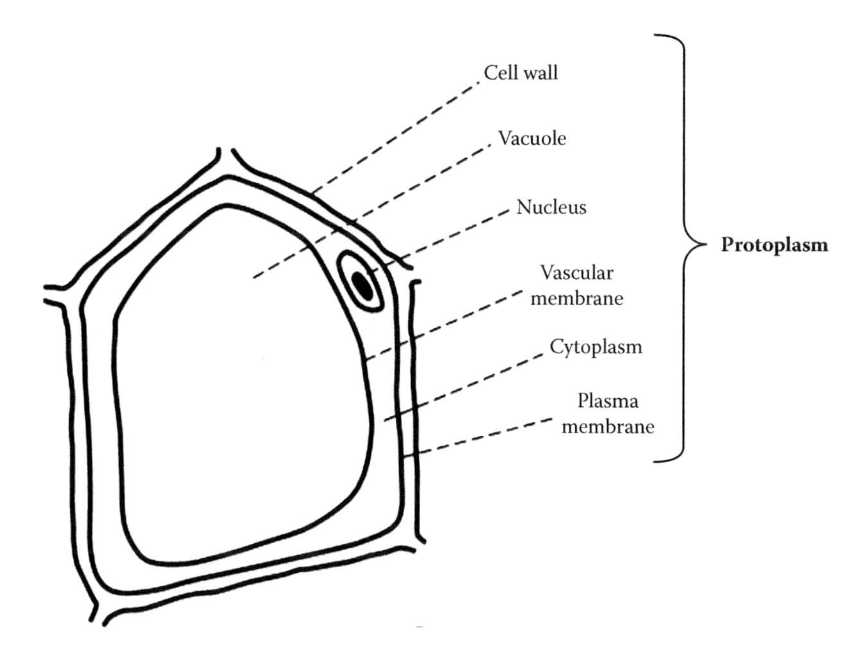

FIGURE 9.1 Cellular structure of plant-based food material.

cell are called the protoplast and constitute a living unit, distinct from the inert walls and inclusions (Iglesias and Chirife 1982), as shown in Figure 9.1. However, in food processing research, researchers consider only three or four components, as needed. Vacuoles, cytoplasm, extracellular space, and the cell wall are the main physiological components in plant-based food material which predominantly consists of water. These four components have been considered by food researchers with a special interest in investigating post-harvest quality assurance, for instance, investigating core water (Cho et al. 2008; Clark et al. 1998; Melado-Herreros et al. 2013), internal browning (Cho et al. 2008; Clark and Burmeister 1999; Gonzalez et al. 2001) and microstructural heterogeneity (Defraeye et al. 2013; Winisdorffer et al. 2015).

In food drying studies, researchers have considered three components in plant tissue: intercellular space, intracellular space and cell wall. The intercellular spaces or environments are those where an unrestricted space has been made by the connection of two or more cells. This unrestricted space contains water and air. The vacuole and cytoplasm together are considered intracellular spaces that contain most of the water in the plant tissue. The proportion of these water environments is different, and these are described in the following section.

9.2.2 Cellular Water Environment (Free and Bound Water) in Plant-Based Food Tissue

The water exists in different spaces within the hygroscopic plant tissue. Food drying research considers three types of water: capillary water, intracellular water and cell wall water as physically bound water. On the other hand, a small amount of water is bound to metabolites (nutrition), which is also referred to as chemically bound water. This compartment should not be targeted for transport due to its importance in maintaining the taste and flavor of the dried food (Kuprianoff 1958). Therefore, transport of physically bound water is discussed in the following, which is the main concern in food processing.

Water residing in the intercellular space is known as capillary water or free water (FW), the water inside the cell is referred to as intracellular water and the water that occupies the fine space inside the cell wall is called the cell wall water (Khan et al. 2016b). These different types of water and their environments are shown in Figure 9.2. Intracellular water and cell wall water are known as physically bound water or simply bound water (Karel and Lund 2003). Based on its mobility, bound water present inside the cells (intracellular water) is referred to as loosely bound water (LBW) while cell wall water is termed strongly bound water (SBW) (Caurie 2011).

Due to the hygroscopic behavior of plant-based food materials, transport of bound water has a great effect on material shrinkage during food drying. Joardder et al. (2015) suggested that migration of free water has no effect on material structure, whereas migration of LBW contributes to cellular shrinkage, pore formation and collapse of the cell. Furthermore, overall food tissue structure is deformed due to the migration of SBW, as shown in Figure 9.3. In contrast, Prothon et al. (2003) showed that migration of the three types of water causes overall tissue shrinkage, cellular shrinkage as well as structural collapse.

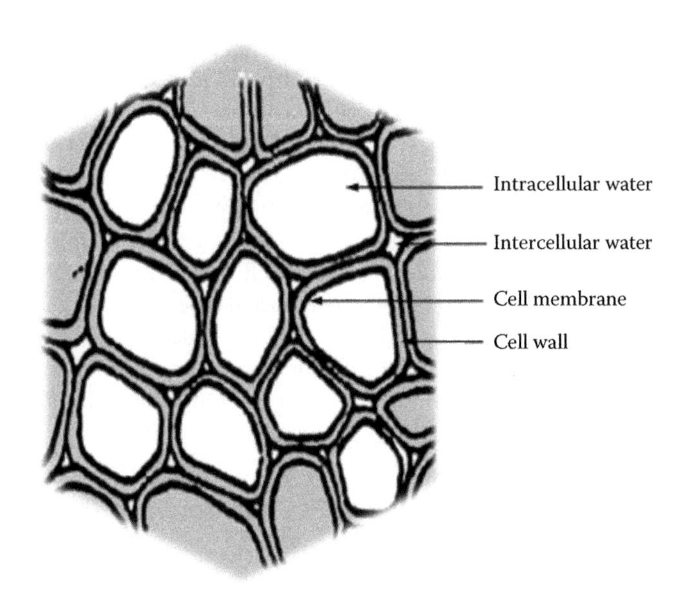

FIGURE 9.2 Cellular structure of plant-based food material for food drying. (From Khan, M.I.H. et al., *Innov. Food Sci. Emerg. Technol.*, 38, 252, 2016b.)

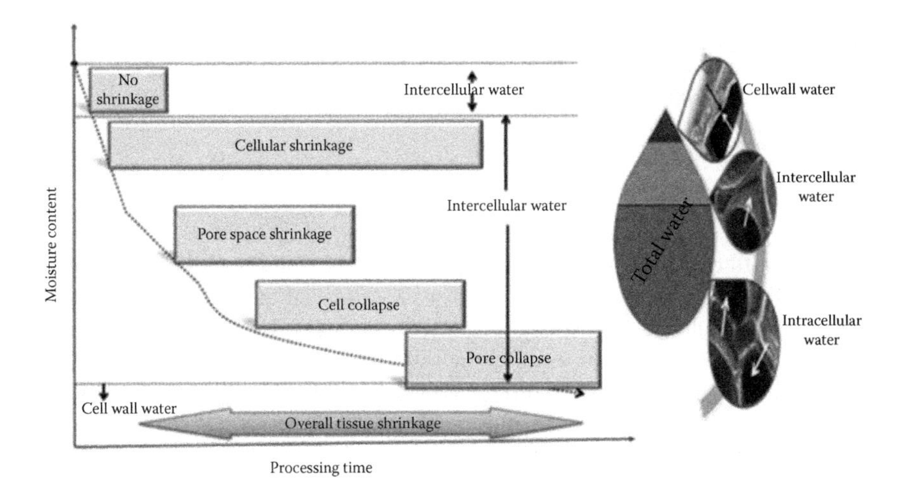

FIGURE 9.3 Water distribution with material shrinkage. (With kind permission from Springer Science+Business Media: *Porosity: Establishing the Relationship between Drying Parameters and Dried Food Quality*, 2015, Joardder, M.U.H., Karim, A., Brown, R.J., and Kumar, C.)

Therefore, unless bound water transport is considered, a food drying study cannot provide a realistic understanding of the heat and mass transfer mechanism as well as material deformation (shrinkage) during drying.

In intermittent drying, the temperature distribution is more uniform compared to continuous convective drying (Lima et al. 2016). Therefore, level and extent of cell rupture might be different; consequently, the bound water might migrate at a different rate from the intracellular space to intercellular spaces. Hence, a better understanding of the migration of bound and free water during intermittent drying is required.

9.3 TECHNIQUES TO INVESTIGATE CELLULAR LEVEL WATER DISTRIBUTION

There are a number of techniques available for investigating the bound and free water in cellular tissue. These are discussed in the following sections.

9.3.1 Nuclear Magnetic Resonance

Nuclear magnetic resonance spectroscopy (NMR) is an established technique for examining the chemical environment of atomic nuclei with an odd number of protons and/or neutrons. Such nuclei have a magnetic moment that precesses in a magnetic field with a characteristic frequency, known as the Lamour frequency. When placed in the presence of a strong magnetic field, a (small) proportion of the nuclei aligns with the magnetic field in an anti-parallel manner similar to a bar-magnet in a magnetic field. The number of spins aligning in this manner is proportional to the strength of the magnetic field, with the end result being a sample with a net magnetization vector that can be studied spectroscopically. The distribution of nuclei between the ground state and the aligned, lower energy state is given by the Boltzmann relation (Abragam 1961; Hanson 2008):

$$\frac{N_{upper}}{N_{lower}} = e^{-\Delta E/kT} = e^{-h\nu/kT} \tag{9.1}$$

where the difference in energy of the two states is ΔE, k is the Boltzmann constant, T is the absolute temperature, h is the Planck's constant and ν is the resonant frequency. For a magnetic field strength of 18.8 tesla (800 MHz for proton), at thermal equilibrium and at room temperature, the ratio of nuclei in the two energy states is 0.999872. This small population difference is utilized by the technique of NMR and is the origin of the low sensitivity of the technique.

Nuclei with a magnetic moment have a specific gyromagnetic ratio, γ, and the resonant frequency, ν_o, is dependent on the magnetic field strength, B_o (Derome 2013), such that

$$\nu_o = \frac{\gamma B_o}{2\pi} \tag{9.2}$$

Application of a short radio frequency (RF) signal matched to the precession frequency of the nucleus of interest results in absorption of energy and coherent precession of the nuclei. The net magnetization vector is moved out of alignment with the magnetic field. The degree of rotation of the magnetization vector is determined by the power of the RF pulse. When the RF signal ceases, the nuclei return to the equilibrium state and their spins lose coherence.

The NMR signal is measured after application of the RF pulse by placing the sample in a coil that is tuned to the resonance frequency of the nuclei of interest. The coherently processing magnetic moment of the nuclei induces a small current in a receiver coil, which decays as the nuclei return to equilibrium. The resulting intensity-time signal can be recorded and is Fourier transformed to yield an intensity–frequency spectrum. The maximum magnetization that can be detected, M_o, is given by

$$M_o = \frac{\gamma^2 h^2 N_s B_o}{4kT} \tag{9.3}$$

where

γ is the nuclear species-specific gyromagnetic ratio
N_s is the number of spins
B_o is the magnetic field strength (Hanson 2008)

It can be seen from this equation that the strength of the recorded signal is proportional to the number of nuclei in the sample volume and the strength of the applied magnetic field. In the presence of a strong external magnetic field, the local magnetic field experienced by a nucleus is influenced by the electrons in its immediate environment. Thus, the molecular structure in which an atom participates can alter the precise resonant frequency of the nucleus. This behaviour in the presence of different electronic environments results in populations of spins corresponding to the electron-shielding experienced by nuclei in each environment. This information is utilized to generate an NMR spectrum and the signal dispersion due to local shielding effects is termed chemical shift.

The return to alignment with the magnetic field is described by a relaxation time constant, T_1 (spin–lattice relaxation), and the loss of coherence is described by a relaxation time constant, T_2 (spin–spin relaxation). When nuclei are in a restricted environment, energy is lost to the surrounding atoms (lattice) and T_1 is shortened, relative to an equivalent nucleus in a less constrained environment. T_2 relaxation results from interaction between the processing spins of the excited nuclei. This form of relaxation can be equal to or shorter than T_1 relaxation, depending on local environments of the nuclei (Derome 2013). An advantage arising from the change in relaxation time constants with environment is the ability to relate the relaxation time to the mobility of the nuclei being studied, with longer relaxation times being associated with more mobile nuclei. Conversely, nuclei that are in a restricted environment will have a shorter relaxation time.

The NMR signal can be spatially localized by making use of the frequency dependence of magnetic field strength. Application of a linear magnetic field gradient across a sample can be achieved by passing a current through 'gradient' coils that are integrated within the static magnetic field of the NMR magnet. The resulting distortion of the B_o magnetic field by the current applied to the gradient coils provides a gradation of the magnetic field that results in a distribution of resonance frequency for a population of nuclei based on their spatial position within the sample (in the direction of the applied gradient). Stepwise variation in the strength of the current enables a corresponding localization of nuclei across the sample. Spatial information in a second dimension is obtained by using phase encoding of the applied RF signal. The slice thickness of the two-dimensional plane created by the imaging sequence is controlled by the application of an orthogonal gradient and RF pulse. The image resulting from Fourier transformation of the acquired data provides a map of the signal intensity throughout the plane of interest. The stepwise technique results in the measurement of signals from discrete regions or voxels. This spatial encoding of the NMR signal forms the basis of magnetic resonance imaging (MRI). Because the signal is proportional to the number of nuclei in each voxel, the reduced size of the effective sample results in a smaller signal. For this reason, only nuclei with the greatest NMR sensitivity (largest gyromagnetic ratio), such as ^{1}H, are suitable for MRI. In biological systems, tissue water provides the greatest concentration of ^{1}H nuclei and therefore the strongest signal in biological samples.

Examining the signal from water in biological samples forms the basis of most MRI studies. The information recorded can range from distribution mapping of the concentration of water to the spatial mapping of relaxation times as a marker of differences in the local environment or tissue water. By selecting imaging sequence timing parameters, the regional intensity of concentration-based images can be filtered according to the relaxation parameters that are influenced by the tissue environment.

Proton nuclear magnetic resonance (^{1}H NMR) relaxometry studies have proven to be valuable in the study of plants and plant-based food materials submitted to stress, reflecting anatomical details of the entire tissue and the water status in particular (Gambhir et al. 2005; Van Der Weerd et al. 2002). ^{1}H NMR relaxometry signals, which are an average over the whole sample, provide information on the water environments within the plant tissue since, as described earlier, the proton signal is dominated by water protons (Van Der Weerd et al. 2001) and the proton NMR signal intensity is directly proportional to the proton density of the tissue (Westbrook and Kaut 1993). The water exchange rates between these compartments are controlled by the water proton relaxation behaviour T_2, which strongly depends on the water mobility in the microscopic environment of the tissue and the strength of the applied magnetic field. The spin–spin T_2 relaxation is the transverse component of the magnetization vector, which exponentially decays towards its equilibrium value after excitation by RF energy. It can be expressed as:

$$M(t) = \sum_{i=1}^{n} A_i e^{-t/T_2^i} \tag{9.4}$$

where

$M(t)$ is the function of relaxation time
A is the relative contribution of sets of protons
T_2 is the relaxation time of water proton
i is the number of the contributing component

To decide whether the T_2 signal contains one or more components, the T_2 signal can be fitted with a mono-exponential model. Fitting a mono-exponential function will result in an erroneous value for T_2 if more than one component is present. In such cases, we need to fit for the additional component(s). The number can be estimated by plotting the natural logarithm of the signal $M(t)$ against time (as shown in Figure 9.4), after which it can be straightforward to see if the function is bi-exponential or tri-exponential, as there will be distinct linear regions of different slopes in the resulting plot (Boulby and Rugg-Gunn 2004). From Figure 9.4, it can be predicted that there are three distinct regions for three different water environment.

However, for optimum accuracy, this model requires a good signal-to-noise ratio (SNR). If SNR is low, the fit to the four (A_1, A_2, T_{21}, T_{22}) parameters for the bi-exponential model or six parameters ($A_1, A_2, A_3, T_{21}, T_{22}, T_{23}$) for the tri-exponential model becomes uncertain, reducing accuracy and precision. In addition, the relaxation times should be within an order of magnitude of each other (and preferably should differ by no more than a factor of 2–3) and the population fractions ideally should not fall below about 15%.

Sometimes it is possible that even more components may be present. In such cases, the approach is to use a statistical test to determine whether or not additional terms need to be added to the fit (Armspach et al. 1991).

For investigating different water environments, such as bound and free water in biological tissue, NMR is a renowned technique and therefore many studies have demonstrated the presence of different tissue environments. NMR has been used to show water compartmentation in a number of animal tissues: lung (Cutillo et al. 1992;

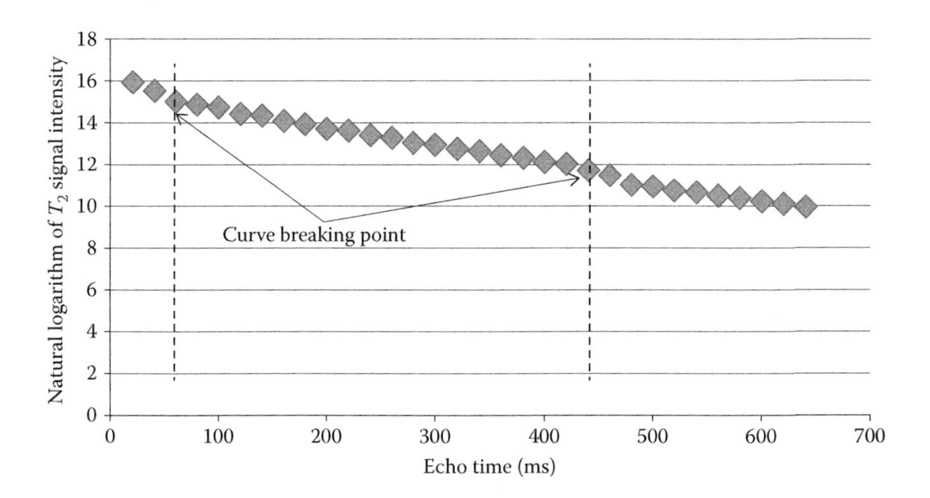

FIGURE 9.4 Natural logarithmic of T_2 signal intensity with echo time.

Sedin et al. 2000), brain (Berenyi et al. 1998; Furuse et al. 1984; Inao et al. 1985; Sulyok et al. 2001; Vajda et al. 1999), liver (Moser et al. 1992; Moser et al. 1996) and red blood cells (Besson et al. 1989). In the case of plant-based food material, T_2 relaxation theories have been applied to the investigation of sugar content in fruit tissue (Delgado-Goni et al. 2013), the quality of fruits and vegetables (Chen et al. 1989; Van de Velde et al. 2016) and the maturity of fruits and vegetables (Chen et al. 1993; Ruan et al. 1999). Literature describing the measurement of free and bound water in plant-based food materials is rare. Hills and Remigereau (1997) investigated T_2 relaxation times for assessing the migration of different types of water during drying and freezing in apple tissue. However, they did not quantify the types of water in that tissue. Gonzalez et al. (2010) studied the effect of high pressure on cell membrane integrity during thermal processing of onion. They investigated the change in T_2 with processing time; however, that study did not report data on the proportion of FW, LBW and SBW in the onion tissue.

Previous NMR experiments have used T_2 relaxometry to study the development of the core water (Cho et al. 2008; Clark et al. 1998; Melado-Herreros et al. 2013), internal browning (Cho et al. 2008; Clark and Burmeister 1999; Gonzalez et al. 2001) and microstructural heterogeneity (Defraeye et al. 2013; Winisdorffer et al. 2015) in apple tissue. However, these studies also did not report the proportion of FW, LBW and SBW.

To investigate the different cellular compartments of water, NMR is a unique tool; however, it has some limitations. The disadvantages of NMR predominantly relate to the low sensitivity of the technique. Due to the low number of nuclei aligned with the magnetic field, significant signal averaging is required to obtain satisfactory SNR. Measurement of water in biological systems provides reasonable signal but when molecules other than water are examined, this low sensitivity becomes a greater problem. The measurement of nuclei other than protons amplifies the problems of low SNR because of their lower gyromagnetic ratio. Measurement of such samples requires either high concentrations of the measured nucleus or longer acquisition times or a combination of both.

NMR measurement is not possible for nuclei that do not have a magnetic moment and even for those that are NMR-sensitive, the technique relies on an environment with a homogeneous magnetic field. At the molecular level, this can be disrupted by paramagnetic nuclei, which reduce the NMR sensitivity. The strong superconducting magnets used in NMR systems are associated with significant costs, being expensive and requiring regular maintenance of cryogen levels.

Despite these disadvantages, the wealth of information available from NMR experiments makes the technique extremely valuable. In addition are the advantages of requiring minimal sample preparation and being a non-destructive technique, allowing repeated measurements of the same sample.

9.3.2 Differential Scanning Calorimetry

Differential scanning calorimetry (DSC) is a thermodynamic tool that determines the temperature and heat flow associated with material transitions as a function of time and temperature. During a change in temperature, DSC measures a heat quantity, which is radiated or absorbed excessively by the sample on the basis of a temperature difference between the sample and the reference material (Haines et al. 1998). As a powerful analytical tool, DSC is capable of elucidating the factors that contribute to the folding and stability of biomolecules. For DSC analysis, it is assumed that all the freezable water that can freeze at a certain temperature in hygroscopic food materials act as a free water with a transition temperature, enthalpy and peaks in DSC curves (Figure 9.5) being the same as pure water (Nakamura et al. 1981). The unfreezable

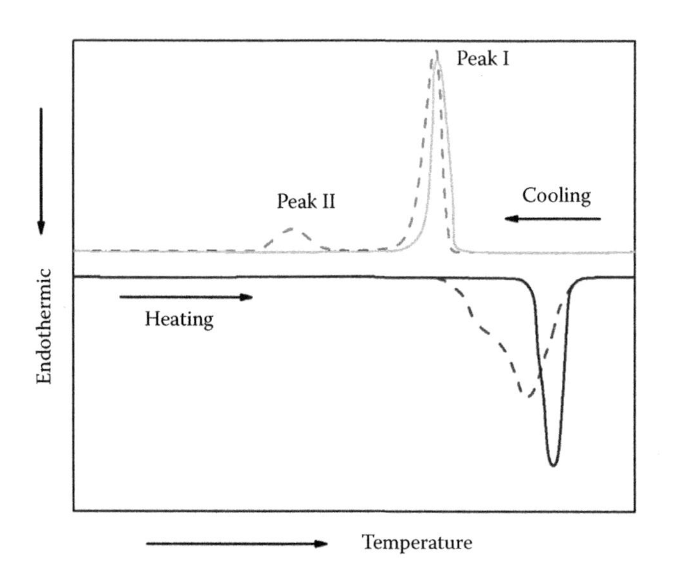

FIGURE 9.5 A typical DSC curve.

water that does not freeze while lowering the temperature up to a specific temperature range (usually –40°C to –60°C) is categorized as bound water, which is restricted by its macromolecular content and the transition is not detected in the first-order transition (Nakamura et al. 1981). In a heat flux DSC, a well-defined heat conduction path of known thermal resistance is used for the heat exchange of the sample to be measured. The primary signal is a temperature difference between a sample and a reference. The instrument determines the temperature and heat flow associated with material phase transitions as functions of time and temperature (Marangoni and Fernanda Peyronel 2014). Some of the physical phenomena studied with DSC are glass transitions, melting profiles, heats of fusion, amount of crystallinity, oxidative stability, curing kinetics, crystallization kinetics and other phase transitions. Using this method, much research has been conducted to find the bound water content in polymers (Hatakeyama et al. 1988, 2012; Ohno et al. 1983), sludge (Lee and Hsu 1995), carboxymethylcellulose (Nakamur et al. 2004), cotton, kapok, linen, jute, wood (Nakamura et al. 1981) and beef (Aktaş et al. 1997; Schwartzberg 1976). The literature on the application of DSC to investigate bound and free water in plant-based materials is very limited; however, some reports describe the investigation of the quality relationship with water distribution (Goñi et al. 2011; Kerch et al. 2012).

9.3.2.1 Mathematical Analysis in DSC

When DSC is used in the scanning mode, the temperature is changed linearly. The heat flow rate, Q, is proportional to the heating rate dT/dt:

$$Q = K\frac{dT}{dt} \tag{9.5}$$

with K as the proportionality factor. In DSC, the differential heat flow rate depends on the differential heat capacity between a pan containing the sample and a reference pan, usually empty, and the heating rate. The measured heat flow in scanning mode is never zero and is made up of three parts:

$$Q(T,t) = Q_o(T) + Q_{cp}(T) + Q_r(T,t) \tag{9.6}$$

where

- Q_o is due to the difference in temperature between the sample and reference positions
- Q_{cp} is due to the difference in heat capacity between sample and reference
- Q_r is the heat flow contribution from the latent heat of transition in the sample

The first and second terms define the baseline and the third term defines the 'peak' of the measured curve, as shown in Figure 9.5.

It can be seen from the DSC curve (Figure 9.5) that there are two peaks: peak I and peak II. These two peaks depend on the enthalpy associated with the change in water phase (Nakamura et al. 1981), and differences in water mobility (Marangoni and Fernanda Peyronel 2014). During the DSC experiment, a certain amount of additional water is needed with the sample to increase the enthalpy of freezing and uses the following equation:

$$W_l = W_1 + W_2 + W_{nf} \quad (9.7)$$

$$W_l = W_m + W_{nf} \quad (9.8)$$

where

W_l is the total weight of water added to the sample

W_1 and W_2 are the weight of water that can be calculated from the enthalpy of peak I and peak II, respectively (Figure 9.5)

W_{nf} is the weight of non-freezing water

W_m is the weight of water that can be calculated from the melting enthalpy

The bound water can be calculated from the following equation:

$$W_b = W_2 + W_{nf} \quad (9.9)$$

Then, the proportion of free and bound water can be calculated from the following equations:

$$\%BW = \left(\frac{W_b}{W_s}\right) \times 100 \quad (9.10)$$

$$FW = 1 - BW \quad (9.11)$$

where

BW is the bound water

FW is the free water

W_s is the weight of sample

9.3.2.2 Basic Arrangement of a DSC

A typical DSC comprises two calorimeters that are assumed to be identical. Figure 9.6 shows a schematic diagram of the inside of the cell and the location of both pans. Sample and reference pans are placed on top of the two platforms projecting from the sensor base. The cell is heated at a linear heating rate. The two calorimeters are part of a bigger unit, the 'sensor' (Danley 2001). The sensor body is made out of constantan (an alloy usually of 55% copper and 45% nickel), consisting of a thick flat base and a pair of raised platforms where the sample pan and reference pan are positioned for the analysis. The thin side (wall) of the sensor creates the thermal resistance. This constantan material provides good thermal conduction for both the sample and the reference. As the temperature of the furnace is changed, heat is transferred from the silver base of the enclosure to the sensor body and thus to the pans. The temperature of the furnace is controlled by the refrigerated cooling system (RCS). Thermocouples on the underside of each platform are used to measure the temperature of the sample and the reference. A third detector is used to measure

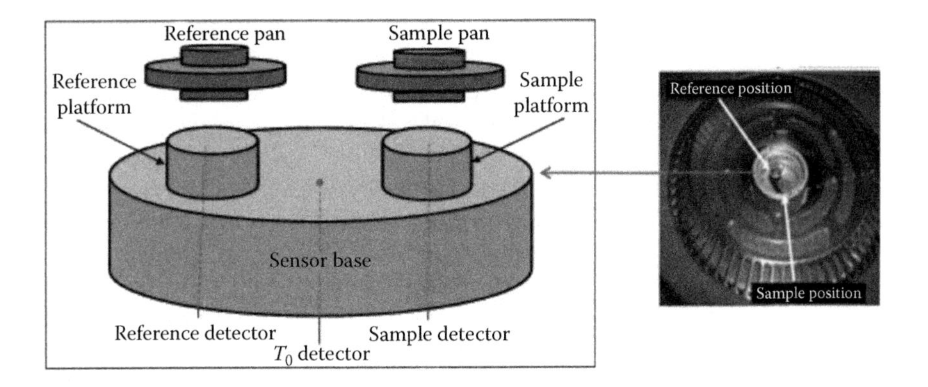

FIGURE 9.6 A typical DSC with different components labelled.

the temperature of the sensor base. Pans are made from gold, copper, aluminium, graphite and platinum or alodine-aluminium (Guide for choosing DSC pans, TA Instrument). The lids can either provide a complete enclosure or can contain a pin hole to release the pressure that builds up inside, due to evaporation of water as the temperature is increased. Most samples can be run in non-hermetically sealed pans either uncovered, or crimped with an aligned or inverted cover. Atmospheric interaction is optimized by using an open (uncovered) pan. Crimped pans improve the thermal contact between the sample, pan and disc, reduce thermal gradients in the sample, minimize spillage and enable retention of the sample for further study. These pans are coated with an inert fluoro-phosphate layer, which gives the pans a slightly yellow or gold colour rather than the typical silver colour of aluminium. This coating renders the pans inert to many chemicals. A schematic diagram of a conventional DSC is given in Figure 9.6.

9.3.2.3 Calibration of DSC

Before doing the experiment, calibration of heat flow is important for maximizing the accuracy of the data. Heat flow calibration can be performed using a sample with a well-characterized physical transformation, for example the melt of a metal such as indium, which occurs in the temperature region of interest for subsequent experiments. During the calibration, the sample should be carefully weighed and loaded into a pan and installed in the calorimeter. An empty pan, similar to the reference position pan, should be chosen and installed on the reference position of the calorimeter. A heating scan can be run at heating rate equal to the rate that will be used in subsequent experiments. The enthalpy of the sample phase transition can be measured by integration of the peak area and compared to known values. The ratio of the measured to the standard value can be used as a multiplier to scale the output of subsequent experiments. Calibration of the sample temperature can be carried out using a series of standards that have well-characterized transitions to enable correction of the sample temperature using a curve fitted to the differences between the measured and correct temperatures for the transitions. The temperature and heat of fusion calibrations ensure that the sample thermocouple reading is correct under the chosen experimental conditions. It involves the use of standard materials, such as high-purity metals, indium and gallium, which are melted at the same heating rate used in the analysis of future samples. High-purity materials are used because they give narrow peaks. This calibration gives three different outputs: the cell constant, the onset slope and the temperature calibration. The cell constant is the ratio between the theoretical heat of fusion and the measured experimental heat of fusion of the standard. The onset slope or thermal resistance is a measure of the temperature drop that occurs in a melting sample in relation to the thermocouple. Theoretically, a standard high-purity sample should melt at a unique temperature. However, as it melts and draws more heat, a temperature difference develops between the sample and the sample thermocouple. The thermal resistance between these two points is calculated as the onset slope of the heat flow versus temperature curve on the low-temperature side of the melting peak. The temperature calibration makes a correction based upon the difference between the observed and the theoretical melting temperature of the high-purity metal used. If only one standard is used, then the calibration shifts the sample temperature by a constant amount. A two-point calibration shifts the temperature with a linear correction (straight line) and projects this correction to temperatures above and below the two calibration points.

9.3.3 Dilatometry

Dilatometry (DIL) is a thermo-analytical technique that can be used for measuring bound water, based on thermal expansion or contraction of material while subjected to a controlled temperature. This method is based on the fact that when water freezes it expands about 9% in volume. In this procedure, the sample and a non-freezing (at the temperatures employed) indicator fluid are entered into the dilatometer as shown in Figure 9.7. The change in the level of the meniscus of the indicator fluid

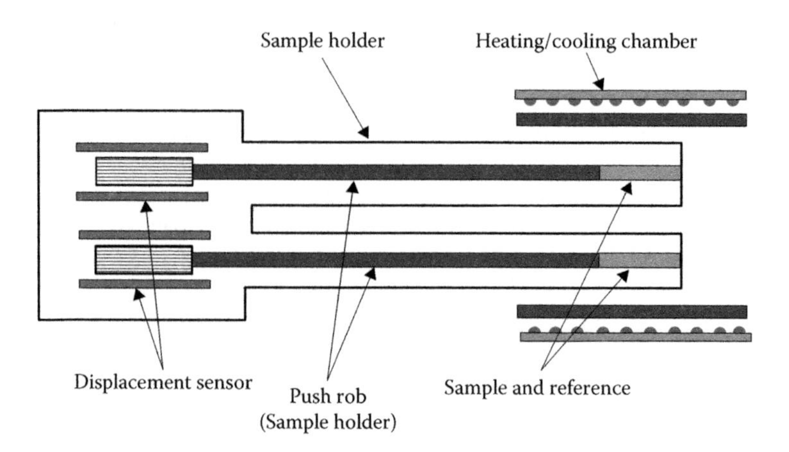

FIGURE 9.7 Schematic of a typical dilatometry.

indicates the change in volume of the freezing sample. Moreover, if a sample of material of known water content is immersed in a liquid with a low freezing point which is immiscible with water and enclosed in a system in which small changes in volume can be measured, the expansion accompanying freezing can be measured and the amount of water which froze can be calculated. From this the amount of unfrozen or bound water can be calculated.

During dilatometry measurements, it can be assumed that the bound portion of water is held by forces greater than those which act to orient water molecules into the crystal lattice of ice.

The main mechanism of dilatometry is the thermal expansion or contraction of material. The thermal expansion or contraction can be calculated from the following equation:

$$\frac{\Delta V}{V} = \exp\left(\int_{T_1}^{T_2} \alpha_V(T)\,dT\right) - 1 \tag{9.12}$$

where

V is the volume of material (m^3)

T_1 and T_2 are the initial and final temperature

α_V is the volumetric thermal expansion coefficient as a function of temperature

The thermal expansion coefficients may be different in different directions (x, y and z). This is due to the anisotropic nature of fruit structure. Consequently, the total volumetric expansion is distributed unequally among the three axes. In such cases, it is necessary to treat the coefficient of thermal expansion as a tensor with up to six independent elements. Thermal expansion generally decreases with increasing bond energy. The volumetric expansion or contraction changes the thermal expansion coefficient. Therefore, the bound water that has strong bonding energy with solid can be readily found. Substantial work has been conducted using this method to study bound water in sludge and soil (Lee 1996; Smith and Vesilind 1995; Wu et al. 1998); however, investigation of bound water in plant tissue is less well studied.

9.3.3.1 Calibration of Dilatometry

Two calibrations are necessary before performing dilatometry measurements for the determination of bound water because these calibration factors limit the accuracy of the determination of the length change. The first of the calibrations can be made for the thermal dilatation behaviour of the sample holder and the push-rods, which is normally accounted for by the baseline of the dilatometer. The second calibration relates to the accuracy of thermocouple that is usually located between the experimental sample and the reference specimen. Much research has been conducted to validate the calibration techniques for the thermal analysis of dilatometry (Liu et al. 2004; Naofumi et al. 2001).

9.3.4 Bioelectrical Impedance Analysis

Bioelectrical impedance analysis (BIA) is a very simple and established technique for measuring body composition. It measures the resistance of tissues to the flow of electrical current. The proportion of different components in tissue can be calculated as the current flows more easily through the parts of the material that are composed mostly of water. Using this method many researchers have examined the morphological behaviour in different biological tissues, for instance, the level of injury due to freeze–thaw cycles in potato (Zhang et al. 1990), estimating the extent of bruising in apples (Cox et al. 1993), assessing the maturity of nectarine (Dejmek and Miyawaki 2002) and examining the effect of drying and freezing–thawing treatments on eggplants, tubers and carrot roots (Wu et al. 2008).

The approach has the advantage that it is simple and non-destructive. It can be used for measuring water distribution in biological tissue. Some studies have used this technique to find the intracellular and extracellular water in animal tissue (Dean et al. 2008; Webber and Dehnel 1968). However, application of this method to investigate bound water in plant-based food tissue is uncommon. Halder et al. (2011) successfully investigated intracellular water in different plant-based food tissues using BIA and found about 78%–96% water present in intracellular space. However, due to the limitations of the method, they did not investigate the proportion of strongly bound water (cell wall water) in plant-based food tissue.

BIA is mainly used for analysing fat proportion in animal tissue. However, in plant tissue, it may not be a sufficiently accurate technique for predicting moisture migration pathways because it cannot detect the position of a component of water because it has limited spatial resolution.

9.4 CELLULAR WATER DISTRIBUTION IN PLANT-BASED FOOD MATERIAL

There are three major cellular environments containing water in plant-based food material. Different cellular environments contain different proportions of water. It is found that the majority of water (80%–96%) in apple tissue is present in the intracellular environment, only 8%–15% water is present in intercellular spaces, and a smaller amount of water (6%–4%) exists in the cell wall environment (Khan et al. 2016b,c), as shown in Figure 9.8. These three types of water have been categorized as LBW, FW and SBW (Khan et al. 2016a).

Halder et al. (2011) investigated intracellular water in different plant-based food materials using the BIA method. They confirmed that depending on the types of fruit and vegetable, about 78%–96% water was present in the intracellular environment.

The water distribution discussed earlier was measured in fresh food samples. However, measurements dealing with the cellular water (re)distribution during drying have not been made because the mechanisms of cellular water transport are unclear. A brief discussion on cellular water distribution during drying is given in the following section.

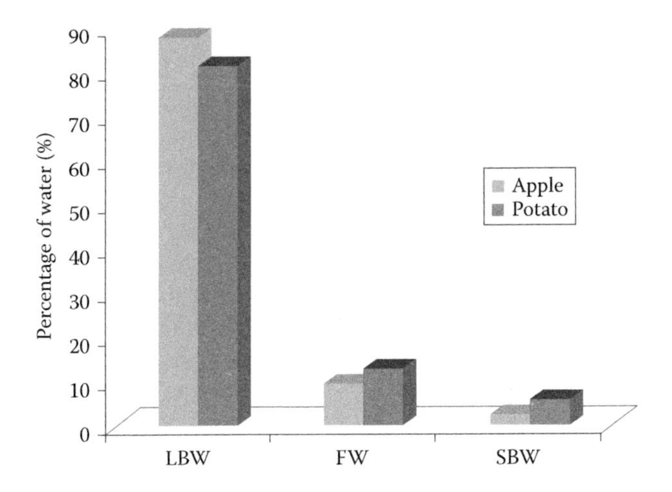

FIGURE 9.8 Cellular water distribution in apple and potato tissue.

9.5 CELLULAR WATER TRANSPORT DURING DRYING

The study and understanding of transport mechanisms of LBW is inadequate and unclear (Feng et al. 2001). The transport mechanisms of free water (FW) and intracellular water may be different in different food materials due to their more diversified transport properties. Free water can migrate from intercellular spaces to the surface by diffusion and convection, where it can be removed by evaporation from the surface to the environment (Srikiatden and Roberts 2007). On the other hand, Turner et al. (1998) assumed that intracellular water transfer was caused by diffusion only, while others postulated a capillary mechanism to characterize intracellular water flow (Peishi and Pei 1989). Feng et al. (2001) hypothesized that a universal driving force that comes from the chemical potential gradient is responsible for migrating intracellular water. It is also hypothesized that intracellular water is removed by progressive vaporization within the solid matrix, followed by diffusion and pressure-driven transport of water vapour through the solid (Datta 2007).

Therefore, the transport mechanism of intracellular water is still a controversial issue for food processing. It is assumed that most of the intracellular water migrates after the collapse of cells, moving to intercellular spaces and becoming equivalent to free water. Halder et al. (2011) investigated the intracellular water transport mechanism during progressive drying in plant-based food material. They found that intracellular water can move following rupture of the cell membrane, the extent of which depends on temperature. They suggested that below 50°C, the cells remain intact; therefore, conversion of the intracellular water to free water remains unchanged with the drying progress. As cells remain intact, the intracellular water (BW) moves to the intercellular space only through micro-capillaries and therefore this migration is very slow. Halder et al. (2011) argued that all of the membranes of the cell collapse at once after reaching a specific temperature during processing.

However, Srikiatden and Roberts (2007) reported that cells may collapse progressively from the surface to the centre.

In intermittent drying, specifically when using intermittent microwave drying, the heat is dissipated from centre to the surface. In convective drying, at the initial stages of drying, intercellular water migrates from surface to the environment through evaporation. Therefore, at the initial stage of drying, microwave heating might not be effective. The intracellular water starts to migrate through rupturing the cell membrane at the second stage of drying, while water that is initially free fully migrates to environment. After the initial heating and water loss, rapid heat energy is needed to damage the cell membranes and therefore microwave energy can be used for more rapid disruption of the cell membrane. The literature also suggests that microwave energy is effective in the final stages of drying, leading to damage to the highly rigid cell membranes, which results in faster moisture transport and reduced material shrinkage (Feng et al. 2012).

9.6 CONCLUSION

This chapter presents cellular level water distribution and its investigation techniques. Techniques such as nuclear magnetic resonance (NMR), differential scanning calorimetry (DSC), dilatometry (DTA) and bioelectrical impedance analysis (BIA) are the main methods available for investigating different cellular levels and environments of water. Among these methods, NMR is the most appropriate technique to investigate water at the cellular level. The proportion of different water environments at the cellular level in plant-based food materials depends on their properties, cell orientation and cell size. Depending on the structural diversity of different plant-based food materials they contain about 70%–95% intracellular water, 8%–15% intercellular water, and very small amount (4%–6%) of cell wall water. During convective drying, the transport mechanisms for these different types of cellular water may be different. The region of intracellular water loss migrates from the sample progressively, starting at the surface and moving to the centre after rupturing of the cell membranes. However, for intermittent drying, the mechanism of water transport may involve loss from centre of the sample followed by progression of the drying front to the surface of the sample. If this is the case, then microwave heating would be less effective during the initial stages of drying.

REFERENCES

Abragam, A. *The Principles of Nuclear Magnetism*. Oxford University Press, UK, 1961.

Aktaş, N., Y. Tülek, and H. Y. Gökalp. Determination of differences in free and bound water contents of beef muscle by DSC under various freezing conditions. *Journal of Thermal Analysis* 50(4) (1997): 617–624.

Armspach, J.-P., D. Gounot, L. Rumbach, and J. Chambron. In vivo determination of multiexponential t 2 relaxation in the brain of patients with multiple sclerosis. *Magnetic Resonance Imaging* 9(1) (1991): 107–113.

Berenyi, E., I. Repa, P. Bogner, T. Doczi, and E. Sulyok. Water content and proton magnetic resonance relaxation times of the brain in newborn rabbits [in English]. *Pediatric Research* 43(3) (March 1998): 421–425.

Besson, J. A., D. N. Wheatley, E. R. Skinner, and M. A. Foster. 1H-NMR relaxation times and water content of red blood cells from chronic alcoholic patients during withdrawal [in English]. *Magnetic Resonance Imaging* 7(3) (May–June 1989): 289–291.

Boulby, P. A. and F. J. Rugg-Gunn. T2: The transverse relaxation time. In *Quantitative MRI of the Brain: Measuring Changes Caused by Disease*, P. Tofts (Ed.), pp. 143–201. West Sussex, England: John Wiley & Sons, Ltd, 2004.

Caurie, M. Bound water: Its definition, estimation and characteristics. *International Journal of Food Science & Technology* 46(5) (2011): 930–934.

Chen, P., M. J. McCarthy, and R. Kauten. NMR for internal quality evaluation of fruits and vegetables. *Transactions of the ASAE* 32(5) (1989): 1747–1753.

Chen, P., M. J. McCarthy, R. Kauten, Y. Sarig, and S. Han. Maturity evaluation of avocados by NMR methods. *Journal of Agricultural Engineering Research* 55(3) (July 1993): 177–187.

Cho, B.-K., W. Chayaprasert, and R. L. Stroshine. Effects of internal browning and watercore on low field (5.4 MHz) proton magnetic resonance measurements of T2 values of whole apples. *Postharvest Biology and Technology* 47(1) (January 2008): 81–89.

Chou, S. K. and K. J. Chua. New hybrid drying technologies for heat sensitive foodstuffs. *Trends in Food Science & Technology* 12(10) (October 2001): 359–369.

Clark, C. J. and D. M. Burmeister. Magnetic resonance imaging of browning development in 'Braeburn' apple during controlled-atmosphere storage under high CO_2. *HortScience* 34(5) (1999): 915–919.

Clark, C. J., J. S. MacFall, and R. L. Bieleski. Loss of watercore from 'Fuji' apple observed by magnetic resonance imaging. *Scientia Horticulturae* 73(4) (April 1998): 213–227.

Cox, M. A., M. I. N. Zhang, and J. H. M. Willison. Apple Bruise assessment through electrical impedance measurements. *Journal of Horticultural Science* 68(3) (January 1993): 393–398.

Cutillo, A. G., A. H. Morris, D. C. Ailion, C. H. Durney, and K. Ganesan. Determination of lung water content and distribution by nuclear magnetic resonance. In *New Aspects on Respiratory Failure*, E. Rügheimer (Ed.), pp. 138–146. Berlin, Germany: Springer, 1992.

Danley, R. L. and P. A. Caulfield. *DSC Baseline Improvements Obtained by a New Heat Flow Measurement Technique*. New Castle, DE: TA Instruments, 2001.

Datta, A. K. Porous media approaches to studying simultaneous heat and mass transfer in food processes. I: Problem formulations. *Journal of Food Engineering* 80(1) (2007): 80–95.

Dean, D. A., T. Ramanathan, D. Machado, and R. Sundararajan. Electrical impedance spectroscopy study of biological tissues. *Journal of Electrostatics* 66(3–4) (March 2008): 165–177.

Defraeye, T., V. Lehmann, D. Gross, C. Holat, E. Herremans, P. Verboven, B. E. Verlinden, and B. M. Nicolai. Application of MRI for tissue characterisation of 'Braeburn' apple. *Postharvest Biology and Technology* 75 (January 2013): 96–105.

Dejmek, P. and O. Miyawaki. Relationship between the electrical and rheological properties of potato tuber tissue after various forms of processing [in English]. *Bioscience, Biotechnology, and Biochemistry* 66(6) (June 2002): 1218–1223.

Delgado-Goni, T., S. Campo, J. Martin-Sitjar, M. E. Cabanas, B. San Segundo, and C. Arus. Assessment of a 1h high-resolution magic angle spinning NMR spectroscopy procedure for free sugars quantification in intact plant tissue [in English]. *Planta* 238(2) (August 2013): 397–413.

Derome, A. E. *Modern NMR Techniques for Chemistry Research*. Elsevier, Amsterdam, Netherlands, 2013.

Feng, H., J. Tang, R. P. Cavalieri, and O. A. Plumb. Heat and mass transport in microwave drying of porous materials in a spouted bed. *AIChE Journal* 47(7) (2001): 1499–1512.

Feng, H., Y. Yin, and J. Tang. Microwave drying of food and agricultural materials: Basics and heat and mass transfer modeling [in English]. *Food Engineering Reviews* 4(2) (June 2012): 89–106.

Furuse, M., T. Gonda, H. Kuchiwaki, N. Hirai, S. Inao, and N. Kageyama. Thermal analysis on the state of free and bound water in normal and edematous brains. In *Recent Progress in the Study and Therapy of Brain Edema*, K. G. Go and A. Baethmann (Eds.), pp. 293–298. Boston, MA: Springer US, 1984.

Gambhir, P. N., Y. J. Choi, D. C. Slaughter, J. F. Thompson, and M. J. McCarthy. Proton spin–spin relaxation time of peel and flesh of navel orange varieties exposed to freezing temperature. *Journal of the Science of Food and Agriculture* 85(14) (2005): 2482–2486.

Goñi, O., C. Fernandez-Caballero, M. T. Sanchez-Ballesta, M. I. Escribano, and C. Merodio. Water status and quality improvement in high-CO_2 treated table grapes. *Food Chemistry* 128(1) (September 2011): 34–39.

Gonzalez, J. J., R. C. Valle, S. Bobroff, W. V. Biasi, E. J. Mitcham, and M. J. McCarthy. Detection and monitoring of internal browning development in 'Fuji' apples using MRI. *Postharvest Biology and Technology* 22(2) (May 2001): 179–188.

Gonzalez, M. E., D. M. Barrett, M. J. McCarthy, F. J. Vergeldt, E. Gerkema, A. M. Matser, and H. Van As. (1)H-NMR study of the impact of high pressure and thermal processing on cell membrane integrity of onions [in English]. *Journal of Food Science* 75(7) (September 2010): E417–E425.

Gunasekaran, S. Pulsed microwave-vacuum drying of food materials. *Drying Technology* 17(3) (March 1999): 395–412.

Gustavsson, J., C. Cederberg, and U. L. F. Sonesson. *Global Food Losses and Food Waste-Extent, Causes and Prevention*. Rome, Italy: Food and Agriculture organization of United nations, 2011.

Haines, P. J., M. Reading, and F. W. Wilburn. Differential thermal analysis and differential scanning calorimetry. In *Handbook of Thermal Analysis and Calorimetry*, M. E. Brown and P. K. Gallagher (Eds.), Vol. 1, pp. 279–361. Amsterdam, The Netherlands: Elsevier B.V. 1998.

Halder, A., A. K. Datta, and R. M. Spanswick. Water transport in cellular tissues during thermal processing. *AIChE Journal* 57(9) (2011): 2574–2588.

Hanson, L. G. Is quantum mechanics necessary for understanding magnetic resonance? *Concepts in Magnetic Resonance Part A* 32(5) (2008): 329–340.

Hatakeyama, T., K. Nakamura, and H. Hatakeyama. Determination of bound water content in polymers by DTA, DSC and TG. *Thermochimica Acta* 123 (Jaunary 1988): 153–161.

Hatakeyama, T., M. Tanaka, A. Kishi, and H. Hatakeyama. Comparison of measurement techniques for the identification of bound water restrained by polymers. *Thermochimica Acta* 532 (March 2012): 159–163.

Hills, B. P. and B. Remigereau. NMR studies of changes in subcellular water compartmentation in parenchyma apple tissue during drying and freezing. *International Journal of Food Science & Technology* 32(1) (1997): 51–61.

Iglesias, H. A. and J. Chirife (Eds.). *Handbook of Food Isotherms: Water Sorption Parameters for Food and Food Components*, pp. 336–343. Waltham, MA: Academic Press, 1982.

Ilker, R. and A. S. Szczesniak. Structural and chemical bases for texture of plant foodstuffs. *Journal of Texture Studies* 21(1) (1990): 1–36.

Inao, S., H. Kuchiwaki, N. Hirai, S. Takada, N. Kageyama, M. Furuse, and T. Gonda. Dynamics of tissue water content, free and bound components, associated with ischemic brain edema. In *Brain Edema: Proceedings of the Sixth International Symposium*, November 7–10, 1984, Tokyo, Japan, Y. Inaba, I. Klatzo and M. Spatz (Eds.), pp. 360–366. Berlin, Germany: Springer, 1985.

Jackman, R. L. and D. W. Stanley. Perspectives in the textural evaluation of plant foods. *Trends in Food Science & Technology* 6(6) (June 1995): 187–194.

Joardder, M. U. H., A. Karim, R. J. Brown, and C. Kumar. *Porosity : Establishing the Relationship between Drying Parameters and Dried Food Quality*. Springer, Berlin, Germany, 2015.

Karel, M. and D. B. Lund. *Physical Principles of Food Preservation*. New York: CRC Press, 2003.

Karim, M. A. and M. N. A. Hawlader. Mathematical modelling and experimental investigation of tropical fruits drying. *International Journal of Heat and Mass Transfer* 48(23–24) (November 2005): 4914–4925.

Kerch, G., A. Glonin, J. Zicans, and R. M. Meri. A DSC study of the effect of ascorbic acid on bound water content and distribution in chitosan-enriched bread rolls during storage. *Journal of Thermal Analysis and Calorimetry* 108(1) (2012): 73–78.

Khan, M. I. H., M. U. H. Joardder, C. Kumar, and M. A. Karim. Multiphase porous media modelling: A novel approach to predicting food processing performance. *Critical Reviews in Food Science and Nutrition* (2016a) doi.org/10.1080/10408398.2016.1197881.

Khan, M. I. H., C. Kumar, M. U. H. Joardder, and M. A. Karim. Determination of appropriate effective diffusivity for different food materials. *Drying Technology* 35(3) (2017): 335–346.

Khan, M. I. H., R. M. Wellard, S. A. Nagy, M. U. H. Joardder, and M. A. Karim. Investigation of bound and free water in plant-based food material using NMR T 2 relaxometry. *Innovative Food Science & Emerging Technologies* 38 (2016b): 252–261.

Khan, M. I. H., R. M. Wellard, N. D. Pham, and M. A. Karim. Investigation of cellular level of water in plant-based food material. In *The 20th International Drying Symposium*, Gifu, Japan, 2016c.

Kumar, C., M. U. H. Joardder, T. W. Farrell, and A. Karim. Multiphase porous media model for intermittent microwave convective drying (IMCD) of food. *International Journal of Thermal Science* 104 (2016a): 304–314.

Kumar, C., M. U. H. Joardder, T. W. Farrell, G. J. Millar, and M. A. Karim. A mathematical model for intermittent microwave convective (IMCD) drying of food materials. *Drying Technology* 34(8) (2015): 962–973.

Kumar, C., M. U. H. Joardder, A. Karim, G. J. Millar, and Z. Amin. Temperature redistribution modeling during intermittent microwave convective heating. *Procedia Engineering* 90 (2014a): 544–549.

Kumar, C., A. Karim, M. U. H. Joardder, and G. J. Miller. Modeling heat and mass transfer process during convection drying of fruit. In *The Fourth International Conference on Computational Methods*, Gold Coast, Queensland, Australia, November 25–28, 2012a.

Kumar, C., A. Karim, S. C. Saha, M. U. H. Joardder, R. J. Brown, and D. Biswas. Multiphysics modeling of convective drying of food materials. In *Proceedings of the Global Engineering, Science and Technology Conference*, Dhaka, Bangladesh, December 28–29, 2012b.

Kumar, C., M. A. Karim, and M. U. H. Joardder. Intermittent drying of food products: A critical review. *Journal of Food Engineering* 121(0) (2014b): 48–57.

Kumar, C., G. J. Millar, and M. A. Karim. Effective diffusivity and evaporative cooling in convective drying of food material. *Drying Technology* 33(2) (2016b): 227–237.

Kumar, C., S. C. Saha, E. Sauret, A. Karim, and Y. T. Gu. Mathematical modeling of heat and mass transfer during intermittent microwave-convective drying (IMCD) of food materials. In *Australasian Heat and Mass Transfer Conference 2016*, Brisbane, Queensland, Australia, 2016c.

Kuprianoff, J. Bound and free water in foods. In *Fundamental Aspects of the Dehydration of Foodstuffs*, J. Kuprianoff (Ed.), pp. 14–23. London, U.K.: Society of Chemical Industry, 1958.

Lee, D. J. Interpretation of bound water data measured via dilatometric technique. *Water Research* 30(9) (September 1996): 2230–2232.

Lee, D. J. and Y. H. Hsu. Measurement of bound water in sludges: A comparative study. *Water Environment Research* 67(3) (1995): 310–317.

Lima, D. A. B., J. M. P. Q. Delgado, S. F. Neto, and C. M. R. Franco. Intermittent drying: Fundamentals, modeling and applications. In *Drying and Energy Technologies*, pp. 19–41. Springer International Publishing, Berlin, Germany, 2016.

Liu, Y. C., F. Sommer, and E. J. Mittemeijer. Calibration of the differential dilatometric measurement signal upon heating and cooling; Thermal expansion of pure iron. *Thermochimica Acta* 413(1–2) (April 2004): 215–225.

Marangoni, A. G. and M. Fernanda Peyronel. Differential scanning calorimetry. AOCS Lipid Library, Urbana, IL, 2014.

Melado-Herreros, A., M.-A. Muñoz-García, A. Blanco, J. Val, M. Encarnación Fernández-Valle, and P. Barreiro. Assessment of watercore development in apples with MRI: Effect of fruit location in the canopy. *Postharvest Biology and Technology* 86 (December 2013): 125–133.

Moser, E., P. Holzmueller, and G. Gomiscek. Liver tissue characterization by in vitro NMR: Tissue handling and biological variation. *Magnetic Resonance in Medicine* 24(2) (April 1992): 213–220.

Moser, E., P. Holzmueller, and M. Krssak. Improved estimation of tissue hydration and bound water fraction in rat liver tissue. *Magnetic Resonance Materials in Physics, Biology and Medicine* 4(1) (1996): 55–59.

Nakamur, K., Y. Minagaw, T. Hatakeyam, and H. Hatakeyama. DSC studies on bound water in carboxymethylcellulose–polylysine complexes. *Thermochimica Acta* 416(1–2) (June 2004): 135–140.

Nakamura, K., T. Hatakeyama, and H. Hatakeyama. Studies on bound water of cellulose by differential scanning calorimetry. *Textile Research Journal* 51(9) (September 1981): 607–613.

Naofumi, Y., A. Remi, and O. Masahiro. A calibration method for measuring thermal expansions with a push-rod dilatometer. *Measurement Science and Technology* 12(12) (2001): 2121.

Ohno, H., M. Shibayama, and E. Tsuchida. DSC analyses of bound water in the microdomains of interpolymer complexes. *Die Makromolekulare Chemie* 184(5) (1983): 1017–1024.

Peishi, C. and D. C. T. Pei. A mathematical model of drying processes. *International Journal of Heat and Mass Transfer* 32(2) (1989): 297–310.

Prothon, F., L. Ahrne, and I. Sjoholm. Mechanisms and prevention of plant tissue collapse during dehydration: A critical review. *Critical Reviews in Food Science and Nutrition* 43(4) (2003): 447–479.

Ruan, R. R., P. L. Chen, and S. Almaer. Nondestructive analysis of sweet corn maturity using NMR. *HortScience* 34(2) (April 1999): 319–321.

Schwartzberg, H. G. Effective heat capacities for the freezing and thawing of food. *Journal of Food Science* 41(1) (1976): 152–156.

Sedin, G., P. Bogner, E. Berenyi, I. Repa, Z. Nyul, and E. Sulyok. Lung water and proton magnetic resonance relaxation in preterm and term rabbit pups: Their relation to tissue hyaluronan. *Pediatric Research* 48(4) (October 2000): 554–559.

Smith, J. K. and P. A. Vesilind. Dilatometric measurement of bound water in wastewater sludge. *Water Research* 29(12) (December 1995): 2621–2626.

Srikiatden, J. and J. S. Roberts. Moisture transfer in solid food materials: A review of mechanisms, models, and measurements. *International Journal of Food Properties* 10(4) (2007): 739–777.

Sulyok, E., Z. Nyúl, P. Bogner, E. Berényi, I. Repa, Z. Vajda, T. Dóczi, and G. Sedin. Brain water and proton magnetic resonance relaxation in preterm and term rabbit pups: Their relation to tissue hyaluronan. *Neonatology* 79(1) (2001): 67–72.

Turner, I. W., J. R. Puiggali, and W. Jomaa. 5th UK national heat transfer conferencea numerical investigation of combined microwave and convective drying of a hygroscopic porous material: A study based on pine wood. *Chemical Engineering Research and Design* 76(2) (February 1998): 193–209.

Vajda, Z., E. Berenyi, P. Bogner, I. Repa, T. Doczi, and E. Sulyok. Brain adaptation to water loading in rabbits as assessed by NMR relaxometry [in English]. *Pediatric Research* 46(4) (October 1999): 450–454.

Van de Velde, F., M. H. Grace, D. Esposito, M. É. Pirovani, and M. A. Lila. Quantitative comparison of phytochemical profile, antioxidant, and anti-inflammatory properties of blackberry fruits adapted to Argentina. *Journal of Food Composition and Analysis* 47 (April 2016): 82–91.

Van Der Weerd, L., M. M. A. E. Claessens, C. Efdé, and H. Van As. Nuclear magnetic resonanceimaging of membrane permeability changes in plants during osmoticstress. *Plant, Cell and Environment* 25(11) (2002): 1539–1549.

Van Der Weerd, L., M. M. A. E. Claessens, T. Ruttink, F. J. Vergeldt, T. J. Schaafsma, and H. Van As. Quantitative NMR microscopy of osmotic stress responses in maize and pearl millet. *Journal of Experimental Botany* 52(365) (December 2001): 2333–2343.

Waldron, K. W., A. C. Smith, A. J. Parr, A. Ng, and M. L. Parker. New approaches to understanding and controlling cell separation in relation to fruit and vegetable texture. *Trends in Food Science and Technology* 8(7) (June 1997): 213–221.

Webber, H. H. and P. A. Dehnel. Water balance of whole animal, muscle tissue, and muscle cells in the prosobranch gastropod, *Acmaea scutum*. *Journal of Experimental Zoology* 168(3) (1968): 327–335.

Westbrook, C. and Kaut, C. *MRI in Practice*. Oxford, U.K.: Blackwell Scientific, 1993.

Winisdorffer, G., M. Musse, S. Quellec, M.-F. Devaux, M. Lahaye, and F. Mariette. MRI investigation of subcellular water compartmentalization and gas distribution in apples. *Magnetic Resonance Imaging* 33(5) (June 2015): 671–680.

Wu, C. C., C. Huang, and D. J. Lee. Bound water content and water binding strength on sludge flocs. *Water Research* 32(3) (March 1998): 900–904.

Wu, L., Y. Ogawa, and A. Tagawa. Electrical impedance spectroscopy analysis of eggplant pulp and effects of drying and freezing–thawing treatments on its impedance characteristics. *Journal of Food Engineering* 87(2) (July 2008): 274–280.

Zhang, M. I. N., Stout, D. G., and Willison J. H. M. Electrical impedance analysis in plant tissues: Symplasmic resistance and membrane capacitance in the hayden model. *Journal of Experimental Botany* 41(224) (1990): 371–380.

10 Multi-Scale Modelling in Intermittent Drying

M.M. Rahman, Chandan Kumar and M. Azharul Karim

CONTENTS

10.1 INTRODUCTION

Thermal drying of food material is a basic operation in numerous industrial processes and is characterized as a highly energy-intensive process (Rahman et al., 2016). However, traditional drying systems (i.e., convection drying) often have several disadvantages, including high energy consumption and poor food quality (Kumar et al., 2015). To improve energy efficiency and food quality, intermittent drying of food materials was introduced (Kumar et al., 2016a). In this method, the drying conditions are changed with time. This is done in many ways, such as by varying the temperature, air flow, pressure and humidity. Researchers have proved that intermittent drying is more energy efficient than continuous drying (Kumar et al., 2014a). Moreover, they have also found that better-quality dried food can be obtained by the intermittent drying method.

The important sources of essential dietary nutrients such as vitamins, minerals and fibre are fruits and vegetables. However, they are considered perishable commodities as their moisture content is more than 80% (Paull, 1999; Kumar et al., 2014a). A common technique for storing perishable items is to maintain low temperature to preserve their nutritional value. However, it is difficult to maintain a low temperature throughout the food chain, as it is energy intensive and time consuming. This is particularly problematic for post-harvest management of fruits and vegetables in developing countries where the distribution chain cannot maintain the required low level of temperatures, thus ruining essential nutritional values of perishable foods (Kumar et al., 2016b). An effective answer to this problem is drying. To promote food security of over 20% of the world's perishable crops, drying is a traditional choice (Sagar and Suresh Kumar, 2010). Drying of fruits and vegetables enhances storage stability, minimizes packaging requirements and reduces transport weight and cost (Rahman et al., 2015). The main purpose of drying is to remove moisture from a product, thus ensuring its prolonged life. Minimizing operational cost is also crucial in the drying process. Food is a multi-component system, composed of a complex integration of different chemical and physical components. Foods are mainly composed of proteins and carbohydrate polymers, and these elements work as structure builders (Mitchell, 1998). Moreover, there are complex relationships between the macroscopic and the microscopic properties of food materials. The explanation of this macro–micro property relationship cannot be explained by means of bulk experimental methods (Limbach and Kremer, 2006). The application of mathematical modelling has been proven to explain the complicated process (Baschnagel et al., 2000; Attig et al., 2004). A good mathematical model can help understand the transport process inside food materials during drying (Kumar et al., 2016c). Drying of porous materials like food is one of the most complex problems in the engineering field and is also multi-scale in nature (Mujumdar, 2007). Multi-scale simulation methods can serve as an effective approach and can extend the vision of macro–micro relationship in the food drying process. A multi-scale modelling technique can establish the relationships between the macroscopic properties and the microstructure of food materials during drying. This chapter will discuss the application of the multi-scale modelling approach in the intermittent drying of food materials.

10.2 MULTI-SCALE TRANSPORT PHENOMENA

Researchers have found 85%–95% of water is inside the cells of most food materials and is known as intracellular water (Halder et al., 2011). The remaining water is found in the intercellular spaces and cell wall. Three main transport mechanisms dominate the internal moisture transfer during drying: (1) transport of bound water, (2) transport of free water and (3) transport of vapour. The flow of the liquid water occurs due to capillary forces, the transport of water vapour occurs through diffusion and the transport of bound water takes place by desorption and diffusion processes (Joardder et al., 2015a).

TABLE 10.1
Water Transport Mechanism inside Food Materials during Dehydration

Moisture Transfer during Drying		Scale	References
	Vapour	Macroscale	Lewis (1921), Engels et al. (1986)
		Microscale	Nieto et al. (2001)
Diffusion	Surface	Macroscale	Zogzas et al. (1994), Jayasundera et al. (2009)
	Liquid	Macroscale	Islam et al. (2003)
		Microscale	Shi and Le Maguer (2002), Nieto et al. (2001)
Capillary		Macroscale	Sankat and Castaigne (2004)
Evaporation condensation		Macroscale	Bhandari and Howes (1999)
Hydraulic flow		Macroscale	Sankat and Castaigne (2004)
Molecular diffusion		Microscale	Karel and Saguy (1991), Aguilera et al. (2003)
Knudsen flow		Microscale	Sagara (2001)
		Macroscale	Rossello et al. (1992)
Mutual diffusion		Macroscale	Sano and Yamamoto (1993)
Slip flow		Macroscale	Waananen et al. (1993)

Water flux inside plant tissue generally follows three pathways during drying: intercellular, wall to wall and cell to cell. Intercellular transport refers to water vapour movement through intercellular spaces. Wall-to-wall transport refers to the capillary water flow through the cell wall. Cell-to-cell transport refers to the water flux through the vacuoles, the cytoplasm and the cell membranes. In tissue with high moisture content, liquid flows due to dominating capillary forces. The amount of liquid in the pores decreases with decreasing moisture content. A gas phase is built up, which decreases the moisture content and causes a decrease in liquid permeability. Gradually, mass transfer is taken over by vapour diffusion in a porous structure with increasing vapour diffusion. However, there are different types of mass transfer mechanisms that dominate at different scales in food materials, presented in Table 10.1.

Water transport during hot-air drying leads to the evolution of porosity. Moreover, shrinkage of the food material towards the centre of the sample compensates for void volume. Therefore, understanding the phenomenon of structural deformation during heat and mass transfer is important for designing the drying process. Also, the process parameters and the transfer mechanisms influence the structural changes and the water transfer mechanisms (Rastogi et al., 2002; Ziaiifar et al., 2008). Pore formation is prevented by shrinkage. The degree of structural change depends on the distribution of water within the food tissue (Alamar et al., 2008). Water can be classified on the basis of its spatial distribution: intercellular water and intracellular water. Intracellular water remains inside the cells, while intercellular water remains in the capillaries. Considering the structure–property relationship and water migration mechanisms, several ways of predicting drying kinetics and deformation have been reported in the literature.

10.3 MULTI-SCALE MODELLING APPROACH

There are two basic approaches used to model the drying phenomena of food materials: macroscopic continuum approach and microscopic (cellular) approach. The macroscopic properties and the tissue strength of food materials are greatly influenced by the cell size and shape, turgor pressure, as well as the presence of intercellular space (Konstankiewicz and Zdunek, 2001). The concept of multi-scale modelling addresses some of these limitations, such as the absence of equilibrium at the microscopic scale or the consideration of product variability at the industrial scale. The concept of multi-scale modelling in drying is illustrated in Figure 10.1.

We will consider both macroscale and microscale model formulations in this chapter.

10.3.1 MACROSCALE APPROACH

In the macroscale approach, the tissue is considered as homogeneous, and the modelling is carried out based on the lumped properties of cell walls and pores (Veraverbeke et al., 2003; Nguyen et al., 2006). In the microscale modelling approach, the tissue is considered as heterogeneous, and the complex cellular structure is represented by a geometric model. Most of the works regarding the drying of food materials are based on the classical continuum approach, where the transport properties are considered as a lumped single value throughout the tissue. Various macroscopic continuum approaches for modelling intermittent drying, along with their governing equations and solution techniques, are discussed in more detail in Chapter 8.

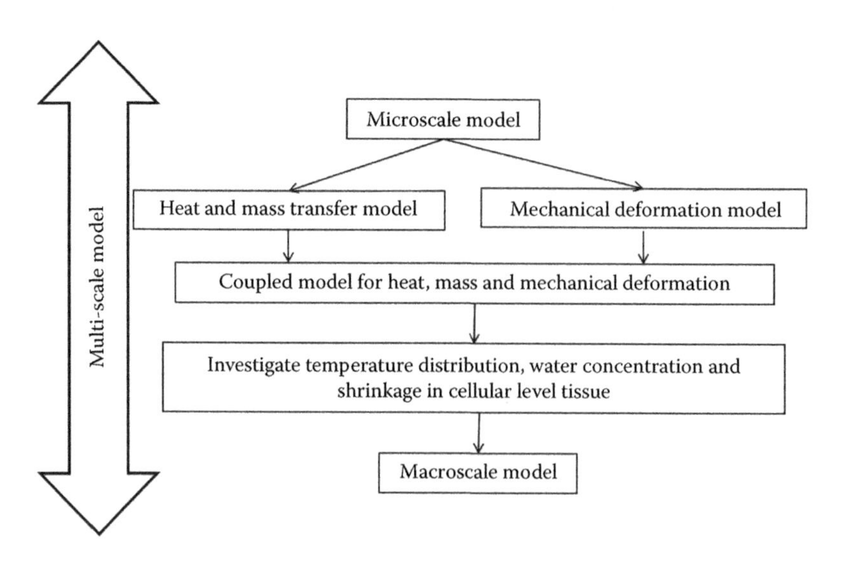

FIGURE 10.1 Concept of multi-scale modelling in drying technology. (From Rahman, M.M. et al., *Crit. Rev. Food Sci. Nutr.*, doi: 10.1080/10408398.2016.1227299, 2016.)

10.3.2 MICROSCALE APPROACH

The microscale model is an effective approach for understanding better the influence of drying conditions on the tissue. The main drawback of the macroscale model is that it does not provide a platform to investigate the microstructural behaviours of the tissue during drying, which can be overcome by micro-scale modelling. The microscale model can describe the deformation of the cell wall and the local distribution of water at the cellular level. For predicting the drying process of the entire sample, the microscale model can be extended and expressed as a multi-scale model. This multi-scale model can provide detailed information of the microstructural features, combining the continuum type models.

10.3.3 RELATIONSHIP BETWEEN MICROSCALE AND MACROSCALE MODELS

Basically, a multi-scale model is a hierarchical combination of a number of submodels that describes the process phenomena at a different spatial scale (Aregawi et al., 2014). The main challenge of formulating the multi-scale model is to establish a logical relation between the scales (Perré and Rémond, 2006; Perré et al., 2012). Drying of food materials is a multi-scale phenomenon, and a better approach to formulate the drying model is to use a multi-scale model (Ho et al., 2013). For bridging the macroscale and the microscale models, an effective approach was proposed, which is known as 'homogenization'. It is a powerful mathematical tool, and the macroscopic properties can be predicted from the microscopic description by using this tool. The homogenization process includes the governing equation. The apparent diffusion can be calculated by solving the cellular diffusion equation. A sample of homogenization approach is illustrated in Figure 10.2 (Ho et al., 2010).

Recently, van der Sman et al. (2012) developed a multi-scale model with a serial coupling technique (Esveld et al., 2012a,b). The model can predict the dynamics of the moisture diffusion in food materials at cellular level. To develop the model, the researchers used an X-ray microtomography image to get information about the cellular network and the pores. The model considers the local diffusivity through the moisture sorption and the pores. These parameters were incorporated in the macroscale model for predicting the dynamic moisture profile of crackers.

Although researchers have always shown more interest in the macroscopic scale, the microscopic transport phenomenon is also an important factor in the drying process (Aguilera, 2005; Fanta et al., 2014). For simulation of the drying process, numerous mathematical equations have been proposed (Whitaker, 1977, 1998). Later, several researchers modified the equations and assumptions and proposed new models considering internal evaporation and pressure-driven transport (Ni and Datta, 1999). Some researchers have considered internal over-pressure and bound water diffusion (Hills and Remigereau, 1997), while others have considered the evolution of porosity in their model (Ni et al., 1999; Datta, 2007). Usually, microscale models are formulated to calculate the tissue properties of the sample more accurately. The heterogeneous multi-scale method (HMM) can be a suitable option for

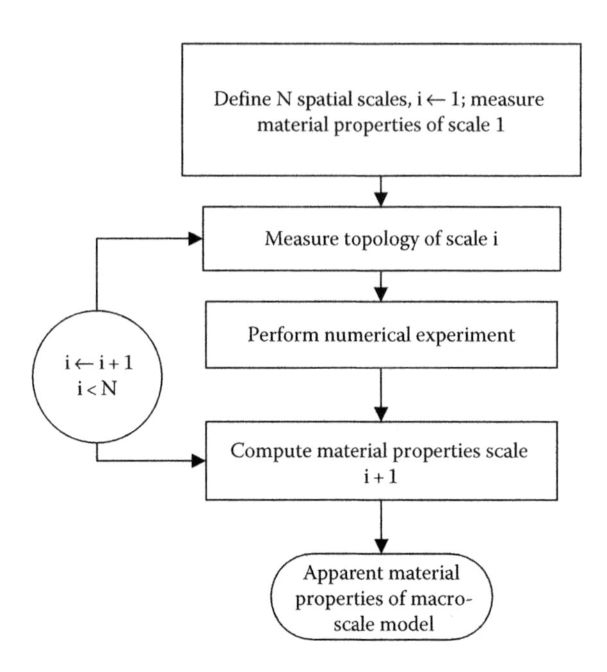

FIGURE 10.2 A sample of homogenization process. (From Ho, Q.T. et al., *J. Exp. Bot.*, 61(8), 2071, 2010.)

coupling both the micro- and macroscales, as the constitutive relation relies on many variables (E et al., 2007). The HMM can be initialized by discretization of macroscale equations. The finite element method (FEM) is an efficient option in this regard. The framework for the heterogeneous multi-scale modelling is illustrated in Figure 10.3.

If a microscopic system is known, the state variable u can be written as

$$f(u,b) = 0 \tag{10.1}$$

where b is the set of auxiliary conditions, such as initial and boundary conditions for the problem. Here, the microscopic details of u are not of interest; rather, the macroscopic state of the system is of interest, which is represented by U in the following macroscopic equation:

$$F(U,D) = 0 \tag{10.2}$$

where D represents the macroscopic data that are necessary for the model to be completed. Considering the compression operator Q that maps u and U, and R is any operator that reconstructs u from U, we can find that,

$$Qu = U \quad RU = u \tag{10.3}$$

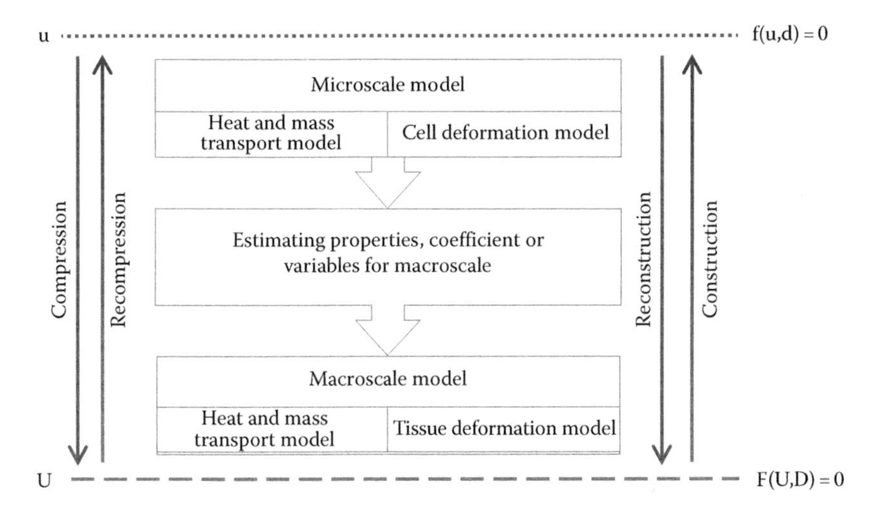

FIGURE 10.3 Heterogeneous multi-scale method framework. (From Rahman, M.M. et al., *Crit. Rev. Food Sci. Nutr.*, doi: 10.1080/10408398.2016.1227299, 2016.)

where compression and reconstruction operators (U) are similar to the projection and prolongation operators (a and b) used in multi-grid methods. The goal of HMM is to compute U using the abstract form of F and the microscopic model. It consists of two main components: selection of a macroscopic solver and estimation of the missing macroscale data, D, using the microscale model.

10.4 NUMERICAL SOLUTION FOR THE DRYING MODELS

Several advanced numerical methods for the solution of both microscale and multi-scale modellings have been used. This section will discuss the numerical solution techniques that have been used for solving the multi-physics-related problem at various scales.

10.4.1 SPH and DEM Method

The key idea in mesh-free particle methods is that the material is mass-discretized into material points. Some researchers have been working on particle-based numerical models to simulate micromechanics of plant cells (Karunasena et al., 2014). They have developed mesh-free particle methods to simulate the mechanics of both individual plant cells and their response to the external stress. At the same time, there has been significant interest in modelling the mechanical behaviour of tissue. Models consider the cell fluid and the cell wall as the two main features of a plant cell. For developing the cell fluid model, smoothed particle hydrodynamics have been used. A discrete element method (DEM) has been used to model the cell wall. In this model, the biological cell response to mechanical load was assessed

(Karunasena et al., 2014). This model is capable of predicting cellular level shrinkage as a function of moisture content. The authors claimed that the mesh-free nature of the model gives it the capability to handle large deformations, multi-phase interactions and subcellular details, which are the key concerns in multi-scale modelling approach (Karunasena et al., 2014).

10.4.2 Finite Element Method

For solving partial differential equations, the FEM is a flexible and widely used method (Zienkiewicz and Taylor, 2005). In the FEM, the computational domain is subdivided into a number of elements having different shapes and sizes. Every element in the computational domain is interconnected by a certain number of nodal points. An unknown solution is assigned to every element, which is known as a shape function. In the following step, the governing equations are discretized and normalized by a suitable meshing technique. Depending on the time-variable, the system equations will be either ordinary differential equations or algebraic equations. These equations are solved using the finite difference approximation approach. The details of the FEM can be found in several publications (Bathe, 2008; Efendiev and Hou, 2009).

10.4.3 Lattice Boltzmann Method

In order to simulate the microscale model, the lattice Boltzmann method and the molecular dynamics method are also useful. The lattice Boltzmann method differs from traditional approaches (Van der Sman, 2007). Materials and fluids are represented as quasi-particles in the lattice Boltzmann method. These particles maintain the basic law of conservation of energy, momentum and mass while interacting via collision. The rules of the collisions maintain the discretized Boltzmann equation. The discretization of momentum, space and time makes the lattice Boltzmann method different from the other methods.

10.4.4 Molecular Dynamics

In the molecular dynamics method, the momentum of the molecules is computed with the solution of the Newton's equation of motions. The van der Waals, the covalent bond and the electrostatic interaction create a potential field, which is used to compute the forces between the molecules. The evaluation of these potentials is the most computationally intensive step of a molecular dynamics simulation. Molecular dynamics can be considered as a DEM. This method is not used frequently in food processing engineering. This methodology has not yet been used in the drying process.

10.5 SIGNIFICANCE OF GEOMETRY IN MULTI-SCALE MODELLING

A geometrical model is required in order to formulate a mathematical model that will work as a computational domain. The geometrical model can be built by using computer-aided modelling techniques. The geometry at the microscale is more complex. Therefore, it is very challenging to build a geometrical model, especially at the micro level for food materials. Recently, non-destructive imaging techniques have advanced significantly. Microscopic imaging using the scanning electron microscope and transmission electron microscope can capture the exact structure of the food materials at the microscopic level (Frank et al., 2010). The combination of microscopic imaging with computed tomography can provide a detailed explanation of the microstructure of food materials (Moreno-Atanasio et al., 2010; Perré, 2011). The complex geometrical model at the micro level can be built by using a proper image-processing technique. The complex geometrical model can be used as a representative volume element (RVE) to build a realistic multi-scale model. An RVE is defined as the minimum volume over which the properties of a material can be calculated. The RVE has been chosen as a non-hygroscopic material where the solid matrix is considered as the cell surrounded by free water.

The effect of microstructure on transport processes in food drying has been neglected for a long time while developing transport models. But the transport phenomena inside the food materials, together with the physical and biochemical stability, greatly depends on the microstructure. In the case of drying, Fick's first and second laws are used to estimate the rate of mass transfer. In most of the drying models, food has been considered as a solid matrix, ignoring the special features of the cell and intercellular space. Figure 10.4 is a schematic diagram of the conventional RVE used by the researchers.

In practice, plant-based tissue is hygroscopic in nature, where the cells are filled with fluid commonly known as bound water. Therefore, it is necessary to choose an appropriate RVE to develop a model for better prediction. Figure 10.5 is a schematic diagram of the microstructure of cellular tissue.

The actual microstructural geometry for developing a better model is illustrated in Figure 10.6. This geometry will work as a computational domain, which is a necessary requirement for developing a model.

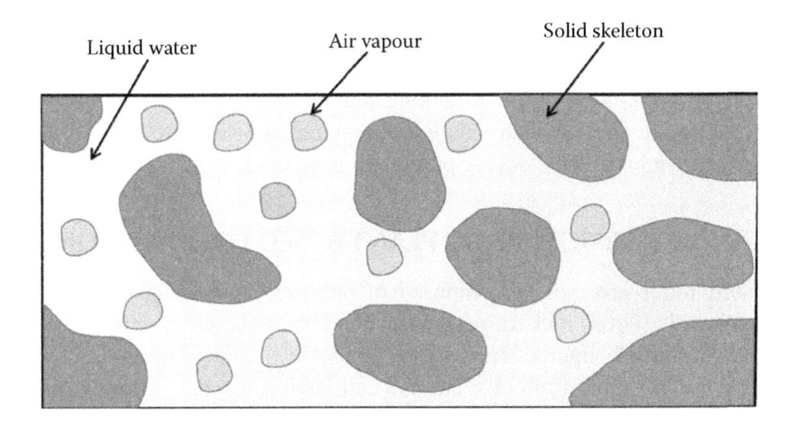

FIGURE 10.4 Conventional RVE used by researchers.

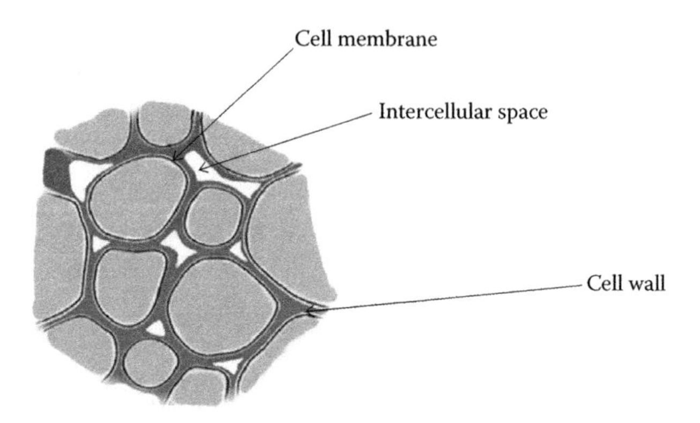

FIGURE 10.5 Cellular structure of plant tissue.

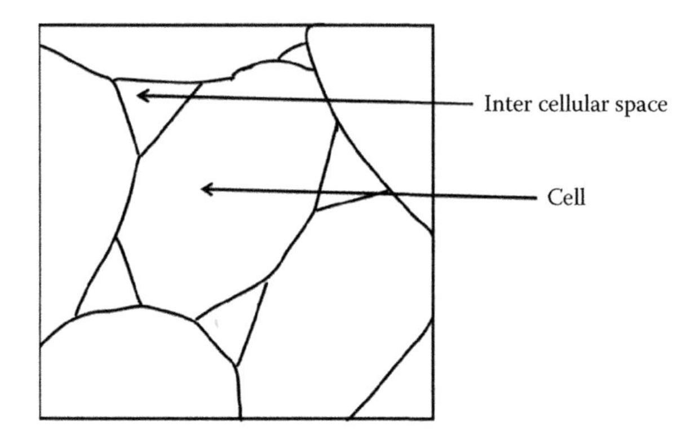

FIGURE 10.6 Schematic diagram of actual geometry of the microstructure of cellular tissue.

In the microscale model, different components of cells are distinguished in terms of position, shape and orientation. It is evident that the cell wall can be a controlling factor in the mass transfer process. Extensive approximation has been made by food engineers for the solution of the diffusion equation in a regular geometry. Moreover, the microstructure of food materials is not regular, and there are limited models to investigate the impact of the drying process in food microstructure.

10.6 MICROSTRUCTURE OF PLANT-BASED FOOD MATERIALS

Plant-based foods are mostly composed of polymers and water but also contain air and minerals (Gross and Kalra, 2002). The food structure complexity is a consequence of multiple interactions of polysaccharides and proteins. Cell walls are formed by the integration of cells, and the cell walls are stabilized with the help of fibres. Cellulose, pectin and hemicellulose are usually the building materials of the cell wall. The water inside the plant cell migrates through pectin, as it is soluble in water (Joardder et al., 2015b).

Fruit tissue can be viewed as a summation of parenchymatic cells and intercellular spaces (Mebatsion et al., 2006). A thin cell wall composed of cellulose and other polysaccharides envelops the cell. The porosity of such cell walls is high and consequently has high permeability to water. Therefore, water retention occurs mainly by surface tension effects. There is also a semi-permeable membrane, called the plasmalemma, inside the cell wall (Aguilera and Stanley, 1999). Each cell contains cytoplasm with a variety of organelles. These organelles play a key role in the metabolic activity of cells in the food materials. Theses organelles are bounded by a membrane known as tonoplast. Inside the tonoplast, there is the vacuole, which has the highest water content of the cell. The membranes are semi-permeable and are responsible for the osmotic effect. The external water potential to the vacuole generates hydrostatic pressure, which ensures that the protoplast is firmly pressed against the cell wall and hence keeps the tissue firm. For this reason, loss of water is accompanied by a loss of internal pressure, while the tissue becomes flaccid. This pressure is known as turgor pressure and plays an important role in the rheology and texture of the tissue.

The structural features at the macro level and the micro level are different. A hierarchical structure of a food material is illustrated in Figure 10.7. The microstructure of food materials can be defined as the spatial arrangement of the cells and pores. The structure of the food materials comprises three basic molecules: polysaccharides,

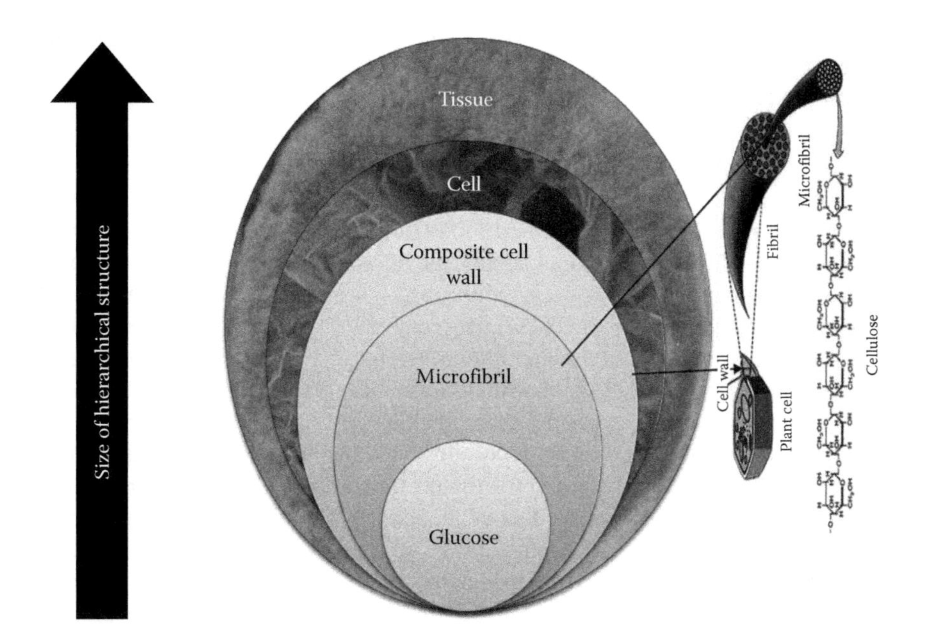

FIGURE 10.7 Illustration of hierarchical structure of cellular-based food materials. (With kind permission from Springer Science+Business Media: *Porosity*, Factors affecting porosity, 2016, pp. 25–46, Joardder, M.U.H., Karim, A., Kumar, C., and Brownet, R.J.)

proteins and lipids. The cellular structure of food can be classified as fibrous, fleshy and encapsulated. The cellular materials are mainly based on glucose.

10.7 RELATIONSHIP BETWEEN MICROSTRUCTURE AND PHYSICAL PROPERTY

Drying environment has a great influence on the microstructure of the food being processed. Food structure is constructed from several compounds and their mutual interactions, which are closely related to quality (Ko and Gunasekaran, 2007; Aguilera and Lillford, 2008; Witek et al., 2010). Drying of food includes modification of microstructures, and this can be regulated by controlling the drying conditions. Novel approaches to the drying process mostly rely on the proper understanding of the architecture and organization of the microstructures.

To design an accurate food processing system, it is imperative to know the structure and the behaviour of the product. Clear knowledge of the relationships between structure and property of the food materials is required for developing a better food drying process (Aguilera et al., 2000). However, the fundamental structure–property relationships are not very clearly understood. As a result, food processing engineers rely on empirical data for designing the drying process. Spectacular advances in materials science during the last decade have emanated from the understanding of the structure of a material, its relation to properties (so-called structure–property relationships) and how to manipulate and control those properties. Material scientists are quite successful in doing this because the microstructures of metals, ceramics and polymers are more or less homogeneous (Limbach and Kremer, 2006). A large collection of data on physical properties of foods including thermal, rheological and mechanical and colour characteristics exists in literature, but understanding their relationship to microstructure is limited (Joardder et al., 2016).

It is suggested that the microstructural changes occurring during drying have a great effect on the texture of the dried food (Bourne, 2002; Rahman, 2008). In the case of industrial food processing systems, the microstructure of plant-based materials breaks down as a result of poor processing environment. This happens because of the loss of turgor and rupture of cellular membranes (Ludikhuyze and Hendrickx, 2001). The sounds produced during biting of fresh fruits or vegetable are the instant release of turgor pressure inside the cell (Vickers and Bourne, 1976). In the case of dried product, this sound comes from the breaking of the cell wall (Duizer, 2001).

Different types of drying have various effects on the microstructure and the product properties. The change in the microstructure that occurs during drying impacts the quality of the food materials (Joardder et al., 2015b). The effect of a combination of hot-air drying and osmotic pre-treatment was evaluated by Prothon et al. (2001). The microstructural studies showed that the quality of the final product was improved by osmotic pre-treatment. Therefore, a proper understanding of the structure–property relationship will enhance the accuracy of the food drying models. Some researchers introduced intermittent microwave drying of food products (Kumar et al., 2014b). Ho et al. (2002) found that intermittent drying can provide higher proportions of ascorbic acid than the other drying methods.

10.8 MODELLING MICROSTRUCTURES OF FOOD MATERIALS

The microstructural information generated by microscopic analysis can be used to understand the transport mechanisms during drying and for assessing the functionality of finished products. A variety of microscopic techniques are available for studying different structures and components of food systems. Building the microstructure of a biological system is the most complex component in the process of formulating the multi-scale modelling. Despite a great deal of advancement in cell biology and fruit science, microstructural characterization and computational modelling of biological systems are still in their infancy. It is not possible to analyze heat and mass transfer at the cellular level by means of FEM due to lack of information on the proper microstructure. SEM imaging provides detailed microstructural information without chemical treatment. The vacuum sample chamber in the SEM limits the direct observation of the dynamic processes of moisture movement (Zadin et al., 2015). Nevertheless, it gives a detailed representation of the food system in a static condition. A suitable image processing technique can extract the geometrical data and provide a clear idea about the microstructure.

Figure 10.8 is a schematic block diagram of the development microstructure from microscopic image.

Transport of water in intercellular spaces and the cell walls network is predicted by using the chemical potential as the driving force for water exchange between different microstructural compartments. The transport properties of the water in the cell wall network can be obtained from the literature.

10.8.1 CALCULATION OF THE GEOMETRICAL PROPERTIES OF THE CELL

The geometrical properties can be calculated from the microscopic image. The geometrical characteristics including the perimeters and the area of the cell are estimated by using Green's theorem (Rahman et al., 2016). The aspect ratio and the orientation of the arrangement of the cells were obtained from the ellipse fitting algorithm. The calculation of the geometrical properties with reference to Green's theorem is discussed.

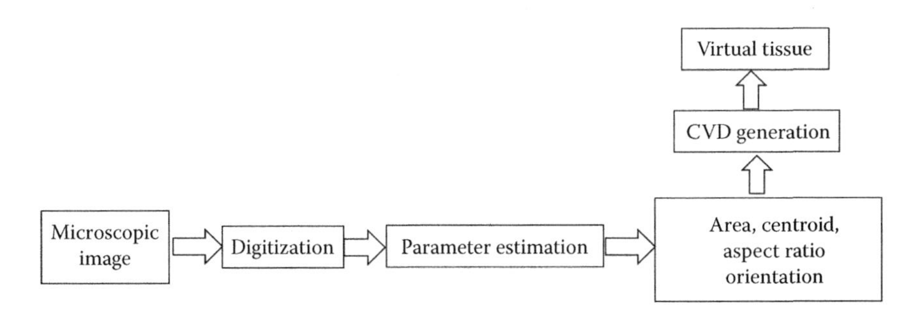

FIGURE 10.8 A step-by-step representation of modelling of cell geometry.

For given two points P(x,y) and Q(x,y), the differentiable functions are as follows:

$$\iint_R \left(\frac{\partial Q}{\partial x} - \frac{\partial P}{\partial y} \right) dx\, dy = \int_B \left(P\, dx + Q\, dy \right) \tag{10.4}$$

where

R is the cell or pore approximated by the polygon

the B is the integral of the boundary

The area of the region has been calculated by the following equation:

$$A = \iint_R dx\, dy. \tag{10.5}$$

Green's transformation of the equation points in anticlockwise direction, and results in

$$A = \frac{1}{2} \sum_{i=1}^{n} x_{i-1} y_i - x_i y_{i-1} \tag{10.6}$$

where n is the number of points on the boundary of the cell, pore or vertices of a Voronoi cell.

By applying Green's transformation, the centroids of the region are found using the following equation:

$$x_c = \iint_R x\, dx\, dy \tag{10.7}$$

$$x_c = \frac{1}{2A} \sum_{i=1}^{n} \left(x_{i-1} - x_i \right) \left(x_{i-1} y_i - x_i y_{i-1} \right) \tag{10.8}$$

$$y_c = \iint_R y\, dx\, dy \tag{10.9}$$

$$y_c = \frac{1}{2A} \sum_{i=1}^{n} \left(y_{i-1} - y_i \right) \left(x_{i-1} y_i - x_i y_{i-1} \right). \tag{10.10}$$

10.8.2 Calculation of the Aspect Ratio and Orientation of the Cells

The aspect ratio and the orientation of the cellular arrangement are calculated using the ellipse fitting algorithm. A discussion of the estimation of the orientation and the aspect ratio is as follows.

For a given number of points, 'n', on the boundary of a given region, the area can be calculated over x, y axes and the plane xy. For this purpose, the following equations have been used:

$$A_{xx} = \iint_R y^2 \, dx \, dy \tag{10.11}$$

$$A_{yy} = \iint_R x^2 \, dx \, dy \tag{10.12}$$

$$A_{xy} = \iint_R xy \, dx \, dy \tag{10.13}$$

$$A_{xx} = \frac{1}{2A}\sum_{i=1}^{n}\left(x_{i+1}y_i - x_i y_{i+1}\right)\left(y_i^2 + y_{i+1}y_i + y_{i+1}^2\right) \tag{10.14}$$

$$A_{yy} = \frac{1}{2A}\sum_{i=1}^{n}\left(x_{i+1}y_i - x_i y_{i+1}\right)\left(x_i^2 + x_{i+1}x_i + x_{i+1}^2\right) \tag{10.15}$$

$$A_{xy} = \frac{1}{2A}\sum_{i=1}^{n}\left(2x_i y_i + x_i y_{i+1} + x_{i+1}y_i + 2x_{i+1}y_{i+1}\right)\left(x_{i+1}y_i - x_i y_{i+1}\right) \tag{10.16}$$

10.8.3 Identifying Cells and Intercellular Space

The cell and the intercellular spaces have distinctive geometrical features. Therefore, it is very important to identify the cell and the intercellular space in the food microstructure. Mebatsion et al. (2006) acquired cell and the intercellular space in the microscopic image by MATLAB® program. For each region, six parameters were calculated, including elongation, convexity, rectangularity, circularity, angularity and roughness. The size and the shape of the cells and the intercellular space in the cellular tissue of an apple can be described by these parameters. The details of the calculation of these parameters can be found in the literature (Pieczywek and Zdunek, 2012). In our proposed method, the regions on the microscopic images have been investigated automatically by the image processing algorithm. The identification process was integrated with the Voronoi tessellation algorithm. The integration approach has made the proposed approach novel as it can detect the cell and the intercellular space.

10.8.4 Generation of Parenchymal Tissue

The formulated geometrical model can be utilized to generate the virtual cellular level tissue. Zadin et al. (2015) compared generated geometries with the original

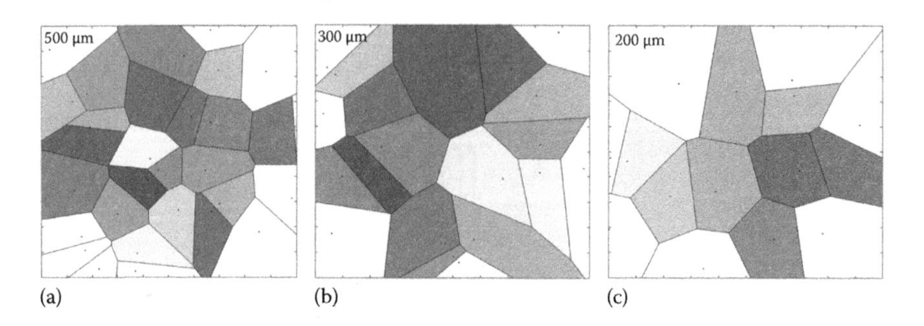

FIGURE 10.9 Initial generation of virtual cell for (a) 500 μm, (b) 300 μm and (c) 200 μm scale.

images obtained using scanning electron microscope. The important comparison features are the size, shape of cells, the orientation of cells and the intercellular space. The initial generation of virtual cell was performed and is shown in Figure 10.9.

For modelling the intercellular space inside the apple tissue, Rahman et al. (2016) modified the initial Voronoi tessellation. The virtual cell has been formulated according to the characteristics of the cellular tissue of the apple. Fresh apple has some porosity, and the porosity inside the cellular tissue is strongly related to the number of intercellular spaces. The intercellular space has a great influence on the diffusivity and the mass transport of water inside the tissue. Using SEM, the distinction between the cell and the intercellular space can be observed. Geometrically, the intercellular spaces are irregular in shape and have high aspect ratio values. The automatic classification of intercellular space inside the apple tissue has been integrated with the Voronoi tessellation method to develop a geometrical model with distinct cell and intercellular spaces. A virtual cell containing intercellular space is presented in Figure 10.10.

The generated cell geometry can be transferred to the COMSOL multi-physics software (COMSOL 5.1, Stockholm, Sweden) via the MATLAB-COMSOL interfacing module. The cell, cell wall and the intercellular spaces can be exported as separate vertices so that different conditions can be applied to each part of the tissue. To minimize computation complexity, only 100 and 150 μm scale geometry was exported to COMSOL, as presented in Figure 10.11.

In the COMSOL multi-physics software, coupled heat and mass transfer (hot-air drying) can be applied to the microstructure. The model was solved using FEMs, where meshing is an important part of the process. Meshing on the formulated geometry performed using the COMSOL mesh generator is presented in Figure 10.12.

Boundary conditions for the cell to cell and cell to intercellular space are also important factors for modelling drying conditions. The cell to cell boundary and the cell to intercellular space boundary are shown in Figure 10.13.

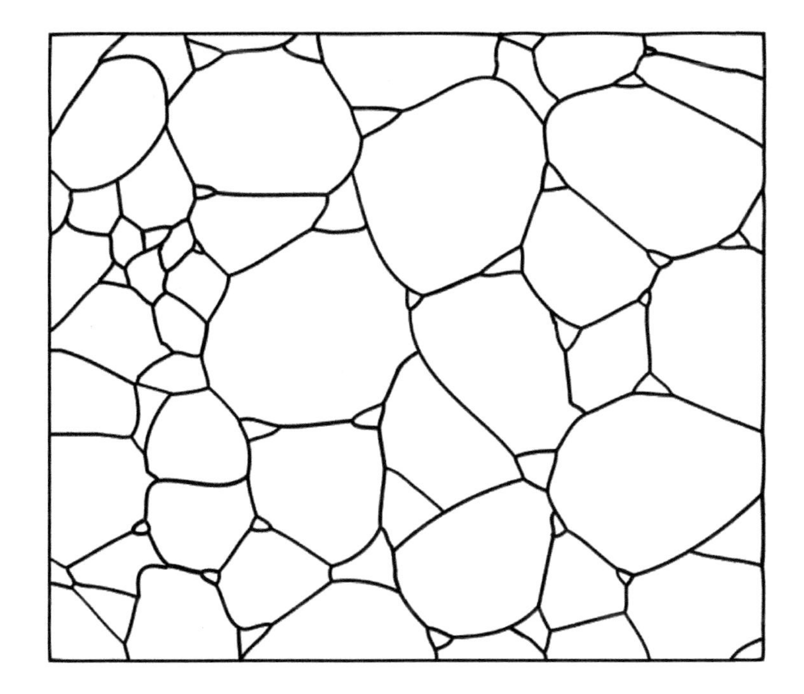

FIGURE 10.10 Virtual cell containing intercellular space.

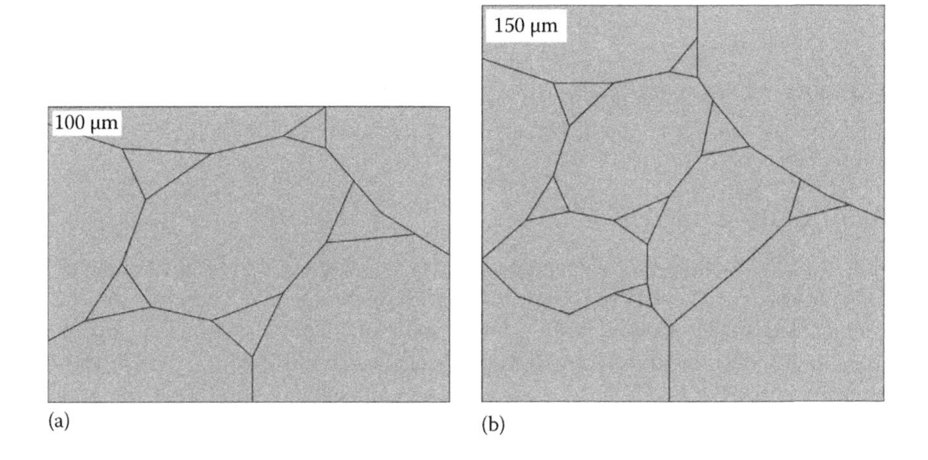

FIGURE 10.11 Virtual cell of apple (a) 100 μm and (b) 150 μm in COMSOL multi-physics environment.

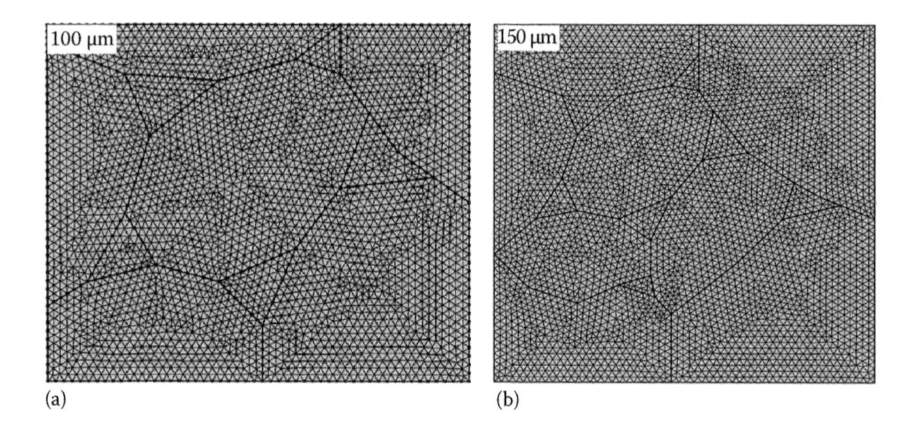

FIGURE 10.12 Generated mesh on microstructure (a) 100 μm and (b) 150 μm.

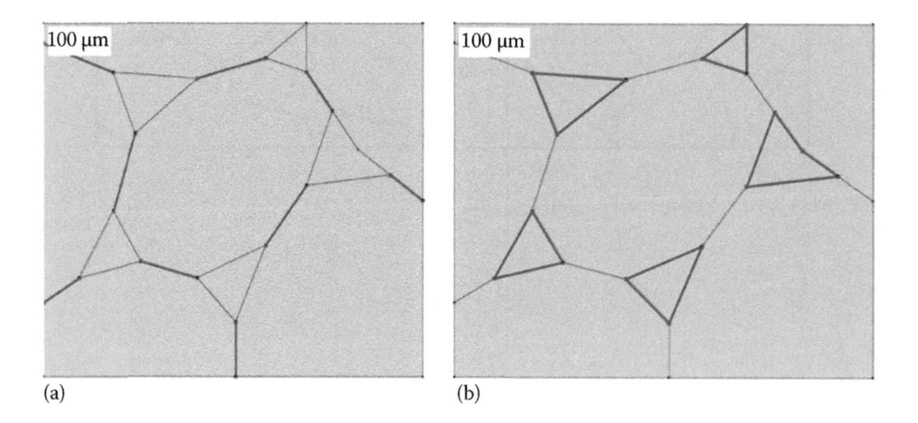

FIGURE 10.13 Boundary inside the virtual tissue (a) cell to cell and (b) cell to intercellular space.

10.9 CONCLUSIONS

Multi-scale modelling is a new approach in the intermittent drying of food material. Food materials are considered as porous hygroscopic materials. Therefore, drying of food materials is considered a multi-phase and multi-scale phenomenon. Advanced modelling and simulation techniques are needed to build a multi-scale model for intermittent food drying. One of the most important challenges in formulating multi-scale modelling is creating the cellular-level geometry. The geometry works as a computational domain for building the mathematical model. This chapter has outlined a unique technique to build the cell-level geometry. This technique is non-destructive and therefore saves computational costs.

The prospect of multi-scale modelling to improve the processes in intermittent food drying technology is highly promising. A multi-scale model will be very useful for optimizing the current drying processes as well as the intermittent drying process. It can be a pathway to an environment-friendly and energy-efficient drying technology.

REFERENCES

Aguilera, J.M. 2005. Why food microstructure? *Journal of Food Engineering* 67 (1):3–11.

Aguilera, J.M., A. Chiralt, and P. Fito. 2003. Food dehydration and product structure. *Trends in Food Science & Technology* 14 (10):432–437.

Aguilera, J.M. and P.J. Lillford. 2008. Structure–property relationships in foods. In *Food Materials Science*, pp. 229–253. Springer-Verlag, New York.

Aguilera, J.M., D.W. Stanley, and K.W. Baker. 2000. New dimensions in microstructure of food products. *Trends in Food Science and Technology* 11 (1):3–9.

Aguilera, J.M. and D.W. Stanley. 1999. *Microstructural Principles of Food Processing and Engineering*. Springer Science & Business Media, Springer-Verlag, New York.

Alamar, M.C., E. Vanstreels, M.L. Oey, E. Moltó, and B.M. Nicolaï. 2008. Micromechanical behaviour of apple tissue in tensile and compression tests: Storage conditions and cultivar effect. *Journal of Food Engineering* 86 (3):324–333.

Aregawi, W.A., M.K. Abera, S.W. Fanta, P. Verboven, and B. Nicolai. 2014. Prediction of water loss and viscoelastic deformation of apple tissue using a multiscale model. *Journal of Physics: Condensed Matter* 26 (46):464111.

Attig, N., K. Binder, H. Grubmuller, and K. Kremer. 2004. *Computational Soft Matter: From Synthetic Polymers to Proteins*. John von Neumann Institute for Computing (NIC), Juelich, Germany.

Baschnagel, J., K. Binder, P. Doruker, A.A. Gusev, O. Hahn, K. Kremer, W.L. Mattice, F. Müller-Plathe, M. Murat, and W. Paul. 2000. Bridging the gap between atomistic and coarse-grained models of polymers: Status and perspectives. In A. Abe (ed.), *Viscoelasticity, Atomistic Models, Statistical Chemistry*, 152: 41–156. Springer, Springer-Verlag, Berlin, Heidelberg.

Bathe, K.J. 2008. *Finite Element Method, Wiley Online Library. Wiley Encyclopedia of Computer Science and Engineering*, pp. 1–12. Massachusetts Institute of Technology, Cambridge, MA.

Bhandari, B.R. and T. Howes. 1999. Implication of glass transition for the drying and stability of dried foods. *Journal of Food Engineering* 40 (1):71–79.

Bouchon, P. and D.L. Pyle. 2005. Modelling oil absorption during post-frying cooling: I: Model development. *Food and Bioproducts Processing* 83 (4):253–260.

Bourne, M. 2002. *Food Texture and Viscosity: Concept and Measurement*, Wiley Encyclopedia of Computer Science and Engineering, pp. 1–12. Massachusetts Institute of Technology, Cambridge, MA.

Datta, A.K. 2007. Porous media approaches to studying simultaneous heat and mass transfer in food processes. I: Problem formulations. *Journal of Food Engineering* 80 (1):80–95.

Duizer, L. 2001. A review of acoustic research for studying the sensory perception of crisp, crunchy and crackly textures. *Trends in Food Science & Technology* 12 (1):17–24.

Efendiev, Y. and T.Y. Hou. 2009. *Multiscale Finite Element Methods:Theory and Applications* (Vol. 4). Springer Science & Business Media, Springer-Verlag, New York.

Engels, C., M. Hendrickx, S. De Samblanx, I. De Gryze, and P. Tobback. 1986. Modelling water diffusion during long-grain rice soaking. *Journal of Food Engineering* 5 (1):55–73.

Esveld, D.C., R.G.M. van Der Sman, G. Van Dalen, J.P.M. Van Duynhoven, and M.B.J. Meinders. 2012a. Effect of morphology on water sorption in cellular solid foods. Part I: Pore scale network model. *Journal of Food Engineering* 109 (2):301–310.

Esveld, D.C., R.G.M. van der Sman, M.M. Witek, C.W. Windt, H. van As, J.P.M. van Duynhoven, and M.B.J. Meinders. 2012b. Effect of morphology on water sorption in cellular solid foods. Part II: Sorption in cereal crackers. *Journal of Food Engineering* 109 (2):311–320.

Fanta, S.W., M.K. Abera, W.A. Aregawi, Q.T. Ho, P. Verboven, J. Carmeliet, and B.M. Nicolai. 2014. Microscale modeling of coupled water transport and mechanical deformation of fruit tissue during dehydration. *Journal of Food Engineering* 124:86–96.

Frank, X., G. Almeida, and P. Perré. 2010. Multiphase flow in the vascular system of wood: From microscopic exploration to 3-D Lattice Boltzmann experiments. *International Journal of Multiphase Flow* 36 (8):599–607.

Gross, R.A. and B. Kalra. 2002. Biodegradable polymers for the environment. *Science* 297 (5582):803–807.

Halder, A., A.K. Datta, and R.M. Spanswick. 2011. Water transport in cellular tissues during thermal processing. *AIChE Journal* 57 (9):2574–2588.

Hills, B.P. and B. Remigereau. 1997. NMR studies of changes in subcellular water compartmentation in parenchyma apple tissue during drying and freezing. *International Journal of Food Science and Technology* 32 (1):51–61.

Ho, Q.T., J. Carmeliet, A.K. Datta, T. Defraeye, M.A. Delele, E. Herremans, L. Opara, H. Ramon, E. Tijskens, and R. van der Sman. 2013. Multiscale modeling in food engineering. *Journal of Food Engineering* 114 (3):279–291.

Ho, Q.T., P. Verboven, B.E. Verlinden, and B.M. Nicolaï. 2010. A model for gas transport in pear fruit at multiple scales. *Journal of Experimental Botany* 61 (8):2071–2081.

Islam, Md.R., J.C. Ho, and A.S. Mujumdar. 2003. Simulation of liquid diffusion-controlled drying of shrinking thin slabs subjected to multiple heat sources. *Drying Technology* 21 (3):413–438.

Jayasundera, M., B. Adhikari, P. Aldred, and A. Ghandi. 2009. Surface modification of spray dried food and emulsion powders with surface-active proteins: A review. *Journal of Food Engineering* 93 (3):266–277.

Joardder, M.U.H., A. Karim, C. Kumar, and R.J. Brown. 2015a. *Porosity: Establishing the Relationship between Drying Parameters and Dried Food Quality*. Springer.

Joardder, M.U.H., A. Karim, C. Kumar, and R.J. Brown. 2016. Factors affecting porosity. *Porosity*, pp. 25–46. Springer International Publishing, New York.

Joardder, M.U.H., C. Kumar, and M.A. Karim. 2015b. Food structure: Its formation and relationships with other properties. *Critical Reviews in Food Science and Nutrition* 57 (6):1190–1205.

Karel, M. and I. Saguy. 1991. Effects of water on diffusion in food systems. In H. Levine, and L. Slade (eds.), *Water Relationships in Foods*, pp. 157–173. Springer, New York.

Karunasena, H., W. Senadeera, Y. Gu, and R.J. Brown. 2014. A coupled SPH-DEM model for micro-scale structural deformations of plant cells during drying. *Applied Mathematical Modelling* 38 (15):3781–3801.

Ko, S. and S. Gunasekaran. 2007. Error correction of confocal microscopy images for in situ food microstructure evaluation. *Journal of Food Engineering* 79 (3):935–944.

Konstankiewicz, K. and A. Zdunek. 2001. Influence of turgor and cell size on the cracking of potato tissue. *International Agrophysics* 15 (1):27–30.

Kumar, C., M.U.H. Joardder, T.W. Farrell, and A. Karim. 2016a. Multiphase porous media model for intermittent microwave convective drying (IMCD) of food. *International Journal of Thermal Science* 104:304–314.

Kumar, C., M.U.H. Joardder, T.W. Farrell, G.J. Millar, and M.A. Karim. 2015. A mathematical model for intermittent microwave convective (IMCD) drying of food materials. *Drying Technology* 34 (8):962–973.

Kumar, C., M.U.H. Joardder, A. Karim, G.J. Millar, and Z. Amin. 2014a. Temperature redistribution modeling during intermittent microwave convective heating. *Procedia Engineering* 90:544–549.

Kumar, C., M.A. Karim, and M.U.H. Joardder. 2014b. Intermittent drying of food products: A critical review. *Journal of Food Engineering* 121:48–57.

Kumar, C., G.J. Millar, and M.A. Karim. 2016b. Effective diffusivity and evaporative cooling in convective drying of food material. *Drying Technology* 33 (2):227–237.

Kumar, C., S.C. Saha, E. Sauret, A. Karim, and Y.T. Gu. 2016c. Mathematical modeling of heat and mass transfer during intermittent microwave-convective drying (IMCD) of food materials. *Australasian Heat and Mass Transfer Conference 2016*, Brisbane, Queensland, Australia.

Lewis, W.K. 1921. The rate of drying of solid materials. *Industrial & Engineering Chemistry* 13 (5):427–432.

Limbach, H.J. and K. Kremer. 2006. Multi-scale modelling of polymers: Perspectives for food materials. *Trends in Food Science & Technology* 17 (5):215–219.

Ludikhuyze, L. and M.E. Hendrickx. 2001. Effects of high pressure on chemical reactions related to food quality. In M.E.G. Hendrickx, and D. Knorr (eds.), *Ultra High Pressure Treatments of Foods*, pp. 167–188. Springer, US.

Mebatsion, H.K., P. Verboven, B.E. Verlinden, Q.T. Ho, T.A. Nguyen, and B.M. Nicolaï. 2006. Microscale modelling of fruit tissue using Voronoi tessellations. *Computers and Electronics in Agriculture* 52 (1):36–48.

Mitchell, J.R. 1998. Water and food macromolecules. Functional Properties of Food Macromolecules 50–76.

Moreno-Atanasio, R., R.A. Williams, and X. Jia. 2010. Combining X-ray microtomography with computer simulation for analysis of granular and porous materials. *Particuology* 8 (2):81–99.

Mujumdar, A.S. 2006. An overview of innovation in industrial drying: Current status and R&D needs. In *Drying of Porous Materials*, pp. 3–18. Springer, Netherlands.

Nguyen, T.A., P. Verboven, N. Scheerlinck, S. Vandewalle, and B.M. Nicola. 2006. Estimation of effective diffusivity of pear tissue and cuticle by means of a numerical water diffusion model. *Journal of Food Engineering* 72 (1):63–72.

Ni, H. and A.K. Datta. 1999. Heat and moisture transfer in baking of potato slabs. *Drying Technology* 17 (10):2069–2092.

Ni, H., A.K. Datta, and K.E. Torrance. 1999. Moisture transport in intensive microwave heating of biomaterials: A multiphase porous media model. *International Journal of Heat and Mass Transfer* 42 (8):1501–1512.

Nieto, A., M.A. Castro, and S.M. Alzamora. 2001. Kinetics of moisture transfer during air drying of blanched and/or osmotically dehydrated mango. *Journal of Food Engineering* 50 (3):175–185.

Paull, R.E. 1999. Effect of temperature and relative humidity on fresh commodity quality. *Postharvest Biology and Technology* 15 (3):263–277.

Perré, P. 2011. A review of modern computational and experimental tools relevant to the field of drying. *Drying Technology* 29 (13):1529–1541.

Perré, P. and R. Rémond. 2006. A dual-scale computational model of kiln wood drying including single board and stack level simulation. *Drying Technology* 24 (9):1069–1074.

Perré, P., R. Rémond, J. Colin, E. Mougel, and G. Almeida. 2012. Energy consumption in the convective drying of timber analyzed by a multiscale computational model. *Drying Technology* 30 (11–12):1136–1146.

Pieczywek, P.M. and A. Zdunek. Automatic classification of cells and intercellular spaces of apple tissue. *Computers and Electronics in Agriculture* 81:72–78.

Prothon, F., L.l.M. Ahrne, T. Funebo, S. Kidman, M. Langton, and I. Sjoholm. 2001. Effects of combined osmotic and microwave dehydration of apple on texture, microstructure and rehydration characteristics. *LWT—Food Science and Technology* 34 (2):95–101.

Rahman, M.M., M.U. Joardder, M.I.H. Khan, D.P. Nghia, and M.A. Karim. 2016. Multi-scale model of food drying: Current status and challenges. *Critical Reviews in Food Science and Nutrition*, doi: 10.1080/10408398.2016.1227299.

Rahman, M.M., S. Mekhilef, R. Saidur, A.G.M. Mustayen Billah, and S.M.A. Rahman. 2016. Mathematical modelling and experimental validation of solar drying of mushrooms. *International Journal of Green Energy* 13 (4):344–351.

Rahman, M.M., A.G.M.B. Mustayen, S. Mekhilef, and R. Saidur. 2015. The optimization of solar drying of grain by using a genetic algorithm. *International Journal of Green Energy* 12 (12):1222–1231.

Rahman, M.S. 2008. Dehydration and microstructure. In C. Ratti (ed.), *Advances in Food Dehydration*, p. 97. CRC Press, Taylor & Francis Group, Boca Raton, FL.

Rastogi, N.K., K.S.M.S. Raghavarao, K. Niranjan, and D. Knorr. 2002. Recent developments in osmotic dehydration: Methods to enhance mass transfer. *Trends in Food Science & Technology* 13 (2):48–59.

Rossello, C., J. Canellas, S. Simal, and A. Berna. 1992. Simple mathematical model to predict the drying rates of potatoes. *Journal of Agricultural and Food Chemistry* 40 (12):2374–2378.

Sagar, V.R. and P. Suresh Kumar. 2010. Recent advances in drying and dehydration of fruits and vegetables: A review. *Journal of Food Science and Technology* 47 (1):15–26.

Sagara, Y. 2001. Structural models related to transport properties for the dried layer of food materials undergoing freeze-drying. *Drying Technology* 19 (2):281–296.

Sankat, C.K. and F. Castaigne. 2004. Foaming and drying behaviour of ripe bananas. *LWT—Food Science and Technology* 37 (5):517–525.

Sano, Y. and S. Yamamoto. 1993. Mutual diffusion coefficient of aqueous sugar solutions. *Journal of Chemical Engineering of Japan* 26 (6):633–636.

Shi, J. and M. Le Maguer. 2002. Osmotic dehydration of foods: Mass transfer and modeling aspects. *Food Reviews International* 18 (4):305–335.

Van der Sman, R. 2013. Modeling cooking of chicken meat in industrial tunnel ovens with the Flory–Rehner theory. *Meat Science* 95 (4):940–957.

Van der Sman, R.G.M. 2007. Soft condensed matter perspective on moisture transport in cooking meat. *AIChE Journal* 53 (11):2986–2995.

Veraverbeke, E.A., P. Verboven, P. Van Oostveldt, and B.M. Nicolaï. 2003. Prediction of moisture loss across the cuticle of apple (*Malus sylvestris* subsp. *mitis* (Wallr.)) during storage: Part 1. Model development and determination of diffusion coefficients. *Postharvest Biology and Technology* 30 (1):75–88.

Vickers, Z. and M.C. Bourne. 1976. A psychoacoustical theory of crispness. *Journal of Food Science* 41 (5):1158–1164. doi: 10.1111/j.1365-2621.1976.tb14407.x.

Waananen, K.M., J.B. Litchfield, and M.R. Okos. 1993. Classification of drying models for porous solids. *Drying Technology* 11 (1):1–40.

Witek, M., W. Węglarz, L. De Jong, G. Van Dalen, J. Blonk, P. Heussen, H. Van As, and J. Van Duynhoven. 2010. The structural and hydration properties of heat-treated rice studied at multiple length scales. *Food Chemistry* 120 (4):1031–1040.

Whitaker, S. 1977. Simultaneous heat, mass, and momentum transfer in porous media: A theory of drying. *Advances in Heat Transfer* 13 (1977): 119–203.

Whitaker, S. 1998. Coupled transport in multiphase systems: A theory of drying. *Advances in Heat Transfer* 31:1–104.

Zadin, V., H. Kasemägi, V. Valdna, S. Vigonski, M. Veske, and A. Aabloo. 2015. Application of multiphysics and multiscale simulations to optimize industrial wood drying kilns. *Applied Mathematics and Computation* 267:465–475.

Ziaiifar, A.M., N. Achir, F. Courtois, I. Trezzani, and G. Trystram. 2008. Review of mechanisms, conditions, and factors involved in the oil uptake phenomenon during the deep-fat frying process. *International Journal of Food Science & Technology* 43 (8):1410–1423.

Zienkiewicz, O.C. and R.L. Taylor. 2005. *The Finite Element Method for Solid and Structural Mechanics*. Butterworth-Heinemann, Burlington, MA.

Zogzas, N.P., Z.B. Maroulis, and D. Marinos-Kouris. 1994. Densities, shrinkage and porosity of some vegetables during air drying. *Drying Technology* 12 (7):1653–1666.

Index

J

K

L

M

N

O

P

V

W

9781138746299